The family Rhinocerotidae has a long and amazing history in North America. From its first appearance about 40 million years ago, it diversified into an incredible array of taxa, with a variety of ecologies that don't resemble any of the five living species. They range from delicate long-legged dog-sized forms, to huge hippo-like forms that apparently lived in rivers and lakes.

This book includes a systematic review of the entire family of North American Rhinocerotidae, with complete descriptions, measurements, and figures of every bone in every species - the first such review in over a century. More importantly, it discusses the biogeographic patterns of rhinos, their evolutionary patterns and paleoecology, and what rhinos tell us about the evolution of North American landscapes and faunas over 35 million years. It is a complete and authoritative volume that will be a reference of interest to a variety of scientists for years to come.

DONALD R. PROTHERO is Chair and Professor of Geology at Occidental College in Los Angeles, California and Lecturer in Geobiology at the California Institute of Technology in Pasadena, California. He is the author or coeditor of 17 books, including textbooks, research volumes, and trade books. He has also published over 200 scientific papers on topics as diverse as mammalian paleontology (specializing in extinct North American rhinoceroses, horses, and camels), magnetic stratigraphy, climatic changes and extinctions of the Cenozoic, micropaleontology and chert petrography. He is on the editorial board of Geology magazine, and serves on the Penrose Conference Committee of the Geological Society of America. In 2000, he was Vice-President of the Pacific Section of the Society for Sedimentary Geology, and from 1998 to 2003 he served as Program Chair for the Society of Vertebrate Paleontology. In 1991 he received the Schuchert Award of the Paleontological Society for the outstanding paleontologist under the age of 40.

The Evolution of North American Rhinoceroses

DONALD R. PROTHERO

California Institute of Technology

CAMBRIDGE
UNIVERSITY PRESS

University Printing House, Cambridge CB2 8BS, United Kingdom

One Liberty Plaza, 20th Floor, New York, NY 10006, USA

477 Williamstown Road, Port Melbourne, VIC 3207, Australia

314-321, 3rd Floor, Plot 3, Splendor Forum, Jasola District Centre, New Delhi-110025, India

79 Anson Road, #06-04/06, Singapore 079906

Cambridge University Press is part of the University of Cambridge.

It furthers the University's mission by disseminating knowledge in the pursuit of education, learning and research at the highest international levels of excellence.

www.cambridge.org
Information on this title: www.cambridge.org/9781108457200

© Cambridge University Press 2005

First published 2005
First paperback edition 2018

A catalogue record for this publication is available from the British Library

Library of Congress Cataloging in Publication data
Prothero, Donald R.
The evolution of North American rhinoceroses/Donald R. Prothero.
p. cm.
Includes bibliographical references (p.).
ISBN 0-521-83240-3 (hb)
1. Rhinoceroses–North America–Evolution. I. Title.
QL737.U63P76 2004
569′.668–dc22 2004051864

ISBN 978-0-521-83240-3 Hardback
ISBN 978-1-108-45720-0 Paperback

This book is dedicated to

Earl Manning

*whose insights and curatorial work
made it all possible*

Contents

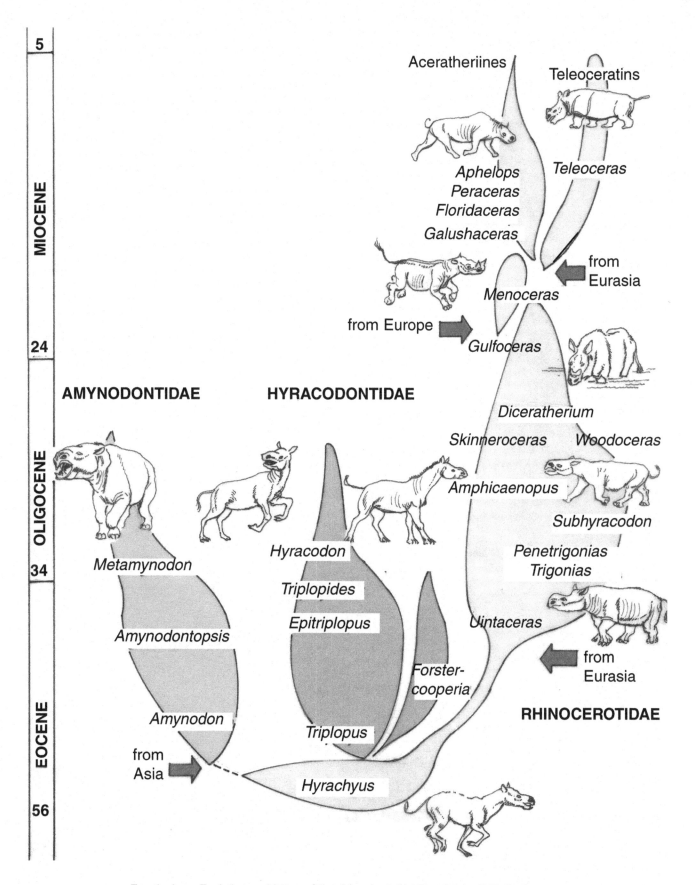

Frontispiece. Evolutionary history of the rhinocerotoids. Drawing by C.R. Prothero.

1. Introduction

Although they are nearly extinct today, rhinoceroses were one of the most widespread and successful groups of mammals on all the northern continents for over 40 million years. They have freely circulated between Eurasia and North America since the middle Eocene, and are known from Africa since the early Miocene. They have also occupied many different modes of life around the world during their long history. There have been sheep-sized running rhinos (*Hyracodon*), short-limbed hippopotamus-like rhinos (*Teleoceras, Metamynodon*), dwarfed rhinoceroses (*Peraceras hessei, Teleoceras meridianum, Gulfoceras westfalli,* and two more new species described in this book), and rhinos with tapir-like prubosces (*Cadurcodon*). The largest land mammal that ever lived (*Paraceratherium*) was also a rhinoceros.

Rhinos occurred in many different shapes and sizes, and some apparently occupied ecological niches that other mammals (such as the hippos in Africa) occupy today. In addition, extinct rhinos had many other characteristics not found in the living forms. For example, most people associate rhinos with their horns, but most fossil rhinos apparently did not have horns. Unlike the bony horns in bovids, antilocaprids, and many other groups, or antlers in cervids, rhino horns are made of agglutinated hair-like fibers, and have no bony core. The only evidence of their existence is a rugose area on the top of the skull, which indicates where the horn attached to the bone. Based on this evidence, most early rhinos were hornless, and horns (as paired horns on the nose) first appeared independently in two different lineages, the Diceratheriinae (*Diceratherium*, early Oligocene) and the Menoceratinae (*Menoceras*, earliest Miocene). Hornless rhinos (such as the Aceratheriinae) continued to live side by side with single-horned rhinos, such as the Teleoceratini, throughout the Miocene. In Eurasia, other horn combinations appeared, such as the tandem horns of the African rhinos, and the huge frontal horn of the elasmotheres.

In North America, rhinoceroses were by far the most common large herbivores on the continent through most of the Cenozoic. After the extinction of the brontotheres in the late Eocene (34 Ma), rhinos were also the largest land mammals on the continent for over 18 million years until the immigration of mastodonts in the middle Miocene (16 Ma). In addition to occupying the large-bodied herbivore niches through most of this time, rhinos also occupied a number of other ecological niches, including the hippopotamus-like forms mentioned above, prehensile-lipped browsers (most of the aceratheriines), and five independent examples of dwarfing (also mentioned above). In some places, such as the Miocene of Nebraska, rhinoceroses occurred in enormous herds, and there are a number of quarries that yield thousands of their bones. Even where they are not common, their bones are typically so large and robust that they are easy to recognize, and can now be identified down to species in many cases. Thus, nearly every important middle to late Eocene, Oligocene, and Miocene locality in North America produces some kind of rhino (even if they are only fragmentary), and it is important when no rhino is known from the fauna. However, rhinos also have some interesting and unusual biogeographic distribution patterns (especially during the Miocene) discussed in Chapter 6.

Given their abundance and interesting ecological and biogeographic patterns, it is surprising that rhinos are not very well studied. Even though they are found in nearly every museum collection, and we have not only teeth, but often skulls and complete skeletons of most of the species, and in some cases, even large quarry samples of single populations, there has been little recent study of them. This is a very unusual situation for fossil mammals, many of which are known only from a few isolated teeth or jaws, and relatively few of which are known from a skull, let alone a complete skeleton. Perhaps it is the sheer size of the specimens, and their enormous abundance that has discouraged workers in the past. The first fossil rhinos were described by Blumenbach (1799), and the first North American rhinos were reported by Leidy in 1850 (Leidy, 1850a, 1850b), yet over the past century, there have been remarkably few systematic monographs on rhinos, mostly on the Eurasian taxa (e.g., Guérin, 1980; Cerdeño, 1992; Antoine, 2002). North American forms have not been studied in detail for decades. In the last century, there were papers by Wood (1927) and Matthew (1932) which gave preliminary phylogenies of some North American rhinos (discussed in detail in Prothero et al., 1986), but no systematic review of all known North American rhinoceroses has ever been attempted. By the middle of the last century, the situation had been so confused that Simpson (1945, p. 257) remarked:

> The human factor in classification is nowhere more evident than in dealing with this superfamily [Rhinocerotoidea]. It is, as mammalian superfamilies go, well known, but what is "known" about it is so inconsistent in places that much of it must be wrong. Some authorities still recognize "genera" (e.g. *Orthocynodon,* an amynodont) that are, beyond much doubt, based on slight individual variation, while others lump together in one genus a whole tribal lineage that must almost certainly include a whole cluster of genera, even if generic lines be drawn as broadly as could be desired (e.g. the supposed European *Dicerorhinus* line). Some of the most competent students (e.g. Matthew) follow very broad lines, emphasize skull and foot characters, and tend to neglect dental mutations; others (e.g. Wood) split the groups into many short, narrow sequences emphasizing minor dental characters and tending to neglect skeletal structure. Much of the published work (aside from that of Matthew, Wood, and some others) is simply incompetent and has not been revised by a properly instructed and judicious student.

Unfortunately, this statement was still accurate for decades after 1945, even though rhinoceros fossils continued to accumulate. The most important new specimens came from the Frick Collection of fossil mammals in the American Museum of Natural History (Galusha, 1975a). From the 1930s until Childs Frick died in 1965, there were full-time crews working most of the important Oligocene and Miocene localities in the western United States, building up enormous collections of fossil mammals. The result was a huge private collection that remained largely unstudied while Frick was alive, and was not even completely curated in the new Frick Wing (with seven floors of storage of fossil mammals, mostly from the Frick Collection) in the mid-1970s. The rhinos occupied an entire floor by themselves, and were first curated by Earl Manning in the early 1970s, and then recurated by myself in 1984. In the process of this curation, Manning developed the first modern concepts of the systematics of North American rhinos (some of his ideas have appeared in Harrison and Manning, 1983; Prothero et al., 1986, 1989; Prothero and Manning, 1987). After Manning left the American Museum in

1981, I continued the curation and study of these collections until 1984. Brief summaries of this work and smaller studies on individual rhinos have appeared elsewhere (Prothero and Sereno, 1982; Prothero et al., 1986, 1989; Prothero and Manning, 1987; Prothero, 1993, 1998; Prothero and Schoch, 2003), but the full study (originally completed in 1984) was not compiled until this volume.

This book seeks to accomplish several tasks. The most important is the documentation of the systematics of all the known North American Rhinocerotidae (Chapter 4), which is now long overdue with the addition of the outstanding material in the Frick Collection. In the process of sorting out the valid species, hundreds of old invalid names (still technically available in the literature) have been synonymized, and a few new species recognized as well (Chapter 3). To do this properly, the patterns of variation within rhino populations need to be assessed, since most the old names are invalid because the authors were taxonomic "splitters" who used typological concepts, and failed to consider variation within populations (Chapter 2). The addition of so much new material, including complete skeletons of most of the species, means that the postcranial anatomy of most North American fossil rhinos is known for the first time. Thus, the huge number of unidentified rhino bones in collections all over the world can now be identified (Chapter 5). Finally, the rhinos have a fascinating history in North America which rivals that of the evolution of the horse for its illustrations of evolutionary and paleoecological principles (Chapter 7), and they are so abundant and widespread that their biogeographic patterns are of considerable interest as well (Chapter 6).

Acknowledgements

I thank many people who helped make this work possible. First, I thank Earl Manning, who introduced me to the rhinos, and whose curation of the Frick rhino collection and insights into their evolution and taxonomy are the basis for most of the ideas in this book. I thank Malcolm McKenna and Richard Tedford for access to the AMNH and F:AM Collections, and for their help and guidance during my graduate school years. I also thank Bruce Hanson, Allen Kihm, and the late Lloyd Tanner for their insights into rhino evolution. I thank Pierre-Olivier Antoine, Kurt Heissig, Spencer Lucas, and Matt Mihlbachler for reviewing the present manuscript. I thank the following people for access to specimens in their care: Mike Voorhies and Bob Hunt (UNSM); Bob Emry and Bob Purdy (USNM); John Ostrom (YPM); Chuck Schaff (MCZ); John A. Wilson (TMM); Phil Bjork (SDSMT); Dave Webb and Bruce MacFadden (FLMNH); Mary Dawson (CM); Howard Hutchison and Pat Holroyd (UCMP); Dave Whistler and Xiaoming Wang (LACM); Bill Turnbull, John Flynn, and Bill Simpson (FMNH); Larry Martin (KU); John Storer (SMNH); and Ted Fremd and Scott Foss (JODA). I also thank the many people of the Frick Laboratory who built the large collections that made this revision possible. This included a long list of collectors over almost 40 years (especially Charles Falkenbach, John Blick, Joe Rak, Ove Kaisen) but especially Ted Galusha and Morris Skinner, not only for making these collections, but for working out and publishing the stratigraphic data behind them (Skinner and Taylor, 1967; Skinner *et al.*, 1968, 1977; Skinner and Johnson, 1984; Galusha and Blick, 1971; Galusha, 1975b). I thank Meg Zepp, Emily CoBabe, and John Alexander for their help in moving and recurating the entire third floor of the Frick Wing during the summer of 1984. I thank Meg Zepp for help with word processing and cataloguing specimens. I thank Steven King for help with spreadsheets, and Patricia Bereck for darkroom work.I thank Matt Mihlbachler for help with the statistical analysis. The beautifully drawn illustrations of rhino foot bones in Chapter 5 are the work of Mark Morita. I thank my wife, Teresa LeVelle, for spending months doing the painstaking Photostudio editing of the photos. I thank Sally Thomas and Jo Bottrill at Cambridge University Press for their help. Most of this research was supported by numerous grants from the National Science Foundation and the Donors of the Petroleum Research Fund of the American Chemical Society. Finally, this work would not have been possible without the love and support of my wife, Teresa LeVelle, and my sons, Erik, Zachary, and Gabriel.

2. Methods

SPECIMENS, MEASUREMENT, AND TERMINOLOGY

This study is based on research begun by Earl Manning in the American Museum of Natural History during the early 1970s (when he originally curated the Frick rhino collection), and then continued by myself in the late 1970s and early 1980s until the present. The AMNH and F:AM rhino collection served as the nucleus for this study, since in most cases, dozens of skulls, jaws, and even skeletons are known for taxa that were once known only from a single jaw or skull. During the late 1970s and 1980s, I also visited the important rhino collections in most of the important museums around North America, including the Museum of Comparative Zoology, Harvard University, Cambridge, Massachusetts; the Yale Peabody Museum of Natural History, New Haven, Connecticut; the Princeton University collection (now at Yale); the National Museum of Natural History, Smithsonian Institution, Washington, D.C.; the Academy of Natural Sciences, Philadelphia, Pennsylvania; the Carnegie Museum of Natural History, Pittsburgh, Pennsylvania; the Florida Museum of Natural History, Gainesville, Florida; the Field Museum of Natural History, Chicago, Illinois; the Saskatchewan Museum of Natural History, Regina, Saskatchewan; the University of Nebraska State Museum, Lincoln, Nebraska; the South Dakota School of Mines and Technology, Rapid City, South Dakota; the University of Kansas Museum of Natural History, Lawrence, Kansas; the Texas Memorial Museum, University of Texas, Austin, Texas; the Colorado Museum of Natural History, Denver, Colorado (now the Denver Museum of Nature and Science); the University of California Museum of Paleontology, Berkeley, California; the Natural History Museum of Los Angeles County, Los Angeles, California; and the collections at John Day Fossil Beds, Oregon.

In each of these collections, a standard series of measurements (see the tables in Chapters 4 and 5) were taken with dial calipers for most of the available specimens. A meter stick and tape-measure were used for longer measurements. Most important specimens were photographed on 35-mm black-and-white film using Nikon cameras, and the descriptions of new specimens were often written in longhand on the spot. In the early 1980s, before personal computers (let alone laptops) and digital calipers, this resulted in an enormous pile of manuscript written on yellow legal pads, and thousands of measurements and sketches on 3" x 5" index cards. Many of these data cards were lost when People Express Airlines (deservedly now out of business) lost my luggage, so I had to return to several museums to remeasure hundreds of specimens. Most of the statistics for these measurements were calculated using Excel spreadsheets.

The landmarks for the measurements of skulls, jaws, and teeth are shown in Figure 2.1. In most cases (such as dental measurements), the landmarks are relatively straightforward, so one can assume that past workers have measured specimens in comparable ways. Whenever possible, I have checked my measurements against those by previous authors such as Wood (1927, 1964) and Tanner (1969, 1975, 1977; Tanner and Martin, 1972), and in most cases, my measurements match theirs within a millimeter or two. In the case of premolar and molar measurements, the landmarks are easy to recognize, since the teeth are nearly rectangular (especially after wear), and the maximum length and width is easy to measure on most specimens. Some variability is introduced on highly worn specimens, because the width of the cheek teeth tends to increase when the tooth is worn down nearly to the base. To a lesser extent, there can be variability of the anteroposterior length of teeth when interstitial or interdental wear takes place. In most cases, if the specimen was extremely worn so that the tooth measurements might be unreliable, or the teeth were damaged or visibly distorted, the measurements were not taken, or notations were made about their reliability.

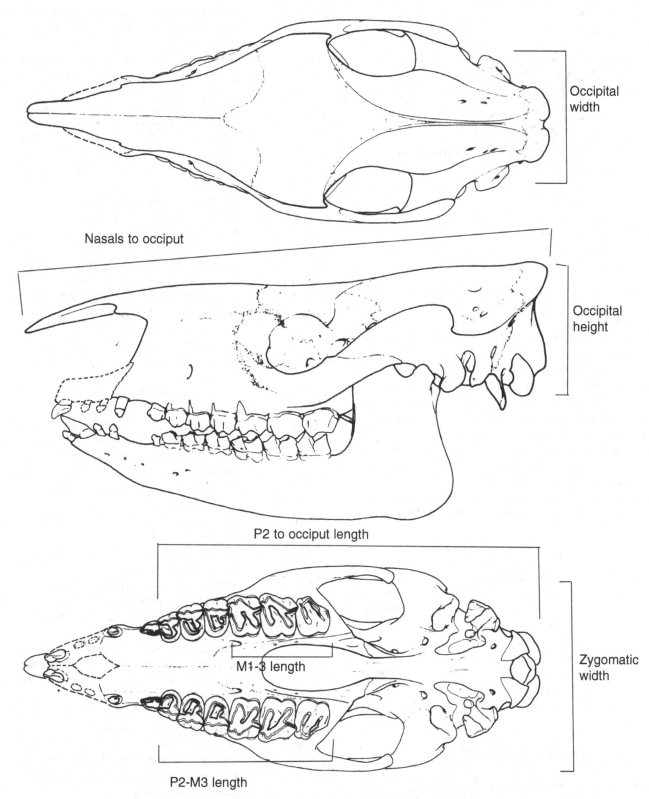

Figure 2.1. Standard measurements of rhinoceros skulls and teeth used in this study. Illustration of *Teletaceras radinskyi* after Hanson (1989). Measurements on the crowns of each tooth are taken at the maximum antero-posterior diameter along the middle of the tooth, and the maximum transverse width measured from the lingual base to the labial base of the tooth. Similar measurements were taken for the lower teeth as well.

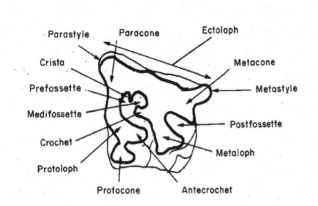

Figure 2.2. Terminology of a typical rhinoceros upper molar (after Prothero *et al.*, 1986).

Anatomical terminology of the skull and skeleton follow Sisson and Grossman (1975) and Scott (1941), and is discussed further in Chapter 5. Dental terminology follows Osborn (1898c, 1904) and is shown in Figure 2.2. Instead of the cumbersome system of superscripts and subscripts for upper and lower teeth, I follow the computer-friendly system of Jepsen (1966) where upper-case letters indicate upper teeth (I1–3 C P1–4 M1–3 for upper incisors, canines, premolars, and molars) and lower-case letters indicate lower teeth (i1–3 c p1–4 m1–3). Biostratigraphic correlations follow the chapters in the Woodburne (2004) volume, as well as other papers cited in the appropriate places. The time scale for the Cenozoic follows Berggren *et al.* (1995) as applied to the North American land mammal "ages" by the chapters in the Janis *et al.* (1998) volume. Other abbreviations are as follows:

dP or dp	deciduous premolars
ht.	height (in tables)
l.f.	local fauna, a stratigraphically and geographic restricted vertebrate assemblage
Ma	millions of years before present
Mc	metacarpal
Mt	metatarsal
m.y.	millions of years as a duration of time
N	sample size
SD	standard deviation

INSTITUTIONAL ABBREVIATIONS

AMNH	Department of Paleontology, American Museum of Natural History, New York
BVM	Buena Vista Museum, Bakersfield, California
CM	Carnegie Museum of Natural History, Pittsburgh, Pennsylvania
DMNH	Denver Museum of Natural History (now Colorado Museum of Nature and Science), Dener, Colorado
F:AM	Frick Collection, Department of Vertebrate Paleontology, American Museum of Natural History, New York
FLMNH	Florida Museum of Natural History, Gainesville, Florida
FMNH	Field Museum of Natural History, Chicago, Illinois
JODA	John Day Fossil Beds National Monument, Oregon
KU	Museum of Natural History, University of Kansas, Lawrence, Kansas
LACM	Natural History Museum of Los Angeles County, Los Angeles, California
LACM (CIT)	California Institute of Technology collection (now housed at the LACM)
LSUMG	Louisiana State University Museum of Geology, Baton Rouge, Louisiana
MCZ	Museum of Comparative Zoology, Harvard University, Cambridge, Massachusetts
MGS	Mississippi Geological Survey, Jackson, Mississippi
MSU	Midwestern State University, Wichita Falls, Texas
OMNH	Sam Noble Oklahoma Museum of Natural History, Norman, Oklahoma
ROM	Royal Ontario Museum, Toronto, Ontario
SDSM	Museum of Geology, South Dakota School of Mines and Technology, Rapid City, South Dakota
SMNH	Saskatchewan Museum of Natural History, Regina, Saskatchewan
TMM	Texas Memorial Museum, University of Texas, Austin, Texas
UCMP	University of California Museum of Paleontology, Berkeley, California
UCR	University of California, Riverside, collection (now housed at UCMP)
UF	Florida Museum of Natural History, University of Florida, Gainesville, Florida
UNSM	University of Nebraska State Museum, Lincoln, Nebraska
UO	University of Oregon Condon Museum of Geology, Eugene, Oregon
USNM	National Museum of Natural History, Smithsonian Institution, Washington, D.C.
UW	University of Washington Burke Museum, Seattle, Washington
YPM	Yale Peabody Museum, New Haven, Connecticut
YPM-PU	Princeton University collection (now housed in the collections of YPM)

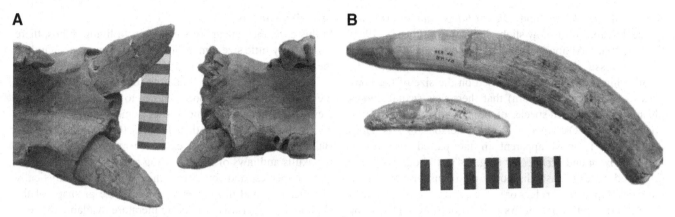

Figure 2.3. Comparison of mandibles with male and female i2 tusks. A. Male (left) and female (right) *Teleoceras fossiger* (AMNH 8391 on left, AMNH 2606 on right). B. Contrast between male and female i2 tusks of *Peraceras superciliosum* from Hottell Ranch, Nebraska (unnumbered UNSM specimens). In general, male i2 tusks are longer, larger, and have no separation between the root and the crown, and are worn much of the length of the crown. Female tusks are shorter, blunter, with a distinct base on the crown separating it from the the root. Scale bar in cm.

A NOTE ON SPELLINGS

According to the *International Code of Zoological Nomenclature* (fourth edition, 1999), adjectival species names must agree with the gender of the genus. Unfortunately, paleontologists are not always well trained in Latin, and have a long tradition of making mistakes in gender of Latinized names, and applying the wrong endings. Generic suffixes such as *-ceros* ("horn" in Greek), *-odus* and *-odon* ("tooth" in Greek), and *-opus* ("foot" in Greek) are masculine, and require a masculine adjectival ending. Generic suffixes such as *-ceras* (another Greek word for "horn") and *-therium* ("beast" in Greek) are neuter, and require neuter adjectival endings in the species name. Thus, the common misspellings such as *Peraceras* "*superciliosus*" and "*profectus*" should be *superciliosum* and *profectum*; *Teleoceras* "*medicornutus*" should be *medicornutum*; *Diceratherium* "*niobrarensis*" should be *niobrarense*; *Brachypotherium* "*americanus*" should be *americanum*, and so on. When species names are transferred to genera of different genders, the adjectival endings must also change, so "*Aceratherium mite*" (neuter) becomes *Subhyracodon mitis* (masculine), and "*Aceratherium platycephalum*" (neuter) become *Amphicaenopus platycephalus* (masculine). Likewise, paleontologists persist in misspellings like "rhinoceratids" and "rhinoceratoids," even in recent publications. However, the stem is the name *Rhinoceros*, so when the suffix is changed, the root is *rhinocero-* (not *rhinocera-*), and the proper spelling is "rhinocerotids" and "rhinocerotoids."

VARIATION DUE TO SEXUAL DIMORPHISM

A persistent question when assessing large sample sizes of rhinos is how much that size variation could be due to sexual dimorphism, and how much is too great to be explained as a population sample from a single species, and indicates differences in species. Fortunately, the sex of many rhino fossils is easy to assess, and there are large samples of both living and extinct rhinoceroses that allow us to assess this problem.

Tusk dimorphism

Osborn (1898c, 1904) first pointed out that the i2 tusks of rhinoceroses are highly dimorphic. In most rhinos, male tusks (Fig. 2.3) are typically long and curved, with a continuous taper of the tooth crown. Female tusks have a distinct base around the crown, and the point of the tusk tends to be shorer and blunter. This pattern is well established for the three living rhinoceros species have lower i2 tusks (*Rhinoceros unicornis, R. sondaicus, Dicerorhinus sumatrensis*), and was confirmed by Voorhies and Stover (1978) when they found probable fetal bones in the pelvic regions of presumed female skeletons of *Teleoceras major* from Ashfall Fossil Bed State Park, Nebraska. Mead (2000) and Mihlbachler (2004) have further documented the size and growth patterns of rhino tusks. For large samples that include lower jaws, this tusk morphology allows for the assessment of sexual dimorphism. Tusks are very important for survival in some living rhinos. For example, Dinerstein (2003) reports that the Indian rhinoceros, *Rhinoceros unicornis*, uses its lower tusks, rather than its horn, as its primary weapon in both intraspecific and interspecific combat.

Horn dimorphism

The second potentially sexually dimorphic character is the horn. Living rhinoceroses show variations in horn dimorphism. *Diceros bicornis* shows no dimorphism (Berger, 1994), while those of *Ceratotherium simum* and *Rhinoceros unicornis* show slight dimorphism (Rachlow and Berger, 1997; Dinerstein, 1991). Data for the rare cryptic forest species *R. sondaicus* and *Dicerorhinus sumatrensis* are sparse, but appear to show slight dimorphism (Pocock, 1945; Groves, 1982). Although the keratinous horn is not preserved in most fossil rhinos (except *Coelodonta antiquitata*, the woolly rhino), it seems clear (based on the size of the bony boss that supported the horn) that there was sexual dimorphism in some horned species of extinct rhinoceroses, as first documented by Osborn (1904) and Peterson (1920). Horn dimorphism is most apparent in the paired-horn rhinos *Diceratherium* and *Menoceras* (Fig. 2.4). In both cases, the presumed males (as established by the i2 tusks) have larger, more developed nasal ridges or bosses, and those of presumed females are faint or absent. As pointed out in Chapter 4, the failure to recognize this dimorphism led to the creation of a number of invalid species that were simply based on hornless females of *Diceratherium* or *Menoceras*. Some dimorphism of the horns may also occur in *Peraceras superciliosum*, where presumed males (based on their i2 tusks) have broad, blunt robust nasals with a highly rugose tip (Fig. 4.32B–D), while those of females are smooth and slender. No dimorphism has been observed in the small horn bosses of *Teleoceras*, however.

Size dimorphism

In this case, the data are less clear-cut. In living rhinos, there is relatively little size dimorphism for most species, including the black rhino (Freeman and King, 1969), the Indian rhino (Dinerstein, 2003), or the Sumatran rhino (Dinerstein, 2003); the Javan rhino is too poorly known to establish its tendencies. Only the white rhino shows some dimorphism, with males about 20% larger than females (Owen-Smith, 1988). Bales (1995) conducted an extensive multivariate study on the skulls and jaws of all five living species, and most of the extinct species, and found that they showed no statistically significant sexual dimorphism in either size or shape of the skull and jaws. Based on all the literature available to them, Prothero and Sereno (1982, p. 16) and Prothero and Manning (1987) argued that most rhinos do not show significant size dimorphism, and so ruled out sexual size differences to account for the dwarf species of rhino from the Texas Gulf Coastal Plain.

Mead (2000) analyzed the large sample of complete articulated skeletons of *Teleoceras major* from Ashfall Fossil Bed State Park, and argued that they do show significant sexual dimorphism. According to Mead (2000), males tend to be significantly larger not only in cranial dimensions, but also in

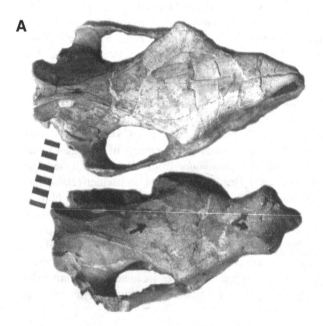

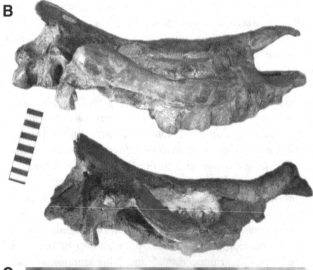

Figure 2.4. Horn dimorphism in *Menoceras barbouri* from Bridgeport Quarry, Nebraska, in dorsal (A), right lateral (B), and anterior (C) views. The darker-colored skull (UNSM 62141) with the horns is a presumed male; the skull with the smooth nasal bones (UNSM 62131) is a presumed female. Note that in this case, the male appears slightly smaller than the female, yet both are fully adult specimens. Scale bar in cm.

Table 2.1. Statistics of samples of presumed male and female skulls and jaws (based on horns and lower incisor tusks) of *Menoceras arikarense* from Agate Springs Quarry. In the upper teeth, the males average slightly larger than females, but in the lower teeth, they are smaller, although none of these differences is statistically significant (P value at 95% confidence level for significance of difference, using a t-test).

	p2-4	m1-3	P2-4	M1-3
Males				
Mean	66.5	96.4	65.0	91.3
Standard Deviation	1.3	2.4	4.5	4.2
Females				
Mean	63.5	93.5	67.2	93.2
Standard Deviation	0.7	5.6	4.3	4.4
P	0.006	0.369	0.725	0.557

limb bone dimensions by 20-29%. He also contended that there are differences in the skulls as well, with males having deeper jaws and more massive mandibular-angular regions. However, this shape difference was not substantiated in the multivariate analysis conducted by Bales (1995) for some of the same taxa. In addition, Mihlbachler (2004) found no consistent evidence of size dimophism in many different quarry samples of both *Teleoceras* and *Aphelops*, and concluded that the Ashfall sample was exceptional among rhinos for its apparent size dimorphism.

To reassess this problem, I measured the same standard variables in large quarry samples of male and females of a number of species, and plotted the data as bivariate plots. Large quarry samples exist for *Diceratherium armatum* and *D. annectens* from the Frick 77 Hill Quarry, Niobrara County, Wyoming. Because males and females of both species are found in the same quarry, we can rule out the possibility that the smaller *D. annectens* is a female of *D. armatum*; there are large female skulls of *D. armatum*, and small male skulls of *D. annectens*, in this and several other quarry samples. Males and females are also abundant in the samples of *Menoceras arikarense* from Agate Springs Quarry, Sioux County, Nebraska, and in several large quarry samples of *M. barbouri*.

Representative data are shown in Table 2.1. In general, there is no consistent trend that can be statistically supported. Measurements of male and female lower jaws (as established by i2 tusks) of *M. arikarense* show that the males tend to be larger than females (but by only 10% at the greatest, and there is overlap), but the male and female skulls (as established by horn bosses) and upper teeth from the same quarry sample show the *opposite* trend—males tend to be smaller than females! Most other plots of quarry samples of *Diceratherium* and *Menoceras* showed no consistent trend. Males and females tended to overlap greatly in body size, or if females were smaller, they were no more than 10% smaller. Mihlbachler (in press) has measured a much larger sample of *Menoceras arikarense* from Agate Springs, and came to the

same conclusion: there was no significant dimorphism in size as measured by teeth, and the females actually did have slightly larger upper cheek teeth. The same could be said for other species that have been sexed. Prothero and Manning (1987, fig. 15, here shown in Fig. 2.5) found no consistent sexual size separation of *Teleoceras medicornutum* from the Barstovian Frick Horse and Mastodon Quarry, Pawnee Creek Formation, Colorado. More germane to the point of this plot, the specimens identified as *T. meridianum* are significantly smaller than even the smallest known females of *T. medicornutum*, so they are distinct dwarfed species, and not sexual dimorphs.

In summary, although as much as 20% size difference in some species due to sexual dimorphism can be documented in a few cases, in most living and fossil rhinoceros species there is either no statistically significant dimorphism, or the females tend to be less than 10% smaller than males. Thus, when assessing the range of size variation of a population to determine whether one or more species are present, these are the guidelines that I will follow.

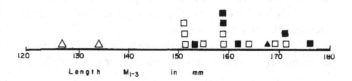

Figure 2.5. Comparison of m1-3 lengths of late Barstovian *Teleoceras*. Solid squares = male *T. medicornutum*, Frick Horse and Mastodon Quarry; open squares = females from same quarry. Solid triangle = type specimen of *T. medicornutum*. Open triangles = *T. meridianum* (after Prothero and Manning, 1987: fig. 15).

VARIATION IN TOOTH CROWN PATTERNS

Molarization of upper premolars

In addition to intrapopulation variation due to sexual dimorphism, another significant source of variation occurs in the cusp and crest patterns of rhinoceros teeth. Traditionally, paleontologists have treated the crown patterns of most mammalian teeth as infallible guides to species-level distinctions, and rarely considered the possibility of variation of tooth crown patterns within a single population. As reviewed in Chapter 3, the paleontologists of the late nineteenth and early twentieth centuries were typological "splitters," who recognized new taxa based on every slight variation in teeth. Consequently, the taxonomy of North American fossil rhinos was grossly oversplit based on the slight difference of the crown pattern of each new specimen.

By the mid-twentieth century, however, paleontologists such as Matthew and Simpson began to think of fossils as parts of living populations, rather than objects that get new names each time they appeared slightly different. Such "population" thinking (and the statistical approach to fossil samples, also introduced to vertebrate paleontology by Simpson and his wife Anne Roe) has now become customary in paleontology, yet because of the slow pace of revision of many groups, the literature is full of invalid taxa created by splitters from over a century ago.

This problem was highlighted in North American rhinos with the discovery and publication of the *Trigonias* Quarry sample from the Chadronian Horsetail Creek Formation in Weld County, Colorado (Gregory and Cook, 1928). In this sample were over a dozen skulls with highly variable crests and cusps on the upper premolars, yet the entire sample was very homogeneous in size and all other features (Fig. 2.6). Gregory and Cook (1928, p. 4) recognized that this suggested that all the specimens were members of a single population, but "for the sake of convenience in describing and cataloguing the material we nevertheless designate the various groups or individuals as variants or 'species' realizing full well that these terms in this instance, and perhaps in many others, merely signify a definable set of characters in certain individuals." Consequently, they recognized six species of *Trigonias* (four of them new) and labeled yet another specimen *?Caenopus premitis* because its upper premolars were more advanced than the other *Trigonias*.

Matthew (1931, 1932), on the other hand, applied essentially modern population concepts to these samples and argued effectively that they could not represent more than one species. His extended discussion (Matthew, 1931, pp. 5–6) of the criteria for species distinctions reads as if it were written very recently. Wood (1931), however, was caught in the middle. He had created many different species within the primitive rhinocerotoids in his 1927 paper, so he was unwilling to discount the importance of minor cusp variations in recognizing species. Still, he reduced Gregory and Cook's (1928) seven species to only three, but then erected another species of *Trigonias*, *T. cooki*, based on minor variations of upper premolars. As detailed in Chapter 3, the same problem occurred with the large, highly variable quarry sample of *Menoceras arikarense* from Agate Springs Quarry when Troxell and Cook split off a number of invalid species, or the large quarry sample of *Teleoceras hicksi* from Wray, Colorado, when Cook and Lane erected multiple species for a single homogeneous population sample (see discussion of each of these species in Chapter 4). The problem tends to be most severe in the more primitive taxa whose upper premolars are not completely molarized (e.g., *Hyracodon, Subhyracodon, Trigonias*), but it also occurs to a lesser degree in taxa (such as *Diceratherium, Menoceras* or *Teleoceras*) with completely molarized upper premolars.

So how do we assess whether variations in cusp morphology are worthy of species recognition? Ideally, a large sample from a single population is required, which is known in only a few instances (e.g., the Ashfall Fossil Bed State Park *Teleoceras major*, which is an instantaneous death assemblage). The next best substitute for a single population is a large quarry sample, which presumably represents individuals from a limited geographic range and span in time (years or at most decades) and approaches a population sample. Such large quarry samples are available for many rhino species, as detailed in Chapter 4. For many oversplit species, we have quarry samples that clearly demonstrate the upper premolar variability in *Trigonias osborni* (the Colorado sample discussed above) and *Subhyracodon occidentalis* (the Harvard Fossil Reserve sample from Goshen County, Wyoming). Lacking a single quarry sample, the best approach is to examine all the specimens from a single restricted stratigraphic level and geographic area (e.g., Prothero, 1996, with *Hyracodon* from the lower Scenic Member of the western Big Badlands). Although there is undoubtedly some time averaging involved, such samples are the best proxy we have for a contemporary population in many instances.

For the variation in molarizing upper premolars, Prothero

Figure 2.6 (opposite page). Variability of upper premolar crests of the *Trigonias* Quarry sample. A. DMNH 884, *T. osborni* "var. secundus." B. DMNH 897, *T. osborni* "var. figginsi." C. DMNH 1025, "*Caenopus premitis*." D. DMNH 881, *T. osborni* "var. figginsi." E. DMNH 1029, *T. "taylori."* F. DMNH 886, *T. "hypostylus."* G. DMNH 414, *T. "precopei."* H. DMNH 878, *T. "preoccidentalis."* (After Gregory and Cook, 1928, Plates III–VI). All of these specimens are about the same size and morphology, and differ only in the details of premolar crests, suggesting that they are a single species, *T. osborni*, and none of these species are valid.

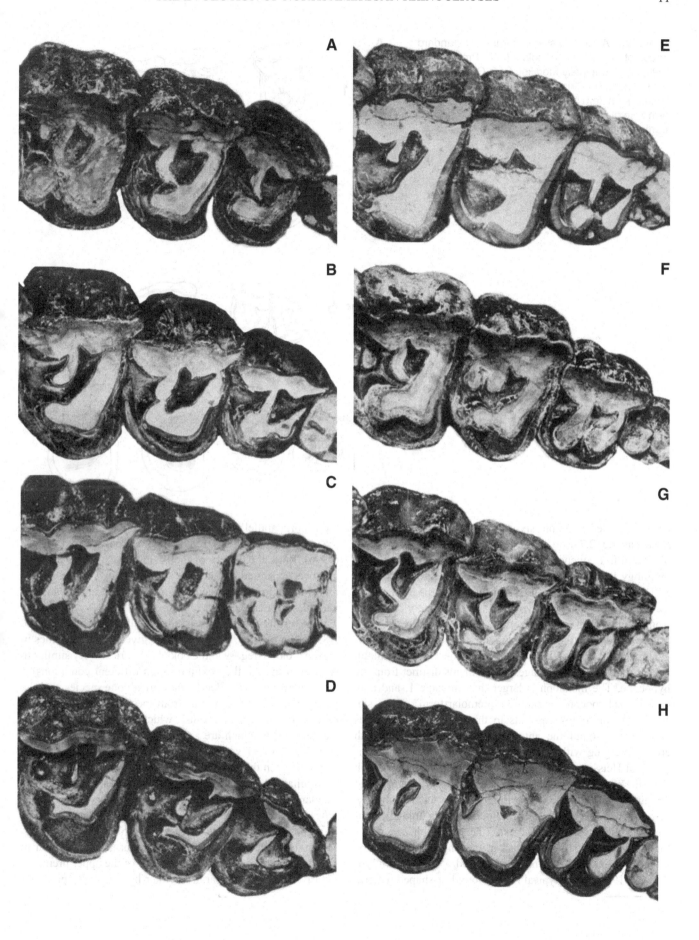

Figure 2.7. A. Coding system for the six standard stages of molarizing premolars (after Prothero, 1996: fig. 2, originally modified from Gregory and Cook, 1928). B. Heissig's (1999) sketches of (from right to left) "premolariform," "submolariform," "semimolariform," and "molariform" upper premolar crests (after Heissig, 1999).

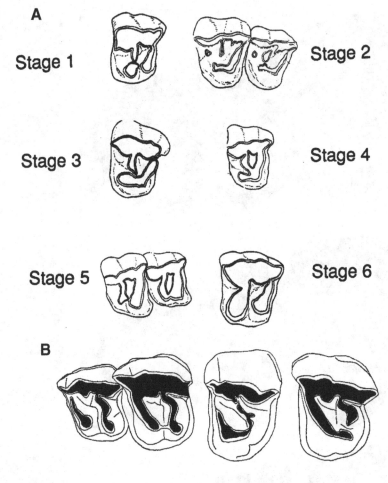

(1996) adopted a simple six-stage descriptive system (here shown as Fig. 2.7A) to codify a large number of samples and plot them in frequency distributions. A slightly different terminology was coined by Heissig (1999; also shown in Fig. 2.7B). In the terminology used here, stage 1 is the least molarized. The protocone and hypocone are completely separated from the ectoloph, and merge with it only in the final wear stages. In stage 2, the protoloph curves around the anterolingual margin of the tooth, but does not connect to the ectoloph except in the final wear stages, and remains distinct from the hypocone. The metaloph is larger than in stage 1, and may contact the hypocone. In stage 3 ("premolariform" of Heissig, 1999), the protoloph connects to the ectoloph, and curves completely around the anterolingual margin of the tooth crown, merging with the hypocone. In stage 4 ("submolariform" of Heissig, 1999), the protoloph contacts the metaloph from the early wear stages, but still has a posterolingual crest remaining in the hypocone position. In stage 5 ("semimolariform" of Heissig, 1999), the protoloph loses its lingual extension, forming two parallel crests (the protoloph and metaloph), which are still in contact at the lingual end. Finally, stage 6 premolars are fully molarized ("molariform" of Heissig, 1999) in the typical rhinocerotoid π-shaped pattern,

with short parallel protoloph and metaloph, and no closure of the valley between them. Following the procedure in Prothero (1996), I plot the pattern of the P2/P3/P4 as a series of three numbers separated by slashes, and this allows the total premolar condition of a specimen to be plotted against another variable, such as size.

The results of coding these specimens are shown in Figure 2.8. As can be seen, the eight skulls given different names in Figure 2.6 by Gregory and Cook (1928) are all very similar in size, even though they exhibit seven different conditions of the upper premolars. Clearly, they are all one species, and the range of premolar conditions (from specimens like "*Trigonias osborni figginsi*," DMNH 881, which is coded 1/2/3, to several specimens which are as advanced as 6/4/4) must represent the range of variation within a population, and differences between these teeth have no systematic significance.

Similar trends are seen in other rhinos as well (Fig. 2.8). The large collection of early Orellan *Subhyracodon occidentalis* from the Lower Nodular Zone of the western Big Badlands (Cottonwood Pass–Big Corral Draw–Sheep Mountain Table area) forms a single normal distribution in size, with eight different combinations of the upper premlars. Clearly, there is no separation either by size or by premolar state, so they cannot be separated into more than one species objectively. This argument is strengthened by examination of the *Subhyracodon* collection from a single quarry, Harvard Fossil Reserve, near Torrington, Goshen County, Wyoming (Schlaikjer, 1936). Of the eight specimens with well-preserved upper premolars, seven combinations are observed, yet this collection again forms a single unimodal distribution in size. Together, these plots show that only a single species, *Subhyracodon occidentalis*, can be recognized for these latest Chadronian–Orellan rhinos, and the many synonyms based on premolar variation (listed in Chapter 4) cannot be substantiated as valid biological species under modern concepts.

This argument also pertains to the large sample of *Diceratherium tridactylum* from the Poleslide Member of the Big Badlands of South Dakota (Fig. 2.8). The 13 skulls, maxillae, and palates with good upper dentitions are slightly larger on average, but they show at least two to three premolar combinations (mostly 6/5/4 and 6/5/5), which overlaps in size and morphology with many specimens of *S. occidentalis* in the same plot. (The species are defined on other criteria, as outlined in Chapter 4.) Once again, the many invalid synonyms of *D. tridactylum* (listed in Chapter 4) cannot be substantiated as meaningful biological species in light of this evidence.

Figure 2.8. Size-frequency distributons of presumed population samples of rhinos, with the P2/P3/P4 of each specimen coded as in Figure 2.7. The type specimens of several *Subhyracodon* species are indicated.

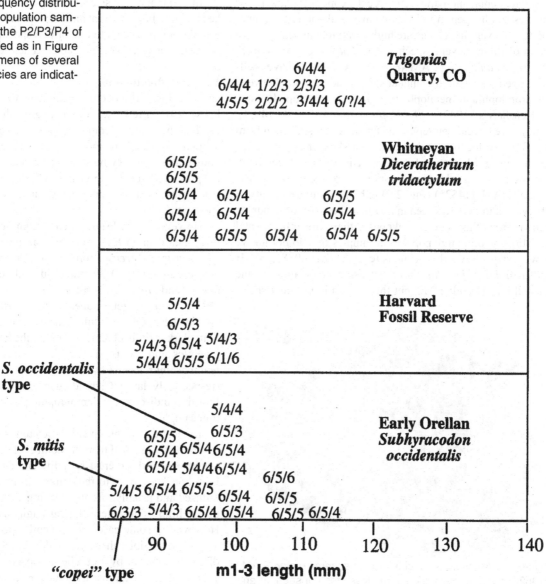

Dental anomalies

Conclusive proof of this argument can be shown by specimens that have different degrees of molarization on the same dentition. For example, F:AM 111871 has a completely molarized (stage 6) P4 on the right side, but a closure between the metaloph and protoloph (stage 5) on the P4 of the left side. If the two maxillae had been separated and described by early paleontologists, the right maxilla would have been called *S. occidentalis*, but the left maxilla would have been called *S. "trigonodus"*! A similar pattern occurs on a specimen (AMNH 32655) of *D. tridactylum*, where the left P3 is stage 5, and the right P3 is stage 6. Clearly, dental anomalies such as these show that the development of the crests on molarizing premolars is not well constrained biologically, and varies from individual to individual in a population, and even on different sides of the same individual in some cases.

Other dental anomalies reinforce the conclusion that peculiarities can happen in individuals in just about any mammal tooth. For example, cristae are highly variable in nearly every taxon of rhino. Several specimens of *Subhyracodon "copei"* or *"trigonodum"* (e.g., AMNH 524, MCZ 2918) have well-developed cristae on the lingual side of the ectoloph between the protoloph and metaloph. The bizarrely developed cristae of Henshaw's (1942) *"Aphelops cristalatus"* (LACM-CIT 2806) have never been replicated on any other specimen from the Tonopah locality, including a nice skull and jaws in the Buena Vista Museum of Natural History in Bakersfield, California, collected by later workers. A palate of *Teleoceras major* (F:AM 114525) (Fig. 2.9) exhibits an unusual pattern of cristae and fossettes seen in no other specimen of hundreds known from this species. Likewise, there are other dental variations within large population samples that do not justify new species. Several specimens (e.g., AMNH 8088, AMNH 1120) of *D. tridactylum* have extensive crenulation of the enamel on M1 and/or M2, but this is seen in no other speci-

mens in this large population sample. A few specimens of *S. occidentalis* or *D. tridactylum* have antecrochets as well, but this character does not consistently appear until species known from the Miocene. Prothero and Manning (1987, fig. 11) illustrated a peculiar skull of *Teleoceras medicornutum* (F:AM 108838) which had an isolated cusp at the posterolingual end of the metalophs of both M3s. Because no other specimen of the large collections of *T. medicornutum* (both from the Trinity River Pit 1, which produced F:AM 108838, and the large samples from elsewhere) shows this feature, it is clearly an individual dental anomaly, not a feature worthy of systematic recognition. There is quite a literature on such dental anomalies (Archer, 1975; Rose and Smith, 1979; Taylor, 1982), and they serve as an object lesson to those typological splitters who would name new taxa on every variation in the tooth crown.

In summary, unless a distinctive tooth crown pattern can be shown to be consistent and widespread within a population sample (not restricted to one or two specimens), it should not be used as a systematic character to define species.

Variation due to ontogeny

Juvenile fossil rhinoceros specimens are very rare and only known from a few taxa. In most cases, they are less robustly ossified than adult specimens, so they break up easily and are seldom fossilized. In most taxa, no juvenile specimens are known at all, so they have never been described. Where they are preserved, most authors have chosen not to measure them, since their measurements are difficult to compare to adult specimens.

However, juvenile specimens are known from a number of taxa. Peterson (1920) reported on juvenile specimens of *Menoceras arikarense* from Agate Springs Quarry. Prothero and Manning (1987) illustrated juvenile dP2-4 of *Peraceras hessei* and *Teleoceras meridianum*. Voorhies and Stover (1978) described fetal bones of *Teleoceras major* apparently still in the uterus of adult females from Ashfall Fossil Bed State Park. Mihlbachler (2003) used the large samples of both juveniles and adults of *Teleoceras proterum* and *Aphelops malacorhinus* from Love bone bed and Mixson's bone bed (respectively latest Clarendonian and early Hemphillian of Florida) to discuss the demographic profiles of these population samples.

In most cases, the juvenile skeleton is simply smaller and less robustly ossified than the adult skeleton, so this has predicated very little interest in individual descriptions. Juvenile limb bones are easy to recognize, because the epiphyses are unfused and usually missing, leaving a rugose area at the end of the limb bone where the cartilaginous cap once existed. However, in some cases, the juvenile specimens are distinct. The most obvious instance of this is the peculiarity of the dP2-4. As shown by a number of taxa (Fig. 2.10), the dP1 is a peculiar tooth, almost triangular in crown view, with long metaloph and protoloph that curve posteromedially and ter-

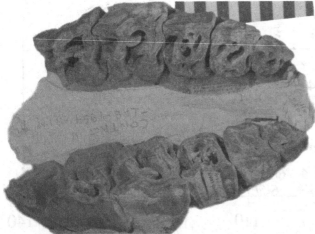

Figure 2.9. Anomalous specimen of *Teleoceras major* (F:AM 114525) showing extra cristae and fossettes not seen on any other individual known in the collections.

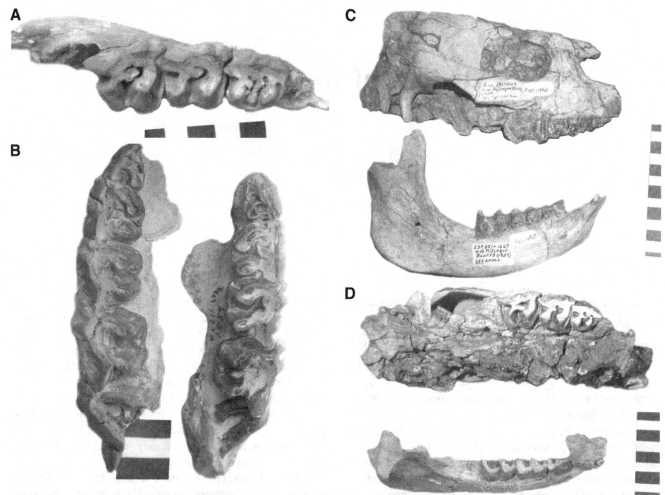

Figure 2.10. Deciduous premolars of representative rhinocerotids, showing the peculiar patterns of the upper dP2-4. A. *Menoceras arikarense*, AMNH 86121, dP2-4, showing a three-lobed dP2 and the molariform dP3 and dP4. B. *Diceratherium annectens*, AMNH 7342, dP1-4-M1, showing a different configuration of the three-lobed dP2. C-D. *Peraceras profectum*, F:AM 114968, crushed juvenile skull and right ramus. The dP1 shows the typical triangular form, with the posteriorly deflected protoloph and metaloph. The dP2 in this taxon is not three-lobed, but more molariform. The dP3 is completely molariform. The deciduous lowers are similar in form to the permanent teeth. Scale bar in cm.

minate in broad tips where the metacone and protocone were expanded. These often have cristae and medifossettes as well. The dP2 is the most peculiar tooth of all. In most taxa, it has a third crest (the "mesoloph" of Antoine, 2002) located between the protoloph and metaloph that makes it look unlike any other rhino tooth. Before this tooth was recognized as a dP2, it was used as the basis for a number of invalid taxa (e.g., "*Hyracodon quadriplicatus*" of Cope, 1873a). And the dP3 and dP4 are also deceptive as well. Typically, they are much more highly molarized (usually stage 6) than the P4 or even the M1, so they can be easily mistaken for an adult M1, which is usually the most worn tooth on a permanent dentition (because it erupts first). However, the best way to avoid this mistake is to examine the color of the enamel. As has been widely observed in many fossil mammals, deciduous enamel tends to be lighter in color than adult enamel (probably because it is thinner), so deciduous teeth can usually be separated from adult teeth on color alone.

The peculiarities of deciduous teeth are particularly apparent on the first deciduous premolar. Wood (1934, p. 263) suggested that the dP1/dp1 of most rhinocerotoids are never replaced, but retained until adulthood (or lost without a replacement tooth). Butler (1952) reviewed this situation, and found that the retention of dP1/dp1 was widespread among perissodactyls. In the Rhinocerotidae, this is also the prevailing condition. Virtually all the teeth in the P1/p1 locus have the crown morphology and light-colored, thinner enamel that suggests they are retained dP1/dp1. However, Tanner and Martin (1972) found that in *Hyracodon*, the lower first cheek tooth is a retained dp1 but the upper dP1 is replaced by a permanent P1. Apparently, the replacement of the upper dP1 is found in tapirs as well, but not in most other perissodactyls.

3. History of Investigations

"Among the most interesting and unexpected paleontological discoveries in America is that of the former existence of the Rhinoceros family upon this continent." (Leidy, 1869)

On December 10, 1850, Joseph Leidy reported the first evidence that rhinoceroses had once lived in North America. At this point in his life, Leidy was 27 and beginning his career as an anatomist and paleontologist (Warren, 1998). He had already described the first North American dinosaurs and a number of other fossils. Geologists from the great government surveys of the west routinely sent him their fossils as the only man in North America qualified to study them. This particular specimen was the fragment of an upper jaw "from Missouri territory" (probably the Big Badlands of South Dakota) collected by Alexander Culbertson of the American Fur Company and Captain Stewart van Vliet. It was given originally to Spencer Fullerton Baird of the Smithsonian Institution, then sent to Leidy to describe. With Vice-President Bridges in the chair of the Academy of Natural Sciences that cold December evening, Leidy got up and briefly displayed his specimen, and gave it the name *Rhinoceros occidentalis*, or "western rhinoceros." (It is now referred to *Subhyracodon*—see Chapter 4). Versatile scientist that he was, he then continued speaking on another topic, his research on the stinging organs of the *Hydra*, a cnidarian related to jellyfish and sea anemones. The published proceedings of this meeting give almost no description of *Rhinoceros occidentalis*, but the name was established nonetheless. Table 3.1 gives a chronological list of all the North American fossil rhino taxa proposed since 1850 to help the reader keep track of this long and confusing story.

Even though rhinos were unknown in North America at the time, Leidy was enough of a comparative anatomist to recognize the distinctive "pi" (π)-shaped crown patterns of the upper molars that are unique to rhinos. He had previously (1847) described the first North American camel, *Poebrotherium wilsoni*. At the December 17, 1850, meeting of the Academy, a week after he described the first North American rhino, he described the first known North American fossil horse, *"Palaeotherium" bairdi* (now placed in *Mesohippus*), along with the first specimens of *Hyracodon*, which he called *Rhinoceros nebraskensis*. Although horses, camels, and rhinos are currently not native to the North American landscape, by 1850 Leidy had evidence that they had all once lived on this continent.

In 1851, Leidy (1851a) gave a slightly more detailed description of *Rhinoceros occidentalis* with some measurements, but still no illustrations of the specimens. Later in 1851, Leidy (1851b) placed both *Rhinoceros occidentalis* and *Rhinoceros nebrascensis* (the species name of the latter misspelled for years with a "c" rather than the "k" that he had originally used) in the European rhino genus *Aceratherium*, because their teeth had cingula unlike those of modern *Rhinoceros*. For many more years, *Aceratherium* would continue to be used as a "wastebasket" genus for hornless fossil rhinos that were not referable to modern *Rhinoceros*.

Finally, in 1852 Leidy gave a full description, discussion, and illustrations of *Rhinoceros occidentalis,* transferring it back from *Aceratherium* (Leidy, 1852a). He illustrated a skull (now USNM 114) collected by David Dale Owen for the Owen Survey of the Northwest Territories as the cotype. This specimen was made the lectotype by Osborn (1898c), because the fragments Leidy originally (1850b) named as types were never illustrated and are now apparently lost. In the same work, Leidy (1852a) also gave a full description and illustration of *Rhinoceros nebrascensis*. Later that year, Leidy (1852b) described *Rhinoceros americanus*, a tooth and partial tooth from "Nebraska territory," and collected by Culbertson (probably from the Big Badlands of South Dakota again). Osborn (1929, fig. 162) figured the specimen, and realized that it is actually a tooth of a brontothere of indeterminate species identity; it is now in the Smithsonian (USNM 167594).

Twice in 1853 Leidy again discussed *Rhinoceros occidentalis* and *Rhinoceros nebrascensis*. One of these descriptions was in his major monograph, *The Ancient Fauna of Nebraska* (Leidy, 1854). Then in 1856, Leidy transferred *Rhinoceros nebrascensis* to his new genus *Hyracodon*, without any generic description, other than that it possessed "a greater number of teeth than any other known member of the *Rhinoceros* family." This refers to the fact that hyracodonts still have the full complement of incisors and canines that rhinocerotids have lost.

Until this point, all of the fossil rhinos known in North America came from the Oligocene beds of the Big Badlands of South Dakota. In 1858, Leidy described two mandibular fragments with deciduous teeth as *Rhinoceros crassus*. They were collected in the Niobrara River Valley, Nebraska, by Ferdinand Hayden on the Warren Expedition. Although the precise origin of these specimens cannot be determined, they are believed to be from the late Barstovian (late middle Miocene) Valentine Formation (Skinner and Johnson, 1984).

As the Civil War was raging, Leidy remained in Philadelphia, and continued teaching anatomy. Although Leidy's three brothers went to war, professors were exempt from the draft, so Leidy served as a pathologist, studying the effects of war wounds and diseases common in the army camps (Warren, 1998). However, he continued to do his paleontological research as well. The California collections of J.D. Whitney came into Leidy's possession in 1865. From these collections, he described some rhinoceros remains found in the "auriferous gravels" and "discovered in association with human and equine remains in Calaveras County, California." This locality, called Chili Gulch (Whitney, 1865), became famous for the controversial "Pliocene" human skull that was immortalized in a Bret Harte poem. Unfortunately, the human skull and horse proved to be a later contamination, but the rhino material is genuine (Wood, 1960). To these rhino specimens Leidy gave the name *Rhinoceros hesperius* (which translates as "western rhinoceros"). Although the type and all the referred material ended up at Harvard (the type specimen is MCZ 9118), the taxon is probably a junior synonym of *Subhyracodon occidentalis* or *Diceratherium tridactylum*. In this same paper, Leidy described an isolated tooth fragment collected by Dr. Benjamin F. Shumard from the Miocene of Washington County, Texas, as *Rhinoceros meridianus*. This specimen was transferred to other genera, such as *Aphelops*, before Prothero and Manning (1987) showed that it is a dwarfed species of *Teleoceras*. In his 1869 monograph, *The Extinct Mammalian Fauna of Dakota and Nebraska*, Leidy reviewed and illustrated all the rhino materials he had seen so far.

In Ferdinand Hayden's 1873 report to the U.S. Geological Survey, Leidy described the Condon Collection from the John Day Formation of Oregon (Leidy, 1873). A partial upper jaw, isolated upper molars, and a mandibular fragment were referred to ?*Rhinoceros hesperius*. A larger species, based on an isolated upper left molar and partial right maxilla from "Alkali Flats" (also in the John Day region) was named *Rhinoceros pacificus* by Leidy in 1872, and further described in the Hayden Report (Leidy, 1873). All of these specimens have since been considered indeterminate by later paleontologists, although they are the oldest names given to John Day rhinos, and most of the specimens are apparently lost as well.

Edward Drinker Cope first began working on North American rhinos in 1873, with collections that had been made in the Pawnee Creek area of northeastern Colorado (Galbreath, 1953). On the top of a butte in the late Barstovian Pawnee Creek Formation, Cope found a beautiful complete skull and jaws that he named *Aphelops megalodus*. From the Orellan Cedar Creek beds below the Pawnee Creek Formation, Cope described many other new fossils. He proposed the species *Hyracodon quadriplicatus* for a maxilla with four deciduous premolars that were larger than any other hyracodont then known. (Like many later workers, Cope was misled by the peculiar pattern of rhino deciduous premolars, and thought they were a new species—see Chapter 2.) In that same year, Cope (1873b) proposed *Hyracodon arcidens* for fragments from two or three individuals from the Cedar Creek beds. (Both of these specimens are indeterminate or invalid junior synonyms, and appear to be lost as well—Prothero, 1996.) Later, Cope (1875a) described *Aceratherium mite* from the same beds in Colorado. *Aceratherium mite* was based on a nearly edentulous mandible (the type specimen), two palates, and a number of other skeletal fragments believed to be associated with the type mandible. Cope characterized *Aceratherium mite* mainly by its smaller size and short mandibular symphysis.

Two years later, Cope described fossils that had been collected on his trips to New Mexico. He had received a lower jaw collected by Dr. H.C. Yarrow near the town of Santa Clara (now thought to be from the Clarendonian Pojoaque Member of the Tesuque Formation—Galusha and Blick, 1971), which he called *Aphelops jemezanus*. This specimen, like so many other fragmentary specimens described by Leidy, Cope, and Marsh, is too incomplete to determine to which taxon it belonged.

Not to be outdone, his rival Othniel Charles Marsh (1870) described a rhino tooth from the Miocene of New Jersey as *Rhinoceros matutinus*. In 1873, he briefly described two new rhinos from the John Day region, *Rhinoceros annectens* and *R. oregonensis*. The latter came from the Barstovian Mascall Formation, and is too incomplete to be considered diagnostic, but the former is still a valid species. In 1875, Marsh described *Diceratherium armatum* based on an excellent skull (YPM 10003) that he had received from the Arikareean Turtle Cove Member of the John Day Formation, Oregon. This skull clearly showed the paired nasal flanges that supported paired

TABLE 3.1. Chronological list of North American rhinocerotid taxa

AUTHOR	NAME	PRESENT STATUS (REVISER)
Leidy, 1850a	*Rhinoceros occidentalis*	*Subhyracodon occidentalis* (Brandt, 1878)
Leidy, 1850b	*Rhinoceros nebraskensis*	*Hyracodon nebraskensis* (Leidy, 1856)
Leidy, 1852b	*Rhinoceros americanus*	Brontothere (Osborn, 1929)
Leidy, 1856	*Hyracodon nebrascensis*	*Hyracodon nebraskensis*
Leidy, 1858	*Rhinoceros crassus*	Indeterminate (this volume)
Leidy, 1865	*Rhinoceros meridianus*	*Teleoceras meridianus* (Prothero and Manning, 1987)
Leidy, 1865	*Rhinoceros hesperius*	*Subhyracodon occidentalis* (this volume)
Leidy, 1871	*Rhinoceros pacificus*	Indeterminate (Peterson, 1920)
Cope, 1873a	*Aceratherium megalodum*	*Aphelops megalodus* (Cope, 1874b)
Cope, 1873a	*Hyracodon quadriplicatus*	*Subhyracodon occidentalis* (Wood and Wood, 1937)
Marsh, 1870	*Rhinoceros matutinus*	*Diceratherium matutinum* (Simpson, 1930)
Marsh, 1873	*Rhinoceros oregonensis*	Indeterminate (Peterson, 1920)
Marsh, 1873	*Rhinoceros annectens*	*Diceratherium annectens* (Peterson, 1920)
Cope, 1874a	*Aceratherium mite*	*Subhyracodon mitis* (this volume)
Cope, 1874b	*Aphelops megalodus*	*Aphelops megalodus*
Cope, 1875a	*Aphelops jemezanus*	Indeterminate (this volume)
Marsh, 1875	*Diceratherium armatum*	*Diceratherium armatum*
Marsh, 1875	*Diceratherium nanum*	*Diceratherium annectens* (Peterson, 1920)
Brandt, 1878	*Subhyracodon*	*Subhyracodon*
Cope, 1878	*Aphelops malacorhinus*	*Aphelops malacorhinus*
Cope, 1878	*Aphelops fossiger*	*Teleoceras fossiger* (Osborn, 1904)
Cope, 1879a	*Anchisodon quadriplicatus*	*Subhyracodon occidentalis* (this volume)
Cope, 1879a	*Anchisodon tubifer*	Indeterminate (this volume)
Cope, 1879a	*Anchisodon truquianus*	*Diceratherium armatum* (this volume)
Cope, 1880a	*Peraceras superciliosum*	*Peraceras superciliosum*
Cope, 1880a	*Caenopus* (for *Aceratherium*)	*Subhyracodon* (Brandt, 1878)
Cope, 1885	*Aceratherium pumilum*	Indeterminate (this volume)
Leidy, 1885	*Rhinoceros proterus*	*Teleoceras proterum* (this volume)
Leidy, 1886	*Eusyodon maximus*	Indeterminate (this volume)
Marsh, 1887	*Aceratherium acutum*	*Teleoceras fossiger* (Osborn, 1904)
Leidy, 1890	*Rhinoceros longipes*	*Aphelops malacorhinus* (Leidy and Lucas, 1896)
Cope, 1891b	*Caenopus simplicidens*	Indeterminate (Osborn, 1898c)
Osborn, 1893	*Aceratherium tridactylum*	*Diceratherium tridactylum* (Osborn, 1923)
Hatcher, 1893	*Diceratherium proavitum*	*Diceratherium tridactylum* (Osborn and Wortman, 1894)
Hatcher, 1894a	*Teleoceras major*	*Teleoceras major*
Osborn and Wortman, 1894	*Aceratherium platycephalum*	*Amphicaenopus platycephalus* (Wood, 1927)
Osborn and Wortman, 1894	*Aceratherium trigonodum*	*Subhyracodon occidentalis* (this volume)
Osborn, 1898c	*Leptaceratherium trigonodum*	*Subhyracodon occidentalis* (this volume)
Osborn, 1898c	*Aceratherium copei*	*Subhyracodon mitis* (this volume)
Matthew, 1899	*Aceratherium profectum*	*Peraceras profectum* (Prothero and Manning, 1987)
Lucas, 1900	*Trigonias osborni*	*Trigonias osborni*
Douglass, 1903	*Aphelops ceratorhinus*	*Peraceras profectum* (this volume)
Osborn, 1904	*Teleoceras medicornutus*	*Teleoceras medicornutum*
Osborn, 1904	*Aphelops planiceps*	*Teleoceras medicornutum* (Matthew, 1918)
Osborn, 1904	*Aphelops brachyodus*	*?Peraceras profectum* (this volume)
Osborn, 1904	*Caenopus persistens*	Indeterminate (this volume)
Osborn, 1905	*Teleoceras bicornutus*	*Lapsus calami* for *T. medicornutum* (this volume)
Barbour, 1906	*Diceratherium arikarense*	*Menoceras arikarense* (Troxell, 1921d)
Peterson, 1906c	*Diceratherium cooki*	*Menoceras arikarense* (Tanner, 1969)
Peterson, 1906c	*Diceratherium niobrarensis*	*Diceratherium niobrarense*
Douglass, 1908	*Aphelops montanus*	*Peraceras profectum* (this volume)
Lambe, 1908	*Aceratherium exiguum*	*Subhyracodon occidentalis* (this volume)
Loomis, 1908	*Diceratherium schiffi*	*Menoceras arikarense* (Peterson, 1920)
Loomis, 1908	*Aceratherium stigeri*	*Menoceras arikarense* (Peterson, 1920)
Loomis, 1908	*Diceratherium aberrans*	*Menoceras arikarense* (Peterson, 1920)
Loomis, 1908	*Diceratherium petersoni*	*Diceratherium niobrarense* (Peterson, 1920)
Cook, 1908	*Aceratherium egregius*	*Menoceras arikarense* (Peterson, 1920)
Cook, 1909	*Metacoenopus egregius*	*Menoceras arikarense* (Peterson, 1920)

Olcott, 1909	*Teleoceras minor*	*Teleoceras meridianum* (this volume)
Cook, 1912a	*Diceratherium loomisi*	*Menoceras arikarense* (Peterson, 1920)
Cook, 1912c	*Epiaphelops virgasectus*	*Diceratherium niobrarense* (Peterson, 1920)
Matthew, 1918	*Peraceras troxelli*	*Peraceras superciliosum* (this volume)
Peterson, 1920	*Diceratherium gregorii*	*Diceratherium niobrarense* (this volume)
Peterson, 1920	*Caenopus dakotensis*	*Penetrigonias dakotensis* (this volume)
Troxell, 1921c	*Caenopus trigonodus allus*	*Subhyracodon occidentalis* (this volume)
Troxell, 1921c	*Caenopus tridactylus metalophus*	*Diceratherium tridactylum* (Scott, 1941)
Troxell, 1921c	*Caenopus tridactylus avus*	*Diceratherium tridactylum* (Scott, 1941)
Troxell, 1921c	*C. platycephalus nanolophus*	*Trigonias osborni* (this volume)
Troxell, 1921d	*Menoceras* for *D. cooki*	*Menoceras arikarense* (Tanner, 1969)
Troxell, 1921d	*Diceratherium labiatum*	*Menoceras arikarense* (this volume)
Troxell, 1921d	*Diceratherium cuspidatum*	*Menoceras arikarense* (this volume)
Troxell, 1921d	*Metacaenopus egrerius*	*Menoceras arikarense* (this volume)
Troxell, 1921d	*Metacaenopus petersoni*	*Menoceras arikarense* (this volume)
Troxell, 1921d	*Metacaenopus gregorii*	*Diceratherium niobrarense* (this volume)
Troxell, 1921d	*Metacaenopus niobrarensis*	*Diceratherium niobrarense* (this volume)
Freudenberg, 1922	*Teleoceras felicis*	Indeterminate (this volume)
Osborn, 1923	*Trigonias tetradactylum*	*Trigonias osborni* (this volume)
Matthew, 1924	*Aphelops malacorhinus mutilus*	*Aphelops mutilus* (Matthew, 1932)
Cook, 1927	*Teleoceras hicksi*	*Teleoceras hicksi*
Lane, 1927	*Paraphelops rooksensis*	*Teleoceras hicksi* (this volume)
Wood, 1927	*Amphicaenopus platycephalus*	*Amphicaenopus platycephalus*
Wood, 1927	*Trigonias wellsi*	*Trigonias wellsi*
Wood, 1927	*Trigonias paucidens*	*Trigonias osborni* (Scott, 1941)
Wood, 1927	*Trigonias gregoryi*	*Trigonias wellsi* (this volume)
Wood, 1927	*Subhyracodon metalophus*	*Subhyracodon occidentalis* (this volume)
Wood, 1927	*Subhyracodon gidleyi*	*Subhyracodon occidentalis* (this volume)
Wood, 1927	*Diceratherium avus*	*Diceratherium tridactylum* (Scott, 1941)
Gregory and Cook, 1928	*Trigonias hypostylus*	*Trigonias osborni* (Scott, 1941)
Gregory and Cook, 1928	*Trigonias precopei*	*Trigonias osborni* (Scott, 1941)
Gregory and Cook, 1928	*Trigonias preoccidentalis*	*Trigonias osborni* (Scott, 1941)
Gregory and Cook, 1928	*Trigonias taylori*	*Trigonias osborni* (Scott, 1941)
Gregory and Cook, 1928	*Trigonias osborni figginsi*	*Trigonias osborni* (Scott, 1941)
Gregory and Cook, 1928	*Trigonias osborni secundus*	*Trigonias osborni* (Scott, 1941)
Gregory and Cook, 1928	*?Caenopus premitis*	*Trigonias osborni* (Prothero and Manning, 1987)
Cook, 1930	*Paraphelops yumensis*	*Teleoceras hicksi* (this volume)
Cook, 1930	*Peraceras ponderis*	*Teleoceras hicksi* (this volume)
Cook, 1930	*A. malacorhinus longinaris*	*Aphelops mutilus* (this volume)
Cook, 1930	*T. (Mesoceras) thomsoni*	*Teleoceras medicornutum* (this volume)
Wood, 1931	*Trigonias cooki*	*Trigonias osborni* (Scott, 1941)
Wood, 1931	*Trigonias osborni osborni*	*Trigonias osborni* (Scott, 1941)
Stock, 1933	*Subhyracodon kewi*	*Subhyracodon kewi*
Figgins, 1934a	*Procaenopus premitis*	*Trigonias osborni* (Prothero and Manning, 1987)
Schlaikjer, 1935	*Caenopus yoderensis*	Indeterminate (this volume)
Schlaikjer, 1935	*Subhyracodon woodi*	Indeterminate (this volume)
Wood and Wood, 1937	*Caenopus premitis*	*Diceratherium annectens* (Albright, 1999b)
Wood, 1964	*Floridaceras whitei*	*Floridaceras whitei*
Wood, 1964	*Diceratherium barbouri*	*Menoceras barbouri* (Tanner, 1969)
Tanner, 1967	*Aphelops kimballensis*	*Aphelops mutilus* (this volume)
Tanner, 1969	*Menoceras marslandensis*	*Menoceras barbouri* (this volume)
Stevens, 1969	*Moschoedestes delahoensis*	*Menoceras barbouri* (this volume)
Tanner, 1972	*Menoceras falkenbachi*	*Menoceras barbouri* (this volume)
Tanner, 1975	*Teleoceras schultzi*	*Teleoceras fossiger* (Voorhies, 1990)
Tanner and Martin, 1976	*Penetrigonias hudsoni*	*Penetrigonias dakotensis* (this volume)
Dalquest and Mooser, 1980	*Teleoceras ocotensis*	*Teleoceras hicksi* (this volume)
Russell, 1982	*Subhyracodon sagittatus*	*Penetrigonias sagittatus* (this volume)
Kihm, 1987	*Trigonias yoderensis*	*Penetrigonias dakotensis* (this volume)
Kihm, 1987	*Trigonias sp.*	Indeterminate (this volume)
Prothero and Manning, 1987	*Peraceras hessei*	*Peraceras hessei*
Hanson, 1989	*Teletaceras radinskyi*	*Teletaceras radinskyi*
Albright, 1999b	*Gulfoceras westfalli*	*Gulfoceras westfalli*

horns, and was the first horned rhinoceros ever described from North America. In the same paper, he described another John Day rhino, *Diceratherium nanum,* about half the size of *Diceratherium armatum.*

Cope continued to describe new specimens, publishing dozens of papers every year during the 1870s and 1880s. In 1878, he described *Aphelops malacorhinus,* based on a fine skull collected by R.S. Hill from the "Loup Fork beds" (now considered to be from the upper Clarendonian Ogallala Formation) of Kansas. In the same year, he described *Aphelops fossiger* from early Hemphillian parts of the "Loup Fork beds" in Kansas. This species was transferred to *Teleoceras* by Osborn (1904).

In 1879, Cope placed a number of fragmentary specimens from the John Day region in his new genus *Anchisodon,* which included his Colorado taxon *Hyracodon quadriplicatus,* and two new species (*Anchisodon tubifer* and *Aceratherium truquianum*). Neither his new genus *Anchisodon* nor any of these species is today considered valid. *Anchisodon tubifer,* based on an indeterminate M2 that was never figured, was referred to *Aceratherium* by Merriam and Sinclair (1907), and then to *Caenopus* by Matthew (1909), but has never been adequately discussed. According to Merriam and Sinclair (1907), the type was originally in the Cope Collection at the American Museum, and now appears to be lost. Based on its size (as given by Cope) and locality, it may be a junior synonym of Marsh's taxon *Diceratherium armatum.* Because the type is lost, and was never adequately described or figured, I consider *Anchisodon tubifer* to be a *nomen nudum.* *Aceratherium truquianum,* on the other hand, came from the "Truckee beds" (lower John Day Formation). It was based on a symphysis and a jaw fragment which were the size and morphology of *Diceratherium armatum* (Peterson, 1920), and is probably also a junior synonym of Marsh's taxon.

In 1880 Cope described a beautiful skull collected by R.H. Hazard from the "Loup Fork beds" of the Republican River, Kansas, as *Peraceras superciliosum* (Cope, 1880a). In another paper later that year, he published his first phylogeny of the rhinos that were known at the time. From an ancestor like the Bridgerian *Triplopus,* Cope proposed that the rhino lineage went through *Caenopus* (as he had renamed "*Aceratherium*" *mite* in this same paper) and through *Aphelops.* At this point, Cope (1880b) thought rhinos branched into two lineages. One lineage led to the Asian rhinos; the other led to the woolly and African rhinos from a *Peraceras*-like ancestor (see Prothero *et al.,* 1986, for a discussion of early rhino phylogenies).

Meanwhile, the discovery of American rhinos caused quite a stir in Europe. At first scientists refused to believe Leidy, but after 20 years, they began to take notice of the American specimens. The German scientist J.F. Brandt (1878) briefly reviewed Cope's and Leidy's discoveries in an obscure Russian monograph written in Latin. He proposed a subgenus of *Aceratherium,* which he named *Subhyracodon,* for *Aceratherium mite, A. occidentalis,* and *A. quadriplicatus*

(these were all the North American rhinos then known that were not referred to *Aphelops*). This subgeneric name was not discovered by North American paleontologists until 1909, at which time Cope's 1880 genus *Caenopus* had become entrenched in the minds of most paleontologists. For a while, there were suggestions that Brandt's 1878 genus *Subhyracodon* should be suppressed (e.g., Wood and Wood, 1937), because of the unfortunate implication that *Subhyracodon* was related to hyracodonts, not rhinocerotids. But several paleontologists (e.g., Wood, 1927; Scott, 1941) established *Subhyracodon* as the oldest valid supraspecific taxon for "*Rhinoceros*" *occidentalis,* and this genus is now more familiar to modern paleontologists than "*Caenopus*" (although the latter name still crops up in works that borrow from outdated sources).

Cope continued to publish on fossils (including North American rhinos) until he died in 1897. In 1885, he described *Aceratherium pumilum,* based on a small edentulous symphysis from the Cypress Hills Oligocene of Canada. This specimen was referred to *Caenopus mitis* (now *Subhyracodon mitis*) by Wood (1927) and Russell (1934), but it could also pertain to *Subhyracodon occidentalis* based on its size, so it is also indeterminate. Cope (1891a) mentioned *Caenopus occidentalis, C. pumilus,* and *C. mitis,* and in his final paper on rhinos, Cope (1891b) named his last new species *Caenopus simplicidens* based on another indeterminate specimen.

After the early 1870s, Leidy had nearly abandoned paleontology because of the competition and warfare between the two great rivals, Cope and Marsh. However, in the 1880s, some new specimens were sent to him. In 1884, he first mentioned some fossils from Florida collected by Dr. J.C. Neal in a bed of clay (known as the Mixson's, or Mixon's, bone bed, from the early Hemphillian Alachua Clay, Levy County, Florida—Harrison and Manning, 1983). Leidy recognized that there was a rhino of "shorter stature" in this material, and in 1885 he described the rhino bones, which were "mangled together in greatest confusion, badly fractured, but not water worn," as *Rhinoceros proterus.* Hay (1902) referred these specimens to *Teleoceras,* and that generic assignment is still valid. In 1886, Leidy described a large broken tusk from this locality as "an extinct boar from Florida," and named it *Eusyodon maximus.* He later (Leidy, 1890) realized that it was actually a tusk of a rhino, but it is too broken to tell to which currently accepted taxon it belongs. In that same year, he described some additional foot bones from the Mixson collection that were not short and stubby like *Rhinoceros proterus,* so he called them *Rhinoceros longipes.* A few years later, Leidy and Lucas (1896) synonymized this taxon with *Aphelops malacorhinus.*

Up to this point, the literature of North American vertebrate paleontology (and rhino taxonomy) was dominated by its three founding fathers, Leidy, Cope, and Marsh. With the deaths of Leidy in 1891, Cope in 1897, and Marsh in 1899, a new generation of paleontologists entered the scene, and came to dominate the science during the new century. The two most important figures of the next generation were Henry Fairfield Osborn and William Berryman Scott. While still Princeton undergraduates, Scott and Osborn took collecting trips to the Wild West in 1877 and 1878, and published their results in 1887 and following years. Osborn (1893) described *Aceratherium tridactylum* from the Whitneyan (lower Oligocene) Poleslide Member of the Brule Formation in the Big Badlands of South Dakota, as the largest of the White River "aceratheria." It was the first (and still one of the few) fossil rhino taxon whose type is a nearly complete skeleton, and not a tooth or jaw fragment or partial skull. In 1894, Osborn and Wortman again reviewed the White River faunas, and named two new species of *Aceratherium, A. trigonodum* and *A. platycephalum.*

In 1894, John Bell Hatcher described a complete skull of a new rhino from the "Loup Fork beds" in Turtle Canyon of the Niobrara River (now thought to be from the early Clarendonian Cap Rock Member of the Ash Hollow Formation—Skinner and Johnson, 1984). This specimen had a tiny horn boss on its nasals, so he gave it the name *Teleoceras major.* Although several species now referred to *Teleoceras* had already been named and placed in other genera, this specimen was the first that showed the nasal horn rugosities. Hatcher (1894b) also described *Diceratherium proavitum* from the "White River beds of South Dakota." He distinguished it from contemporaneous *Aceratherium tridactylum* by its distinct pair of nasal horns. In 1894, Osborn and Wortman reviewed the White River perissodactyls, and synonymized Hatcher's *Diceratherium proavitum* with *Aceratherium tridactylum.* For many years, *"Aceratherium" tridactylum* was referred to *Subhyracodon,* but in this volume, it is considered *Diceratherium tridactylum.*

Prior to 1898, all of these fossil rhinos had been described in short papers, often with inadequate descriptions, illustrations, or comparisons to other named forms. The first major monographic review of the North American rhinoceroses was Osborn's memoir, *The Extinct Rhinoceroses* (Osborn, 1898c). In this work, he not only reviewed the taxonomy of most of the fossil rhinos then known, but also provided one of the first extensive discussions of fossil rhino anatomy and phylogeny, and the first adequate comparative figures and descriptions of American fossil rhinos. Osborn (1898c) erected a new species of *Aceratherium, A. copei,* and a new genus, *Leptaceratherium,* for *A. trigonodum. Leptaceratherium* was distinguished from *Aceratherium* by its retention of the upper canine, although later workers (Wood, 1927; Scott, 1941) found this character to be highly variable and thus unreliable for generic distinction.

The beginning of the twentieth century saw Lucas (1900) propose the taxon *Trigonias osborni* for a Chadronian rhinoceros that had clearly retained all its incisors, but still had the blade-like first upper incisor and incisiform upper canines that characterize the Rhinocerotidae. A few years later,

Osborn (1904) reviewed all the known species of Miocene rhinoceroses, and was surprised to find a primitive-looking form he named *Caenopus persistens,* apparently from the Miocene of Colorado.

Before the twentieth century, virtually all of the North American mid-Tertiary rhinoceroses were collected from either the White River badlands or the John Day beds of Oregon. In 1904, the Agate Springs quarries were discovered in western Nebraska, and two papers were published describing the new paired-horned rhinoceros found by the thousands in these quarries. On June 15, 1906, Erwin H. Barbour described it in a Nebraska Geological Survey publication and named it *Diceratherium arikarense* (Barbour, 1906). Apparently unaware of Barbour's work, two months later (August 15, 1906), Olof Peterson of the Carnegie Museum named the same rhino *Diceratherium cooki,* and a larger rhino *Diceratherium niobrarensis* [*sic*]. Barbour's priority was ignored by Peterson (1920), who incorrectly (Peterson, 1920, footnote 24) cited the date of Barbour's publication as December 14, 1906. This was actually the date of Barbour's second publication on the Agate rhino. This misunderstanding remained uncorrected until 1969, when Lloyd Tanner straightened out the priority and referred *"Diceratherium" arikarense* (= *"D. cooki"*) to *Menoceras.* But most museum labels and many popular books still have their Agate *Menoceras* incorrectly labeled *"Diceratherium cooki."*

The Agate Springs rhinos then became the subjects of a series of papers by Harold Cook and Frederic Loomis, who coined a new genus and species for just about every variant in their collections. Loomis (1908) reviewed the known species of *Diceratherium,* and coined three new species, *D. petersoni, D. schiffi,* and *D. aberrans,* and a new species of *Aceratherium, A. stigeri.* Almost all of these specimens are just *Menoceras arikarense* with slight variations in their teeth. Cook (1908) described *Aceratherium egregius* (misspelled *egrerius* by Loomis, Peterson, and others) from a quarry four miles west of Agate. Then in 1909, Cook placed it in his new genus *Metacaenopus,* since he felt it was an advanced form of Oligocene *Caenopus.* In 1912, Cook described *Diceratherium loomisi* from Agate Springs quarry, and shortly thereafter Peterson (1920) synonymized most of these species with *D. niobrarense* or *D. cooki,* since they were all clearly based on specimens that were different due to sexual, individual, or ontogenetic variation. Cook (1912b, 1912c) then named another rhino from near Agate (*Epiaphelops virgasectus*), which came from "a Harrison Formation channel." According to Tedford (pers. commun. to E. Manning, 1978), this locality may be the same as the Morava Ranch Quarry, Box Butte County, Nebraska, which is in the latest Arikareean Harrison Formation. *Epiaphelops virgasectus* was based on a lower jaw that still retains the deciduous lower first premolar, a tooth that is variably retained or lost in many rhinos, and thus not adequate justification for a new genus. Cook (1912b, 1912c) compared *Epiaphelops virgasectus* only to *Caenopus* and *Aphelops*, but (surprisingly) not to *Diceratherium*.

Several other rhino-bearing localities besides Agate Springs Quarry were also described in the first two decades of the twentieth century. Lambe (1908) monographed the Chadronian faunas of the Cypress Hills, Saskatchewan, and named several new species of rhinos and hyracodonts. In his 1908 paper, he recognized *A. mite, A. occidentale, L. trigonodon* [*sic*], and a new species, *Aceratherium exiguum*. Surprisingly, he did not discuss the validity of Cope's species from the same region, *Aceratherium pumilum*.

In 1918, one of the most competent paleontologists of the early twentieth century, William Diller Matthew, weighed in about the Miocene rhinos. In his monograph on the Snake Creek faunas of Sioux County in western Nebraska, he set up the first clear definitions of the best-known Miocene rhino genera, *Teleoceras, Aphelops*, and *Peraceras*. He finally assigned *Aphelops fossiger* and several other species to Hatcher's *Teleoceras*, reduced the number of species of *Aphelops*, and recognized a new species of *Peraceras, P. troxelli*. In many ways, our modern concepts of these genera closely follow those of Matthew (1918), with some slight modifications.

After Osborn's (1898c) monograph, the second great monograph on North American Oligocene-Miocene rhinos was Olof A. Peterson's (1920) *The American Diceratheres*. In this work, he reviewed most of the forms known from the John Day and Agate regions. Peterson synonymized most of the invalid taxa proposed earlier by Cook, Loomis, and others, and gave one of the first complete descriptions and bone-by-bone illustrations of the postcranial skeleton of any North American rhinoceros. Peterson (1920) also described a new species, *Diceratherium gregorii*, from the "Miocene Lower Rosebud beds near Rosebud Indian Agency, South Dakota." In contrast to the careful work shown in the rest of the volume, Peterson (1920) also used the name *Caenopus dakotensis* in a table without any description, diagnosis, or discussion whatsoever.

In 1921, Edward Leffingwell Troxell published a series of four papers in the *American Journal of Science*. In the first (1921a), he proposed a new species of *Metamynodon, M. rex*. In the second (1921b), he proposed three new species of *Hyracodon*. In the third (1921c), he discussed the position of *Caenopus* as the ancestral rhinoceros and created the following new taxa: *C. trigonodus allus, C. tridactylus metalophus, C. tridactylus avus*, and *C. platycephalus nanolophus*. Needless to say, all of these subspecies are invalid and based on slight dental variation in primitive rhinos (see Chapter 2). In the final paper (1921d), Troxell reviewed *Diceratherium*, creating two new species, *D. lobatum* and *D. cuspidatum*, and referred the species *niobrarensis, petersoni, egregius* (misspelled "*egrerius*" yet again), and *gregorii* to Cook's previously discredited genus *Metacaenopus*. The only good result of all of this oversplitting by Troxell was the creation of the genus *Menoceras* (type species = *M. cooki*) for the small

Agate rhino with the knob-like horn cores. This genus languished as an obscure subgenus of *Diceratherium* until Tanner (1969) correctly showed that it was a valid genus and had little to do with true *Diceratherium*.

The dominant figure in early twentieth-century North American rhinoceros paleontology was Horace Elmer Wood II. He first published on his lifelong interest in fossil rhinos in 1926, with the naming of yet another species of *Hyracodon*. In 1927 he published his dissertation, a major review of early Tertiary rhinos and hyracodonts, which was a landmark comparable to the monographs of Osborn (1898c) and Peterson (1920). In this work, he reviewed the Eocene ceratomorphs, and created quite a few taxa that have since been synonymized or rearranged by Radinsky (1967). Wood (1927) erected three new species of *Trigonias: T. wellsi, T. paucidens*, and *T. gregoryi*, based on slight variations in premolars, and straightened out much of the synonymy of the species of *Caenopus*. He synonymized *Leptaceratherium* and most of the unnecessary species of *Subhyracodon*, although he raised *Subhyracodon tridactylus metalophus* to specific rank and created a new species, *Subhyracodon gidleyi*, based on one aberrant skull. Wood (1927) questioned the validity of Troxell's (1921d) *Menoceras*, but retained it as "doubtfully distinct" from *Diceratherium*. He raised *D. tridactylum avus* to specific rank. The one new genus created by Wood (1927) that remains valid is *Amphicaenopus*, for Osborn and Wortman's (1894) *Aceratherium platycephalum*. Finally, Wood (1927) presented the first detailed branching diagram of rhinoceros relationships. Most previous workers were deliberately vague in specifying their phylogenetic hypotheses. Although Wood (1927) was a typological oversplitter (as were nearly all of his contemporaries), he was one of the first North American rhino paleontologists to do a thorough comparative study of all the known forms. To his credit, this thoroughness resulted in his sinking more invalid taxa than he created. In 1929, Wood wrote a postscript, describing some new specimens, but fortunately describing no new taxa.

In 1928, the discovery of the *Trigonias* Quarry in the Chadronian Horsetail Creek Formation of Weld County, Colorado, created a problem for the typological oversplitting of rhino taxa. For decades, paleontologists had assumed that every small variation in the crests and cusps of rhino premolars was worthy of systematic recognition and the grounds for new species. But the enormous variability of this single quarry sample was a source of taxonomic controversy. Gregory and Cook (1928) persisted in the old typological practice by recognizing six species of *Trigonias* (four new) and calling one specimen ?*Caenopus premitis*. Matthew's (1931, 1932) reaction to this sample, on the other hand, was surprisingly modern. His 1930 discussion of the typological species concept reads as if it were written today. Matthew correctly realized that, on ecological grounds, such a large number of species in one quarry sample (and probably from a single population) was unlikely, and thus suggested that they were all the same species. Wood (1931) took an intermediate position. He synonymized three species, but created another new one, *Trigonias cooki*. J.D. Figgins (1934a, 1934b), the director of the Colorado Museum of Natural History (which had collected the specimens) agreed that most of the *Trigonias* species were the same, but raised ?*Caenopus premitis* to a new genus, *Procaenopus premitis*. The full implications of the premolar crest variability of the *Trigonias* Quarry sample apparently caused several paleontologists to question the meaning of premolar characters, but none took the next logical step—to synonymize all the superfluous species of *Subhyracodon* and *Hyracodon*, which were also based on similar subtle premolar differences.

In 1927, Harold Cook further burdened the literature with additional names when the large collection of late Hemphillian rhinos from near Wray, Colorado, was discovered and described. Cook (1927) first named this rhino *Teleoceras hicksi*, and this is still the senior synonym for most large late Hemphillian *Teleoceras*. However, the variability of the large sample again led to excessive oversplitting. Lane (1927) called some specimens *Paraphelops rooksensis*, and Cook (1930) then named *Paraphelops yumensis* and *Peraceras ponderis*, all from specimens that were just variants of *Teleoceras hicksi*. In the same paper, Cook gave a large Wray *Aphelops* specimen the name *A. malacorhinus longinaris* (it is merely another *A. mutilus*). Cook also described a Barstovian specimen from the "lower Snake Creek beds" (now the Olcott Formation) in Sinclair Draw, Sioux County, Nebraska, as *Teleoceras (Mesoceras) thomsoni*. This specimen is just a *T. medicornutum* with extraordinarily long nasals and premaxillae, but nonetheless "*Mesoceras thomsoni* Quarry" came to be an important locality in the Sheep Creek-Snake Creek beds (Skinner *et al.*, 1977).

At the end of his brilliant career, Matthew (1931, 1932) again visited the systematics of the rhinoceroses. Among his last works (posthumously published after he died in 1930) were a number of observations that were surprisingly modern, and clarified a lot of the confusion about most rhino taxa, including a more broadly drawn phylogeny than that presented by Wood (1927) or anyone else before him. However, Matthew died before he could do a thorough systematic revision of the North American rhinoceroses. Had he lived, the longstanding confusion about rhino systematics might not have waited another 70 years to be cleared up.

After 1932, the pace of new discoveries and publications diminished considerably over the next 70 years. Stock (1933) described *Subhyracodon kewi* from the late Whitneyan or earliest Arikareean Kew Quarry in the Sespe Formation, Las Posas Hills, Ventura County, California (Prothero *et al.*, 1996). Schlaikjer (1935) described fossils from the early Chadronian Yoder Formation in Goshen Hole, Wyoming, including the new rhinos *Caenopus yoderensis* and *Subhyracodon woodi*. Wood and Wood (1937) reviewed the

nomenclature of *Caenopus*, and restricted it to a few specimens: *C. dakotensis, C. mitis, C. yoderensis,* and *C. premitis*. The impetus for this review was a maxilla from the "Oligocene" of Texas, which they referred to *Caenopus* cf. *premitis*. Prothero and Manning (1987) argued that it is a specimen of *Menoceras arikarense*, and is not even Oligocene in age, but Albright (1999b) has shown that it is a specimen of *Diceratherium annectens*. In the White River monographs, Scott (1941) acknowledged the problem of the variability of the upper premolars of rhinos, and reduced most of the invalid species of *Trigonias* to synonymy, but unfortunately was unwilling to sink *Caenopus* or reduce the species of *Subhyracodon* to synonymy. His greatest contribution, however, was a detailed osteological description of all the well-known White River rhinocerotoids, including *Metamynodon, Hyracodon, Trigonias, Subhyracodon,* and some foot bones referred to *Caenopus*. As a result of his excellent descriptions (and Bruce Horsfall's fine illustrations), the postcranial anatomy of Oligocene rhinoceroses is better described than that of any other North American rhino except Agate *Menoceras arikarense* (described by Peterson, 1920). Indeed, the postcranial anatomy of most *Aphelops, Peraceras,* and *Teleoceras* species has yet to be described, and will be detailed in this volume (Chapter 5).

Horace Wood published a number of general review papers on rhinoceroses between 1941 and 1960, but did little or no taxonomy in them. No new rhinos were described for almost 30 years until the discovery of the Hemingfordian Thomas Farm locality in Gilchrist County, Florida. Wood (1964) named the large rhino from this locality *Floridaceras whitei*, and called the paired-horned rhino from Thomas Farm *Diceratherium barbouri* (he still did not recognize Troxell's (1921d) distinction between *Diceratherium* and *Menoceras*). After a 40-year career, Horace Wood then ceased to publish on rhinoceroses, and died in 1975.

Lloyd Tanner was practically the only paleontologist working on North American rhinocerotids through most of the 1960s and 1970s, and into the 1980s. In 1969, Tanner described *Menoceras marslandensis* from the early Hemingfordian Marsland Formation of Nebraska, and clarified the differences between *Menoceras* and *Diceratherium,* and Barbour's priority for the trivial name *arikarense* over *cooki*. In 1967, he described *Aphelops kimballensis* from an unusually large *Aphelops malacorhinus* skull from what was then thought to be the "Kimballian" or what Schultz thought was latest Hemphillian Cambridge local fauna of Nebraska. In 1975, Tanner described *Teleoceras schultzi* for a specimen of *Teleoceras fossiger,* again mistakenly thinking its large size and "Kimballian" age justified distinguishing it from other known species. If Tanner had not been misled by Schultz into thinking that the "Kimballian" was late Hemphillian, rather than early Hemphillian (Breyer, 1981; Tedford *et al.*, 1987), he would have compared *Aphelops* "*kimballensis*" with *A. malacorhinus,* and *T.* "*schultzi*" with

T. fossiger, and perhaps not created invalid new taxa. Tanner and Martin (1976) established the taxon *Penetrigonias hudsoni* based on three small upper premolars from the Chadron Formation. Even though the type is inadequate and the species is invalid, it is the first generic name proposed that can be used to associate the small rhinos formerly called "*Caenopus*" *dakotensis* and "*Caenopus*" *sagittatus*.

Russell (1982) described the rhinos from the Chadronian of Cypress Hills, Saskatchewan. Like previous authors, he was confused over the persistence of invalid concepts of "*Caenopus*," and the scrambled taxonomy of White River rhinos. He named one new taxon, *Subhyracodon sagittatus*, which is here transferred to *Penetrigonias*. Hanson (1989) described several skulls of a new rhino from the late Uintan or Duchesnean Hancock Quarry in the Clarno Formation of Oregon, which proved to be one of the most primitive rhinocerotids known from North America. He named this animal *Teletaceras radinskyi*, in memory of the late Leonard Radinsky, who died in 1986 after a long career working with ceratomorphs.

This brings us to the 1970s and 1980s, and the curation of the Frick Collection, which completely revolutionized our studies of North American Rhinocerotidae. Where once there were a few isolated teeth, skulls, and jaws known for most taxa, now we have dozens to hundreds of skulls and jaws, and complete skeletons for many taxa. Such an enormous sample demands that modern concepts of statistics, population and ontogenetic variation, and stratigraphy be applied to the North American Rhinocerotidae, so that the clutter of hundreds of valid and invalid names can be reassessed. This task began with the curatorial and systematic work of Earl Manning in the 1970s, and I continued this research in the 1980s. Overviews of much of our work have already appeared (Prothero *et al.,* 1986, 1989; Prothero and Manning, 1987; Prothero, 1993, 1998), but this volume is the first detailed documentation of this research.

Although the long litany of over a hundred rhino taxa that have been proposed is not easy reading, it is important for the reader to realize how much our concepts of North American rhinoceros taxonomy have changed, and how many names and invalid taxa have to be dealt with. The detailed justification for most of the taxonomic assignments shown in Table 3.1 are given in Chapter 4.

4. Systematics

Class Mammalia Linnaeus, 1758
Order Perissodactyla Owen, 1848
Suborder Ceratomorpha Wood, 1937
Superfamily Rhinocerotoidea Owen, 1845
Family Rhinocerotidae Owen, 1845

The systematic relationships of the North American rhinocerotids was presented by Prothero et al. (1986) and Prothero (1998), and shown here as Figure 4.1. Slightly different phylogenies were presented by Cerdeño (1995) and Antoine (2002), but since those focus largely on Eurasian rhinos, they will not be considered here. In the section below, the detailed generic- and specific-level taxonomy of all the North American Rhinocerotidae is documented, based on the methods and principles discussed in Chapters 2 and the history discussed in Chapter 3. Comparisons and measurements of the postcranial anatomy of most of these taxa are given in Chapter 5.

Uintaceras Holbrook and Lucas, 1997

Forstercooperia Radinsky, 1967 (in part)
Forstercooperia Lucas et al., 1981 (in part)
Forstercooperia Lucas and Sobus, 1989 (in part)

Figure 4.2

Type species. *Uintaceras radinskyi* Holbrook and Lucas, 1997
Included species: Type species only.
Known distribution: Late middle Eocene (Uintan), Utah and Wyoming.
Generic diagnosis: "Medium-sized rhinocerotoid (M1–3 length = 81-93 mm), distinguished from hyracodontids by tetradactyl manus and non-cursorial limb structure; synapomorphies of amynodontids, including enlarged and labially deflected M3 metastyle, elongated upper molar metalophs, reduced premolars, elongated lower molar talonids and preorbital fossa lacking; distinguished from rhinocerotids by distinct M3 metastyle; most resembling indricotheriine hyracodont *Forstercooperia*, but distinguished by isolated p3–4 entoconids, relatively tall maxillaries contributing to short, high rostrum, nasal incision above P1, orbits

above M1–2, high braincase, and large sagittal crest; differing from all other rhinocerotoids in possessing buccolingually compressed upper incisors with triangular profile" (from Holbrook and Lucas, 1997, p. 384).

Uintaceras radinskyi Holbrook and Lucas, 1997

Forstercooperia? *grandis* Radinsky, 1967
Forstercooperia grandis Lucas et al., 1981 (in part)
Forstercooperia grandis Lucas and Sobus, 1989 (in part)

Holotype: CM 10004, fragmentary skull and dentition, most of postcranial skeleton, from the late Uintan Myton Pocket, Uinta Basin, Utah.
Hypodigm: Given by Holbrook and Lucas (1997, p. 385).
Known distribution: Same as for genus.
Diagnosis: Same as for genus.
Description: A full description of all the known elements was given by Holbrook and Lucas (1997).
Discussion: These specimens had originally been referred to *Forstercooperia grandis* by Radinsky (1967) and also by Lucas et al. (1981) and Lucas and Sobus (1989). This material represents a very primitive Uintan rhinocerotoid about the size of *Forstercooperia*, so the comparison was not surprising. However, Holbrook and Lucas (1997) demonstrated that these specimens have several important characters (outlined in the generic diagnosis above) that separate them from true *Forstercooperia*, and that they also lack the elongated limbs of hyracodonts. Instead, they have several synapomorphies (especially in the buccodistally compressed upper incisors, and in many parts of the postcranial skeleton) which place them as sister-taxa of the rhinocerotoids. However, *Uintaceras* appears to lack the crucial rhinocerotid synapomorphies of the chisel-like I1 and tusk-like i2 and the loss of the M3 metastyle that Radinsky (1966) and most later workers (e.g., Prothero et al., 1986; Prothero, 1998) used to define the Rhinocerotidae. Holbrook and Lucas (1997, p. 394) chose not to refer *Uintaceras radinskyi* to the Rhinocerotidae and force a redefinition of the family. Instead, they placed it as a sister-taxon to the family until the anterior dentition of *Uintaceras* is better known. Whichever familial assignment is followed, it is still important to mention this specimen in a complete review of the North American Rhinocerotidae.

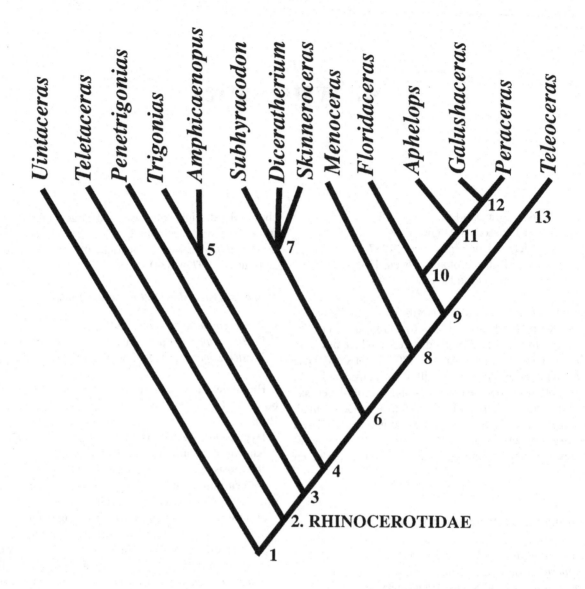

Figure 4.1. Interrelationships of the Rhinocerotidae. Characters at the nodes as follows: 1. Buccolingually compressed upper incisors. 2. I1 chisel-shaped; i2 tusk-shaped; very reduced M3 metacone; M1–2 cristae lost; premaxillary-nasal contact lost; reduced posterior cingulum on M3; shorter posterior ectoloph on M1–2. 3. M1–3 parastyle fold more open; antecrochets enlarged; M3 metacones lost; i3 and lower canine lost in adults; metacone ribs on molars lost; paracone and metacone ribs separate and reduced on premolars; greater hypsodonty; postcondyloid process on ramus; broad ascending ramus on dentary, with straight posterior border; long posteromedially curved process on anterolateral tuberosity of humerus. 4. Broad parasagittal crests; laterally flared lambdoid crests; concave dorsal skull profile; long nasals. 5. Extended occiput; anterodorsally inflected basicranium; long flattened postglenoid process. 6. Third upper and lower incisors, upper canine lost; metacone ribs on P2–4 lost; P2 molarized; mandibular condyle broader, flat-surfaced, and nearly horizontal; distal condyle of humerus more asymmetrical; dorsoventrally compressed posterior articular surface on atlas; postglenoid process faces anteriorly; fifth metacarpal reduced to vestige. 7. DICERATHERIINAE: long, broad supraorbital ridges; paired nasal ridges in males. 8. Strong crochets on molars; I2 lost; reduced sagittal crest; premaxillary further reduced; nasal incision over posterior P2; basicranium shortened relative to palate; upper molar cingula weak or absent; shallow anteroventral notch on atlas. 9. Upper premolars fully bilophodont; overall size increase. 10. ACERATHERIINAE: medial flange of i2 reduced; long diastema posterior to i2; fifth metacarpal enlarged. 11. Premaxillary reduced; I1 lost; nasal incision over anterior P4. 12. Dorsal skull profile flattened. 13. TELEOCERATINI: metapodials shortened; carpals and tarsals compressed dorsoventrally; strong antecrochets; broad zygomatic arches; lateral edges of nasals downturned and thinned, resulting in a U-shaped cross-section; calcaneal tuber elongate; brachycephalic skull; nasal incision retracted to level of anterior P3; p2 lost in some *Teleoceras* (modified from Prothero *et al.*, 1986; Prothero, 1998, Fig. 42.3).

A

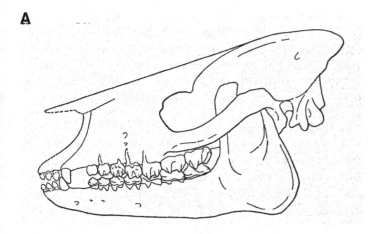

B

C

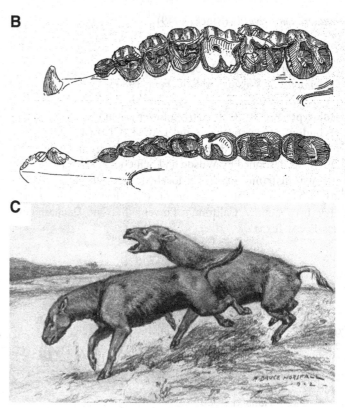

Figure 4.2. *Uintaceras radinskyi* Holbrook and Lucas, 1997. A. Skull of best specimen, UCMP 69722. B. Dentition of UCMP 69722 (A and B after Radinsky, 1967). C. Restoration of the running rhino *Hyracodon*, which closely resembled *Uintaceras*. (After B. Horsfall, in Scott, 1913)

Teletaceras Hanson, 1989

Eotrigonias (?) Stock, 1949 (*non* Wood, 1927)
Eotrigonias Beliaeva, 1959 (*non* Wood, 1927)
Hyracodon Wood, 1963
Pappaceras Radinsky, 1966 (in part) (*non* Wood, 1963)
Juxia Lucas *et al.*, 1981 (in part) (*non* Chow and Chiu, 1964)
?Caenopus sp. Wilson and Schiebout, 1984

Type species: *Teletaceras radinskyi* Hanson, 1989
Included species: Type species, *T. borissiaki* (Beliaeva, 1959), and *T. mortivallis* (Stock, 1949)
Known distribution: Late middle Eocene to late Eocene (Uintan or Duchesnean to Chadronian) of Texas, central Oregon and southern California, and late Eocene of the Maritime Province, Russia.
Diagnosis: "Dental formula = I3/3, C1/1, P4/4 (–3), M3/3. Small rhinocerotids with an I1 and i2 tusk complex characteristic of the family, but not as enlarged as in other incisor-bearing rhinocerotids. Differs from all other rhinocerotids in the possession of an unreduced anterior dental series, sharp crease between the molar parastyles and paracones, more lingually inflected molar metacone axes, and low connection of molar metalophids to protolophids" (after Hanson, 1989, p. 380). Differs from *Penetrigonias* by the presence of a postcanine diastema and a single-rooted p1.

Teletaceras radinskyi Hanson, 1989

"Clarno rhino" Prothero *et al.*, 1986

Figure 4.3

Holotype: UCMP 12900, a nearly complete skull lacking premaxillae, portion of the right zygoma, and occipital crest (Fig. 4.3).
Hypodigm: numerous specimens in the UCMP and UOMNH collections listed by Hanson (1989, p. 381).
Known distribution: Type locality, the ?late Uintan-?Duchesnean (Lucas *et al.*, 2004) UCMP locality V75203, Hancock Quarry, from the uppermost Clarno Formation, Wheeler County, Oregon.
Diagnosis: "Paracone and metacone ribs on P2–4 ectolophs prominent, subequal, and contiguous (not separated by an intervening flat area). Dentition larger and more brachydont than *Teletaceras mortivallis*, smaller than *T. borissiaki*. Mean length M1–3 = 64 mm; m1–3 = 65 mm. Crown height index = 0.66 ± 0.02" (from Hanson, 1989, p. 381).
Discussion: Hanson (1989) provided a thorough description of this new species, and it is not necessary to repeat it here. So far as I know, no additional material of this species has appeared since Hanson's (1989) publication. Prior to its formal description in 1989, this material was well known to many paleontologists, with casts located in many collections, and it was called informally "the Clarno rhino" (e.g., Prothero *et al.*, 1986).

Teletaceras mortivallis (Stock, 1949)

Eotrigonias mortivallis Stock, 1949
Hyracodon mortivallis Wood, 1963
?*Caenopus* sp. Wilson and Schiebout, 1984

Holotype: LACM (CIT) 3564, a lower jaw fragment.

Hypodigm: type specimen and LACM (CIT) 61303, a lower jaw fragment, plus FMNH PM 141, 39, and 55, from the Porvenir l.f. of Texas described by Wilson and Schiebout (1984, p. 32–33).

Known distribution: Type locality, the ?late Duchesnean-?early Chadronian Titus Canyon l.f. (CIT locality 254), Death Valley, Inyo County, California; Porvenir l.f. (late Duchesnean), Trans-Pecos Texas.

Emended diagnosis: Differs from *T. radinskyi* in "its less obtuse metalophid–hypolophid angle, slightly less elongate trigonid, the absence of a distinct p4 entoconid, and smaller size" (Hanson, 1989, p. 392). The mean length of the m1 of *T. radinskyi* is 4 mm longer than that of *T. mortivallis*, and the m2 is even longer in proportion, without overlap in the ranges of tooth measurements.

Discussion: This material was originally described by Stock (1949) as *Eotrigonias mortivallis*. Hanson (1989) demonstrated that the name *Eotrigonias* is no longer available, since Radinsky (1967) transferred its type species, *E. rhinocerinus*, to the hyracodontid *Triplopus*. Instead, Hanson compared it to the Clarno rhino, and concluded that they belong in the same genus.

Wilson and Schiebout (1984, p. 32–3) described several small

Figure 4.3. Composite reconstruction of the skull of *Teletaceras radinskyi*, based primarily on UCMP 12900, with areas reconstructed from UCMP 129001 and UCMP 129039. Scale bar is 5 cm. (Modified from Hanson, 1989).

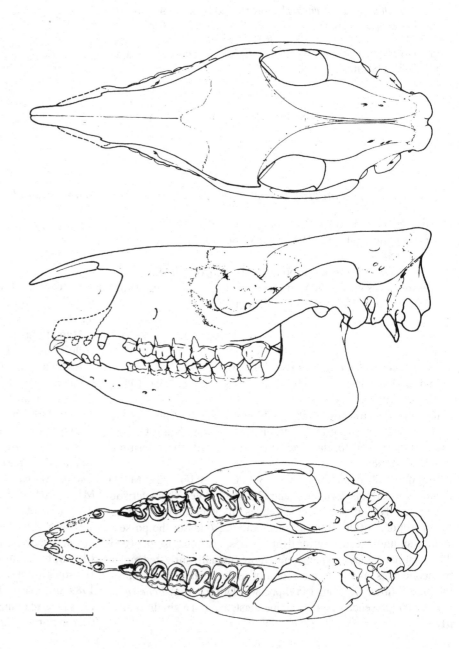

rhinocerotid specimens from the vicinity of the Chambers Tuff in the Vieja Group, and referred them to ?*Caenopus* sp. They correctly realized that the specimens were too small and primitive to belong to any known rhino taxon, and placed them in ?*Caenopus* for lack of a better name. They did not know of the specimens from the Clarno Formation, or they would have compared their material with the Clarno rhino. My own observations of these specimens show that they are clearly referable to *Teletaceras*. Published measurements show that most of them are too small to fall within the size range of *T. radinskyi*, but instead they are smaller forms within the size range of *T. mortivallis*. Unfortunately, they are so poorly preserved that they add little to our knowledge of this species, even though they do extend its geographic range.

Teletaceras borissiaki (Beliaeva, 1959)

Eotrigonias borissiaki Beliaeva, 1959
Pappaceras borissiaki Wood, 1963
Forstercooperia borissiaki Radinsky, 1967
Juxia borissiaki Lucas *et al.*, 1981

Discussion: Although this is not a North American fossil, Hanson (1989, p. 395) pointed out that this material most closely resembles North American *Teletaceras*, and thus belongs in this genus. Hanson (1989) provided a detailed justification of this taxonomic change and the previous systematic confusions, and descriptions of this material, so no further description is required here. This occurrence shows that the first North American rhinocerotids were immigrants from Asia in the Duchesnean.

Penetrigonias Tanner and Martin, 1976

Aceratherium (in part) Osborn, 1898
Caenopus (in part) Peterson, 1920
Caenopus (in part) Schlaikjer, 1935
Caenopus (in part) Russell, 1982
Subhyracodon (in part) Russell, 1982
?*Trigonias* sp. Wilson and Schiebout (1984)
Trigonias (in part) Kihm, 1987

Figures 4.4, 4.5, 4.6; Tables 4.1, 4.2.

Type species: *Penetrigonias dakotensis* (Peterson, 1920)

Figure 4.4. Graph of dimensions of specimens referred to *Penetrigonias*.

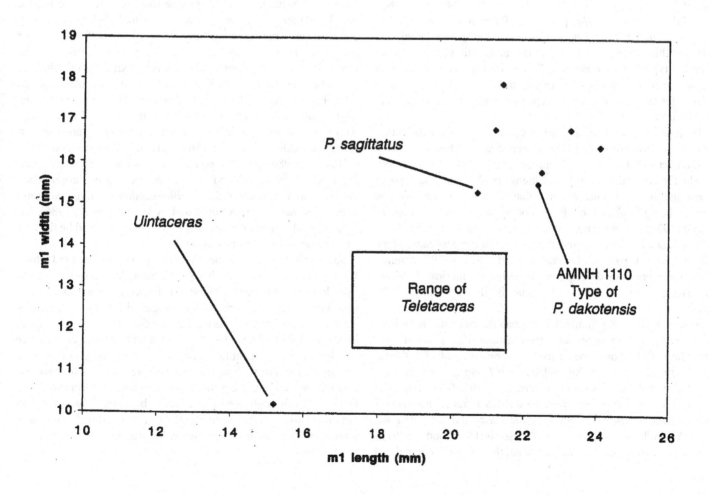

Included species: Type and *P. sagittatus* (Russell, 1982)

Known distribution: Duchesnean, Utah (Lapoint fauna) and Trans-Pecos Texas (Porvenir l.f.); Chadronian, Wyoming, Montana, Nebraska, Colorado, South Dakota, and Saskatchewan; Whitneyan, South Dakota.

Emended diagnosis: Small rhinocerotids (P2–M3 length = 124–135 mm; M1–3 length = 72–83 mm) with very reduced premaxilla still retaining I1–3. Diastema between dP1 and the anterior teeth short. Molarization of premolars and presence of upper canine variable. Larger than any known species of *Teletaceras*, but smaller than any known species of *Subhyracodon, Trigonias,* or *Amphicaenopus*.

Discussion: For over a century, there has been tremendous confusion about the status of smaller rhinocerotids from the White River Group. Osborn (1898c, fig. 39) first described (p. 140) and illustrated a small lower jaw (AMNH 1110) he referred to "*Aceratherium mite*" from the Whitneyan "*Protoceras* beds" of South Dakota (Fig. 4.3). Recognizing that it was much smaller than the type material of "*A. mite*," Peterson (1920, p. 402) gave this jaw the name *Caenopus dakotensis* in a passing reference, but did not formally justify this name. The only indication he gave regarding this name was its presence in the *Protoceras* sandstones. Because this indication was clearly inadequate, Troxell (1921c) placed *Caenopus dakotensis* in synonymy with "*C.*" *tridactylus* from the same deposits. However, Wood (1927, p. 57) asserted that *C. dakotensis* was the jaw figured by Osborn (1898c). According to Wood (1927, p. 57), "as Peterson (1920) explains, *Caenopus dakotensis* is a provisional name, given largely on stratigraphic grounds, to the lower jaw of an animal even smaller than the type of *Caenopus mitis*." I can find no such explanation anywhere in Peterson's monograph, and I cannot imagine how Wood got this information, except by unpublished communications with Peterson.

Despite the fact that the name *Caenopus dakotensis* was a *nomen nudum*, Wood's (1927) reference established it. Thus it became entrenched in the literature (e.g., Wood, 1929; Scott, 1941) for any rhinocerotid specimens smaller than the type of *Caenopus mitis* (despite the fact that the name *Caenopus* was slowly being replaced by its senior synonym *Subhyracodon*). Wood (1929) described additional material from the Big Badlands of South Dakota (horizon uncertain) in the Berlin Museum, which he referred to *Caenopus dakotensis*. Scott (1941, p. 797) summarized these previous descriptions of both the American Museum and Berlin materials, but no additional fossils were known at the time.

Since 1941, several smaller Chadronian rhinocerotids have been recovered, but their systematic status remained confused. Tanner and Martin (1976) described a specimen consisting of a P2–4 from the Chadronian of Nebraska, and named it *Penetrigonias hudsoni*. As the name implies, they compared it to *Trigonias*, plus *Subhyracodon, Amphicaenopus*, as well as hyracodonts, amynodonts, and *Toxotherium*, but for some reason did not notice the obvious similarity to the upper teeth described by Wood (1929) as *Caenopus dakotensis*. Had they done so, they would have realized

that it is identical in size and morphology to that taxon. Nonetheless, they provided the first valid generic name for this taxon.

Russell (1982) described several skulls and jaws from Hunter Quarry in the Chadronian deposits of the Cypress Hills of Saskatchewan. To one nicely preserved skull (Fig. 4.6) he gave the name *Subhyracodon sagitattus*, in reference to its distinctive single sagittal crest. However, he compared it only to larger, more derived species of *Subhyracodon*, and made no comparisons with *Caenopus dakotensis*, or with the recently named (and similarsized) *Penetrigonias hudsoni*. In addition to these skull materials, Russell (1982, p. 52) described several lower jaws as ?*Caenopus* spp., comparing them to *Caenopus mitis*, but failing to note that they are smaller than this taxon, but within the range of *Caenopus dakotensis* (Table 4.1), and also within the size range of his own *Subhyracodon sagittatus*.

Wilson and Schiebout (1984) described some fragmentary jaws and teeth from the Duchesnean Porvenir l.f. that they assigned to ?*Trigonias* sp., even though they noted (p. 36) that they were much smaller than any known specimen of that genus. They noted the similarities with *Penetrigonias hudsoni*, but did not compare their materials to the recently named *Subhyracodon sagittatus*, nor to *Caenopus dakotensis*.

Kihm (1987) described several primitive rhinocerotid specimens from the Yoder Formation of Goshen County, Wyoming. He resurrected Schlaikjer's (1935) name *yoderensis* for these fossils, but placed them in the genus *Trigonias*, even though he noted they were much smaller than any other known *Trigonias* or *Subhyracodon* specimens, and had no diagnostic *Trigonias* features. He briefly mentioned the similarity in size and morphology to Tanner and Martin's (1976) *Penetrigonias*, but chose not to refer this material to their taxon. Even though he resurrected the name *yoderensis*, he did not notice that this material is too large with respect to the type jaw of *Caenopus yoderensis*, and that that specimen might not even be a rhinocerotid, but a hyracodont.

Hanson (1989, p. 396) pointed out that several Chadronian localities have produced small rhinocerotids larger than *Teletaceras*, but smaller than *Subhyracodon* or *Trigonias*, and argued (as was originally suggested by Prothero *et al.*, 1986) that they were all congeneric, and that *Penetrigonias* would be the first valid name assigned to this genus.

Since that time, several new specimens have been found which further confirm this hypothesis, and provide a more complete hypodigm for this genus. These specimens are listed in the sections below, and a few are shown in Figure 4.5 and their measurements given in Table 4.1. As can be seen from Figure 4.3, the specimens referred to *Penetrigonias* form a well-defined size cluster smaller than any specimen now referred to *Subhyracodon* or *Trigonias*, but larger than the materials referred to *Teletaceras*. From these data, it is clear that there is a small rhinocerotid in the Duchesnean through Whitneyan that is between *Teletaceras* and *Subhyracodon* in size. Since it cannot be referred to *Caenopus*, I refer it to *Penetrigonias*, with two valid species, *P. dakotensis* and *P. sagittatus*.

Table 4.1. Cranial and dental measurements of specimens of *Penetrigonias* (in mm)

MEASUREMENT	AMNH 1110 *dakotensis* (type)	F:AM 39577	AMNH 8838	F:AM 105018	F:AM 105019	FMNH P12228	SMNH P1635.2 *sagittatus* (type)	SMNH 23190	UNSM 62049 *hudsoni* (type)	TMM 40807-6
Nasals to occiput							277.0			
Pmax to condyles		290.0								
Width @ zygoma		155.3				148.0	137.5			
Height of occiput										
Width of occiput		98.4								
I1-M3 length										
I2 length										
I2 width										
Diastema length										
P2-M3 length		134.3			128.7		124.9			
M1-3 length		81.6			81.4					
P2 length		15.6			15.0	16.5	16.1			
P2 width		20.0			21.1	19.0	19.4		17.4	
P3 length		19.3			18.3	17.1	17.7		13.2	
P3 width		25.6			24.8	26.3	22.2		18.8	
P4 length		19.3			17.3	17.1	19.2		16.0	
P4 width		27.7			27.1	21.7	23.7		22.5	
M1 length		24.7			25.9	20.0	24.9		15.2	
M1 width		30.6			28.2	30.5	25.1			
M2 length		27.6			26.3	27.4	26.9			
M2 width		32.7			29.3	31.8	27.7			
M3 length		29.2			27.8	28.9	23.3			
M3 width		29.4			27.0	31.5	24.5			
i2 length	12.0									
i2 width	8.0									
diastema length	15.0									
p2-m3 length	125.7			119.4	128.2					127.3
m1-3 length	75.7	79.2		72.3	79.0					79.6
p2 length	16.1			14.5	15.2					12.8
p2 width	8.5			10.2	10.0					7.5
p3 length	18.7			17.3	20.0			18.8		16.4
p3 width	13.4			13.0	12.8			13.8		12.5
p4 length	19.4		22.6	19.5	18.8			19.2		20.0
p4 width	14.7		17.0	13.6	14.7			15.0		13.5
m1 length	22.4	22.5	21.4	21.2	24.1			20.7		23.3
m1 width	15.5	15.8	17.9	16.8	16.4			15.3		16.8
m2 length	25.0	27.4	26.4	25.5	25.3			27.1		25.3
m2 width	16.7	17.0	16.0	18.0	17.0			16.6		18.7
m3 length	27.7	26.7		27.2	29.5			28.5		28.2
m3 width	15.7	16.4		18.0	16.3			14.9		18.2
symphysis to angle	254.0									
ht. coronoid above angle	140.0		150.0							
length symphysis	50.0									
depth jaw below p2	39.0			32.5	27.3					37.3
depth jaw below m2	49.6	42.2	47.7	44.4	41.4					44.6

Penetrigonias dakotensis (Peterson, 1920)

Aceratherium mite (in part) Osborn, 1898
Caenopus dakotensis Peterson, 1920
Penetrigonias hudsoni Tanner and Martin, 1976
?Trigonias sp. Wilson and Schiebout (1984)
Trigonias yoderensis Kihm, 1987 (*non* Schlaijker)

Figure 4.5, Table 4.1

Holotype: AMNH 1110, a lower jaw from the Whitneyan "*Protoceras* beds" of South Dakota (Fig. 4.5A).

Hypodigm: *Duchesnean*: from the Porvenir l.f., Trans-Pecos Texas: TMM 40807-6, mandible (Wilson and Schiebout, 1984, fig. 22); TMM 40807-1, left ramus fragment (juvenile); 40807-5, fragmentary right ramus; 40688-33, left maxilla with P2–M1; FMNH PM87, mandible fragment with m1; TMM 40113-1, mandible with roots of p1–p4, parts of m1–2, roots of m3. From the Lapoint fauna, Uinta Basin, Utah: FMNH PM22411, partial skull.

Chadronian: from UNSM locality Sx-41, Sioux County, Nebraska: UNSM 62049, left P2–4 (type of *P. hudsoni*); from the Yoder Formation, Goshen County, Wyoming: SDSM 6353, 5331, 5337, 6336 (referred to "*Trigonias yoderensis*"); from McCarty's Mountain, Montana: F:AM 105017; from the White River Formation; Ledge Creek and Flagstaff Rim area, Natrona County, Wyoming: F:AM 105018, two rami, 105019, right maxilla and left ramus, 105020, 105021, right ramal fragment, 105022, left ramal fragment, 105023, edentulous symphysis; from the White River Formation, "Government Slide," Beaver Divide, Wyoming: AMNH 14578, left ramus with p4–m3; from the Chadronian Horsetail Creek Member, Logan County, Colorado: AMNH 8838.

Whitneyan: FMNH PM12228, a partial skull, "*Protoceras* beds" of South Dakota (Fig. 4.5B).

White River Group, South Dakota, level unknown or unspecified: AMNH 1173, 39577. Humboldt Museum, Berlin, specimen MB. Ma. 42545.

Diagnosis: *Penetrigonias* specimens with a well-developed chisel-like I2 and small, procumbent tusk-like i2, a large upper canine, and relatively unmolarized premolars.

Description: Although the type specimen of *Penetrigonias dakotensis* has been described elsewhere (Osborn, 1898c; Wood, 1927, 1929; Scott, 1941), our current perspective on rhinos requires additional description. Relatively complete upper teeth and a partial skull were described by Wood (1929), and are redescribed below. In addition, a partial skull from the Whitneyan of South Dakota (FMNH PM12228, Fig. 4.5C–E). shows features not previously described. It has a typically flat rhinocerotid dorsal skull profile (unlike that of *Trigonias*, which is concave). The nasals are arched in cross-section, and have a distinct notch on their lateral margins. The specimen has been badly crushed, so the supraorbital ridges have been bent downward, and the orbits are dorsoventrally flattened. The postorbital portion of the skull is completely missing.

The premaxillae are fairly robust and still have the long posterodorsal portion that includes the maxilla within the nasal incision. The nasal incision is retracted to the level of posterior P2. A chisel-like I1 is present, which is longer anteroposteriorly than the I1 in *Teletaceras* or *Trigonias*, and comparable to the I1 in *Subhyracodon*. There is a short diastema between I1 and I2. The latter tooth is very reduced, with a knob-like crown. Posterior to the premaxillary–maxillary suture is an alveolus for a canine, although these teeth must have been very small. There is a short postcanine diastema.

The tooth crowns are highly worn, indicating that this was a very old individual when it died. No crown pattern can be seen on P4–M3, and the right M1 is missing. P2–3 appear to be fully molarized, with a complete protoloph and metaloph. The lingual cingula are strong on all premolars (where they are not obliterated by wear). The dP1 is present on both sides, and is also very worn. No labial cingula were present on any of the teeth. The anterior premolars are quite reduced relative to the molars.

Little can be said about the crown patterns of M1–2, since they are worn almost to the roots. M3 is typically rhinocerotid, lacking any trace of metacone or metastyle. There are weak anterior and posterior cingula, and a lingual cingulum between the lophs, which do not connect along the lingual end of the protocone or hypocone. The post-pterygoid portion of the skull is also missing.

The teeth of the Berlin skull are less worn than those of FMNH P12228, but the former specimen lacks the anterior dentition or the dorsal part of the skull (seen in the latter specimen). P2–4 in the Berlin skull are completely molarized, although the metaloph and ectoloph are not directly connected. The premolars all have strong lingual cingula and no labial cingula. M1–3 shows the typical rhinocerotid pattern with a distinct anterior cingulum, and a weak posterior cingulum, and no lingual cingula. The posterior cingulum on M3 is very thin and not divergent labially. M2 has distinct crista between the protoloph and metaloph.

The basicranium of the Berlin skull is poorly preserved, although the postglenoid processes are oriented slightly anterolaterally, as in *Trigonias*. The external auditory meatus is open, with the postglenoid and paroccipital processes widely separated. The lambdoid crests are flared about as they are in *Trigonias*, and not as widely as in *Amphicaenopus* or *Subhyracodon*. According to Wood (1939, p. 67), the temporal ridges are widely separated.

The type mandible (AMNH 1110) has a heavy, upturned symphysis that is very narrow laterally. The distance between the i2 tusks is very short. The i2 tusks are both broken at present, although the original figure (Osborn, 1898c, fig. 36) showed that the right i2 had the typical lanceolate rhinocerotid morphology. There is a short diastema between i2 and dp1.

The lower cheek teeth show the stereotyped rhinocerotid crown pattern. They are different, however, in that the lower premolar row length is reduced relative to the molars. This reduction is also seen in the upper premolars. The anterior border of the coronoid is vertical, and the posterior border is anteriorly inclined and only slightly concave. The condyles tilt inward but their trochlear surfaces are very nearly straight (approaching the *Subhyracodon*

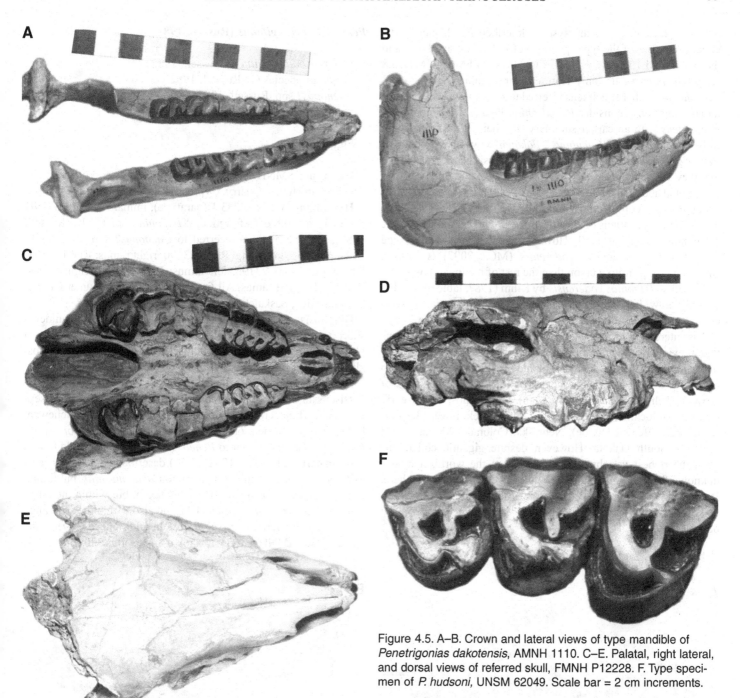

Figure 4.5. A–B. Crown and lateral views of type mandible of *Penetrigonias dakotensis*, AMNH 1110. C–E. Palatal, right lateral, and dorsal views of referred skull, FMNH P12228. F. Type specimen of *P. hudsoni*, UNSM 62049. Scale bar = 2 cm increments.

condition). There is a strong posterior crest behind the condyles that passes from the lateral edge of the trochlea posteromedially to the inside of the ramus. The coronoid fossa is located high on the side of the coronoid process.

A badly broken articulated manus (AMNH 1173) from the upper Poleslide Member (late Whitneyan) of South Dakota may also pertain to *Penetrigonias dakotensis*. The morphology and size are wrong for *Hyracodon*, and it is too small for the other Whitneyan rhinos, *Diceratherium* or *Amphicaenopus*. All four

metacarpals are present in articulation, although Mc2 has been displaced proximally. The Mc5 is long, slender, and fully functional, not very different from the morphology of the Mc5 in *Trigonias*. Most of the carpals are missing.

Discussion: *Penetrigonias dakotensis* is clearly distinct from other rhinocerotids by its smaller size and the condition of the anterior teeth. The molarization of the upper premolars, on the other hand, is highly variable, with relatively primitive conditions found in some specimens, and much more advanced conditions in others. For example, in the system described in Chapter 2, the upper premolars of the type specimen of *P. hudsoni* are 3/4/4, and those of F:AM 105019 are 3/4/5, while those of FMNH PM12228 are 5/6/6. Because the only feature distinguishing the type of *P. hudsoni* from other specimens referred to *P. dakotensis* is this premolar condition, I consider *P. hudsoni* to be a junior synonym of *P. dakotensis*. In an earlier summary (Prothero, 1998), I was not fully convinced that "*Caenopus" dakotensis* was a synonym of *Penetrigonias hudsoni*, but more recent analysis has persuaded me that they are synonyms.

As noted above, the materials Kihm (1987) referred to *Trigonias yoderensis* are completely within the size range (Fig. 4.4, Table 4.1) of other specimens of *Penetrigonias dakotensis*, and have the same morphology as well. However, Schlaikjer's (1935) type specimen of "*Caenopus" yoderensis* (MCZ 2097) is from a smaller, more primitive taxon than the rest of the lower jaw material listed as *Trigonias yoderensis* by Kihm (1987, table 6). As discussed below, it is not even clear whether this material is hyracodont or rhinocerotid, so I consider it to be indeterminate.

It is interesting to note that, even with much larger collections than previously known, *Penetrigonias dakotensis* has a very peculiar pattern of temporal and geographic distribution. It is found in the two principal late Duchesnean localities (the Porvenir l.f. of Texas and the Lapoint fauna of Utah), and in most of the early Chadronian localities of Wyoming (Ledge Creek, lower Flagstaff Rim, Yoder, Beaver Divide), Nebraska, Colorado, Montana, and probably South Dakota. However, despite gigantic collections from the richest vertebrate-bearing beds in the world, it is still unknown from the Orellan in the White River Group or elsewhere. It then reappears as a Lazarus taxon in the Whitneyan of South Dakota, paralleling the pattern seen in *Amphicaenopus* (see below).

Penetrigonias sagittatus (Russell, 1982)

Subhyracodon sagittatus Russell, 1982
Trigonias species C   Russell, 1982
Caenopus? spp. Russell, 1982
Penetrigonias sagittatus Hanson, 1989

Figure 4.6

Holotype: SMNH P1635.2, a nearly complete skull from the Chadronian Hunter Quarry, Cypress Hills, Saskatchewan.

Hypodigm: SMNH P833.1 (paratype), immature skull; SMNH P1635.1, mandible (referred to "*Trigonias* sp. C" by Russell, 1982); and the material referred to *Caenopus*? spp. by Russell, 1982, p. 52-55, including: ROM 23190, right ramus; ROM 23191, left ramus; ROM 23192, left ramus; ROM 23193, right ramus; ROM 23194, left ramus. All from the Chadronian Hunter Quarry, Cypress Hills, Saskatchewan (see Russell, 1982).

Diagnosis: *Penetrigonias* specimens with a small canine, a short postcanine diastema, wide nasals, and apparently a reduced I1 that may have not been very chisel-like (or may have been lost altogether).

Description: See Russell (1982).

Discussion: As discussed above, Russell (1982) described several small rhinocerotids from the Chadronian of Saskatchewan, but he compared them only to *Trigonias* or *Subhyracodon*, not to *Caenopus dakotensis*, nor to *Penetrigonias hudsoni*. In addition to skull materials, Russell (1982, p. 52) described some lower jaws as *Caenopus*? spp., comparing them to *Caenopus mitis*, but failing to note that they are smaller than this taxon, but within the range of *Caenopus dakotensis* (Table 4.1), and also within the size range of his own *Subhyracodon sagittatus*. In addition, his "*Trigonias* sp. C" jaw (SMNH P1635.1, shown in his figs. 20 and 21) is

A

B

C

Figure 4.6. A. Dorsal view; B. right lateral view; and C. palatal view of SMNH P833.1, paratype skull of *Penetrigonias sagittatus*. (After Russell, 1982).

essentially identical to the type specimen of *"Caenopus" dakotensis*, yet Russell (1982) failed to make the appropriate comparison. Now that a much larger sample of *Penetrigonias* is known, it is clear that these specimens should be referred to that genus. They are distinct at the species level primarily in the features of anterior dentition, nasals, and sagittal crest listed above. The adult upper premolars are known only from the type specimen, but they are relatively advanced, coding as 4/5/5 (see Chapter 2).

Trigonias Lucas, 1900

Caenopus (in part) Gregory and Cook, 1928
Procaenopus Figgins, 1934

Figures 4.7, 4.8, Table 4.2

Type species: *Trigonias osborni* Lucas, 1900
Included species: The type species and *T. wellsi* Wood, 1927.
Known distribution: Middle to late Chadronian of the High Plains, Montana, and California.
Diagnosis: Medium-sized (length P1–M3 = 183–258 mm) rhinocerotids retaining all anterior teeth except the lower canine. I1/i2 show the typical rhinocerotid chisel/tusk combination. Upper premolars extremely variable. Nasals very elongate with long slender premaxillae below a large nasal incision. Skull relatively low and saddle-shaped, with low, moderately broad sagittal and lambdoid crests. Posterodorsal portion of skull is constricted just anterior to the flare of the occiput. Manus retains a functional fifth metacarpal.
Description: The genus *Trigonias* was fully described by Scott (1941), so no further description is necessary here. Much of the postcranial skeleton is further described in Chapter 5.
Discussion: As discussed in Chapters 2 and 3, Gregory and Cook (1928) and Wood (1927, 1931) oversplit *Trigonias* on the basis of the highly variable crest pattern on the upper premolars. The evidence from the Colorado *Trigonias* Quarry sample clearly indicates that these variations are all intraspecific and even intrapopulational. Figgins (1934a), Matthew (1931), and Scott (1941) all recognized this fact, and have generally regarded most of the variants seen in the *Trigonias* Quarry sample as *T. osborni*. Scott (1941) retained *T. taylori,* although it does not differ significantly in size from the rest of the quarry sample. The variability of the *Trigonias* sample from Hunter Quarry (Russell, 1982) further corroborates this hypothesis.

The specimen with the most molarized premolars in the quarry, DMNH 1025E, was called ?*Caenopus premitis* by Gregory and Cook (1928). Their own diagram (especially their fig. 1), however, shows that it clearly fits within the range of variability of the rest of the sample. There is no feature in the skull that disallows assignment to *Trigonias*, and it is clearly not a *Subhyracodon*. Nevertheless, the rigid belief in the systematic utility of upper premolar characters was such that Figgins (1934a) created a new genus, *Procaenopus*, for ?*Caenopus premitis*, since he felt that it

was a form intermediate between *Trigonias* and *Caenopus mitis*. Wood (1934), Wood and Wood (1937), and Scott (1941) considered *"premitis"* referable to *Caenopus* and not *Trigonias*. Only Matthew (1931) realized the full implications of the variability of the *Trigonias* Quarry sample for this specimen, and considered it to be an unusually advanced variant of *Trigonias osborni*.

Wood and Wood (1937) referred a maxilla from the Catahoula Formation of Texas to *"Caenopus* cf. *premitis."* Apparently, their belief that the specimen was Oligocene in age prevented them from comparing it with Arikareean taxa, which were then thought to be Miocene in age. Prothero and Manning (1987) argued this specimen is *Menoceras arikarense*, the common Agate Springs rhino, and is almost certainly latest Arikareean, not Oligocene, in age. Albright (1999b) has since shown that this specimen is actually a *Diceratherium annectens*. Thus, the taxon *"Caenopus"* (and *"Procaenopus"*) *premitis* is an artificial composite based on a highly advanced individual of *Trigonias osborni* and a normal *Diceratherium annectens*—neither specimen has anything to do with either *"Caenopus"* or *Subhyracodon*.

Trigonias osborni Lucas, 1900

Caenopus platycephalus nanolophus Troxell, 1921
Trigonias tetradactylum Osborn, 1923
Trigonias paucidens Wood, 1927
Trigonias hypostylus Gregory and Cook, 1928
Trigonias precopei Gregory and Cook, 1928
Trigonias preoccidentalis Gregory and Cook, 1928
Trigonias taylori Gregory and Cook, 1928
Trigonias osborni figginsi Gregory and Cook, 1928
Trigonias osborni secundus Gregory and Cook, 1928
?*Caenopus premitis* Gregory and Cook, 1928
Trigonias cooki Wood, 1931
Trigonias osborni osborni Wood, 1931
Procaenopus premitis Figgins, 1934
Trigonias species A Russell, 1982
Trigonias species B Russell, 1982
Trigonias species D Russell, 1982

Figure 4.7, Table 4.2

Holotype: USNM 3924, partial skull of an old individual with I1–3, C, P1–3, from the "Miocene" of South Dakota (collected by Hatcher). Presumably from the Chadron Formation of the Big Badlands of South Dakota.
Hypodigm: *Chadron Formation, South Dakota and Nebraska:* Lower *Titanotherium* beds, Quinn Draw, South Dakota: type and paratype USNM 4815, a complete left ramus; AMNH 11865, type of *Trigonias paucidens*. AMNH 1138, skull, Hat Creek Basin, Nebraska; AMNH 9792, right ramus with i1–m3, Indian Creek, Shannon County, S.D.; AMNH 12298, left ramus with p1–m3, Indian Creek, South Dakota; AMNH 12389, skull, lower *Titanotherium* beds, South Dakota; AMNH 83331, 85934, 85935,

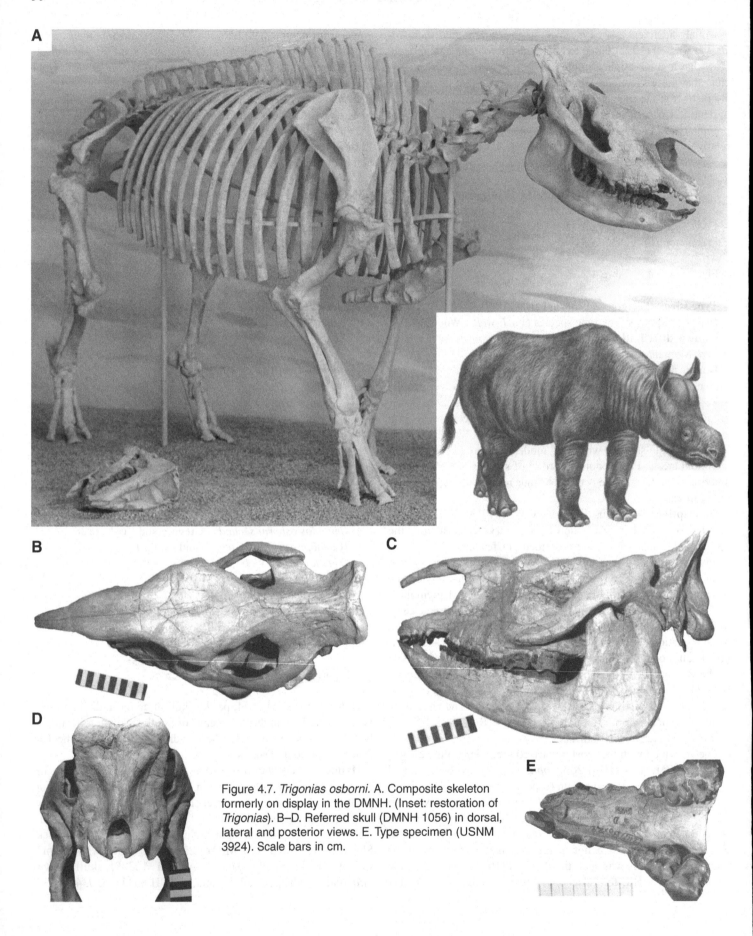

Figure 4.7. *Trigonias osborni.* A. Composite skeleton formerly on display in the DMNH. (Inset: restoration of *Trigonias*). B–D. Referred skull (DMNH 1056) in dorsal, lateral and posterior views. E. Type specimen (USNM 3924). Scale bars in cm.

85936, from *Chadronia* pocket, Sioux County, Nebraska (Harold Cook collection).

Trigonias *Quarry, Horsetail Creek Formation, Weld County, Colorado*: all the *Trigonias* specimens in the Colorado Museum of Nature and Science (listed by Gregory and Cook, 1928) and AMNH 27777, 27781, 27782, 27783, 27784, and 46000.

Pipestone Springs l.f., Jefferson County, Montana: UCMP 113714.

West Easter Lily l.f., Jefferson County, Montana: several specimens in the CM and University of Montana collections (Tabrum *et al.*, 1996, table 3).

Lower Cedarville Flora, Modoc County, California: UCMP 121793.

Hunter Quarry, Cypress Hills, Saskatchewan: *Trigonias* species A, B, and D in the ROM and SMNH collections (listed by Russell, 1982);

Known distribution: Middle Chadronian (as low as 30 feet below Ash G at Flagstaff Rim—see Emry, 1992) to late Chadronian, Colorado, Nebraska, Wyoming, South Dakota, Montana, California, and Saskatchewan.

Diagnosis: Smaller species (P1–M3 length = 180–210 mm) of *Trigonias*. Upper premolar crests extremely variable.

Description: *Trigonias osborni* was thoroughly described by Scott (1941, pp. 776–785).

Discussion: As Matthew (1931, 1932) and Scott (1941) demonstrated, all of the smaller "species" of *Trigonias* are referable to *T. osborni*. The Hunter Quarry *Trigonias* sample (Russell, 1982) confirms the variability of the upper premolars of *Trigonias*, although Russell insisted on separating each minor variant as Gregory and Cook had done over 50 years before. The only useful character is separating species of *Trigonias* is size.

Kihm (1987) referred material from the early Chadronian Yoder Formation of Wyoming (originally named *Subhyracodon woodi* by Schlaikjer, 1935) to *Trigonias*. However, as Kihm (1987) noted, the material is in the same size range as *Subhyracodon*. After re-examination of this material, I cannot tell whether the badly broken lower jaw belongs to either *Trigonias* or *Subhyracodon*. Therefore, I consider Schlaikjer's (1935) taxon *Subhyracodon woodi* to be a *nomen dubium*.

Trigonias wellsi Wood, 1927

Trigonias gregoryi Wood, 1927
Trigonias osborni wellsi Scott, 1941

Figure 4.8

Type specimen: AMNH 13226, a skull without a jaw from an old individual, upper *Titanotherium* beds, Corral Draw, Big Badlands, South Dakota.

Hypodigm: Type and AMNH 13226a (type of *T. gregoryi*), a skull of a young adult and associated parts of skeleton; AMNH 13226b, lower jaw of a third individual; 13226c, d, and e, most of skeleton; AMNH 83329, right ramus with p2–m3, AMNH 83330, skull, AMNH 85933, ramus, all from *Chadronia* Pocket, Sioux County, Nebraska.

Known distribution: Late Chadronian (upper *Titanotherium* beds), Big Badlands of South Dakota, and *Chadronia* Pocket, Sioux County, Nebraska.

Diagnosis: Large *Trigonias* (P1–M3 length = 250–260 mm) with highly variable upper premolar crests.

Description: *Trigonias wellsi* was described by Wood (1927),

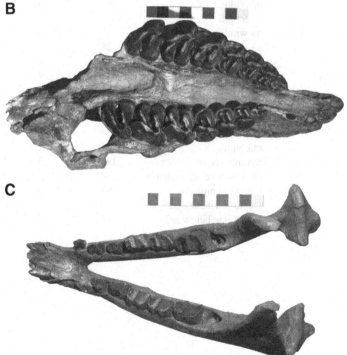

Figure 4.8. *Trigonias wellsi*. A–B. Type skull (AMNH 13226) in lateral and palatal views. C. Referred mandible. Scale bar = 2 cm increments.

Table 4.2. Cranial and dental measurements of *Trigonias* and *Amphicaenopus* (in mm)

MEASUREMENT	*T. osborni*			*T. wellsi*			*A. platycephalus*		
	Mean	SD	N	Mean	SD	N	Mean	SD	N
P2 to occiput	367	22	13	—	—	—	—	—	—
Pmax to condyles	515	16	4	596	2	2	—	—	—
Width @ zygoma	232	29	16	—	—	—	360	—	1
Height of occiput	189	22	13	—	—	—	142	—	1
Width of occiput	121	16	13	—	—	—	—	—	—
I2 length	12	1	4	11	1	2	—	—	—
I2 width	7	1	5	8	1	2	—	—	—
Diastema length	35	8	5	38	6	4	107	—	1
P2-M3 length	176	10	17	220	22	5	219	3	5
M1-3 length	113	8	17	142	2	3	134	5	5
P2 length	24	3	6	28	1	4	25	2	4
P2 width	27	3	6	36	2	4	33	3	4
P3 length	24	3	7	29	1	4	28	1	5
P3 width	26	2	7	46	1	4	42	1	4
P4 length	26	2	10	32	1	4	32	1	5
P4 width	39	1	10	52	2	4	58	1	5
M1 length	33	3	10	44	1	4	41	4	5
M1 width	43	1	10	56	1	4	53	1	4
M2 length	39	2	9	49	1	4	46	1	5
M2 width	48	1	9	62	1	4	57	2	5
M3 length	38	2	9	50	1	4	47	2	6
M3 width	44	1	9	60	1	4	44	4	6
i2 length	13	1	8	13	2	2	29	4	3
i2 width	18	2	8	18	1	2	18	2	3
Diastema length	28	10	7	40	1	2	73	5	3
p2-m3 length	194	4	4	251	2	2	220	6	5
m1-3 length	111	7	17	164	5	2	137	5	4
p2 length	22	2	6	32	1	2	19	—	1
p2 width	15	1	6	21	1	2	15	—	1
p3 length	25	2	6	35	1	2	21	—	1
p3 width	19	1	6	27	1	2	25	—	1
p4 length	28	2	4	36	1	2	33	1	2
p4 width	22	2	4	25	1	2	29	3	2
m1 length	35	3	4	44	1	2	41	1	2
m1 width	26	2	4	28	2	2	33	1	2
m2 length	42	4	4	51	1	2	48	2	3
m2 width	28	2	4	32	1	2	34	2	3
m3 length	45	1	2	—	—	—	57	5	4
m3 width	26	4	2	—	—	—	33	1	4
Symphysis to angle	399	12	3	474	11	2	516	11	3
Ht. coronoid above angle	194	6	3	211	1	2	251	3	2
Length symphysis	83	9	4	97	—	1	18	15	3
Depth jaw below p2	49	6	6	44	1	2	64	13	4
Depth jaw below m2	51	5	4	58	1	2	76	16	5

and since no new material has been discovered, there is nothing additional to report here.

Discussion: All larger late Chadronian *Trigonias* is here referred to *T. wellsi. T. gregoryi* is based on a skull from the same level, and probably the same quarry, as the type of *T. wellsi*. It is the same size as *T. wellsi,* and differs only in its slightly more molarized upper premolars. As is the case for *T. osborni,* upper premolar variation is useless in distinguishing species of *Trigonias*.

Amphicaenopus Wood, 1927

Aceratherium Osborn and Wortman, 1894 (in part) (*non* Kaup)
Caenopus Osborn and Matthew, 1909 (in part) (*non* Cope)
Amphicaenopus Wood, 1927

Type species: *Amphicaenopus platycephalus* (Osborn and Wortman, 1894)
Included species: Type species only.
Known distribution: Chadronian and Whitneyan (but not Orellan), South Dakota and North Dakota.
Diagnosis: Large rhinocerotid (P1–M3 length = 236–238 mm) with broad, dolichocephalic skull and flaring lambdoids. Lower jaw with a nearly circular cross-section. and with large procument i2. P2–3 unmolarized, P4 nearly molariform. Nasals relatively short, with no lateral notches. Anterodorsal portion of premaxilla reduced, allowing maxilla to contact nasal incision. Orbit shifted anteriorly (anterior to rim over M2). Distinguished from *Subhyracodon* and *Diceratherium* by its big, broad skull, cylindrical ramus, and larger size. Much larger than any other Oligocene or early Miocene rhinoceros.

Discussion: This peculiar rhino is very easily distinguished from all other Oligocene rhinos by its large size and hippo-like proportions. Scott (1941, p. 792) commented that *Amphicaenopus* was of doubtful validity until more material was collected, but this is untrue. As Osborn (1904) noted, *Amphicaenopus* shows some resemblances to *"Paracaenopus"* (now *Ronzotherium*) *filholi* in its premolars. Both taxa also have striking similarities in the position and the cross-section of i2. However, the I1 is reduced and pointed in *Ronzotherium filholi* (K. Heissig, pers. comm.).

Amphicaenopus is also peculiar in its stratigraphic distribution. Like *Penetrigonias*, it is known from the Chadronian and Whitneyan of South Dakota, but not from the intervening Orellan (some of the most fossiliferous beds in the world). Instead, the appropriate channel deposits carry an ecologically similar rhino, *Metamynodon planifrons*. Perhaps the two were in competition during the Oligocene, and *Amphicaenopus* was temporarily driven out of South Dakota during the Orellan. Although the Orellan record is one of the best in North America, we have found no evidence of *Metamynodon planifrons* or *Amphicaenopus platycephalus* outside South Dakota, so it is difficult to evaluate this hypothesis.

Amphicaenopus platycephalus (Osborn and Wortman, 1894)

Aceratherium platycephalum Osborn and Wortman, 1894
Aceratherium platycephalum Osborn, 1898
Caenopus platycephalus Osborn and Matthew, 1909
Caenopus platycephalus Troxell, 1921
Amphicaenopus platycephalus Wood, 1927

Figure 4.9, Table 4.2

Holotype: AMNH 542, a skull from the *Protoceras* channels, late Whitneyan of South Dakota.
Hypodigm: *Late Whitneyan "Protoceras beds," Poleslide Member, Brule Formation, South Dakota*: Type and AMNH 540, maxilla with right P1–M3 (paratype); AMMH 545, mandible with left and right i2–m3; AMNH 548, right forelimb (humerus, radius, ulna, Mc2, Mc3); AMNH 1444, mandible; YPM-PU12046, skull; SDSM 407, skull.
Late Chadronian "Titanotherium beds," Chadron Formation, South Dakota: AMNH 1478, skull; AMNH 12453, jaw with left p4–m3, left forelimb with radius, ulna, articulated carpals and metacarpals (no Mc5 known), phalanges.
?Arikareean, "Killdeer" (= Arikaree) Formation, Slope County, South Dakota: (Stone, 1970).
Known distribution: Late Chadronian and late Whitneyan (but not Orellan), South Dakota; ?Arikareean, North Dakota..
Diagnosis: Same as for genus.
Description: Relatively little new material of *Amphicaenopus* has been found since the original descriptions of Osborn (1898c) and Wood (1927). Bjork (1978) reported additional skulls (SDSM 407 and YPM-PU12046) from the Whitneyan of South Dakota that shed important light on the anterior dentition. The upper incisors of the type skull are unknown, but Osborn (1898c) reconstructed them as broad chisels like those of *Subhyracodon*. However, SDSM 407 and YPM-PU12046 have a very small I1, with a conical knob-like crown that shows no sign of wear. The lower tusks (i2) of SDSM 407 are also different from those of the type material of the jaw (e.g., AMNH 545, 1444). In those specimens, the lower tusk is relatively short with a conical crown (typical of female rhinocerotids; see Chapter 2). In SDSM 407, however, the lower tusk is much more robust (typical of male rhinocerotids). All of the lower tusks show wear facets, but because there is no wear on the I1 of SDSM 407, Bjork (1978) suggested that the wear on the lower tusks was caused by the action of a mobile upper lip which drew vegetation across the tusk. This is corroborated by the wear striations on the i2 of SDSM 407, which are nearly horizontal, in contrast to those of *Subhyracodon*, which are caused by wear against the I1. Ringström (1924) postulated a similar scenario for the Asian aceratherine *Chilotherium*, which also lacked an upper incisor to provide wear on the lower tusk. In addition, a healed break on the left tusk of SDSM 407 suggests that *Amphicaenopus* may have used its tusks for intraspecific combat, as do modern rhinos like *Rhinoceros unicornis* (Dinerstein, 2003).

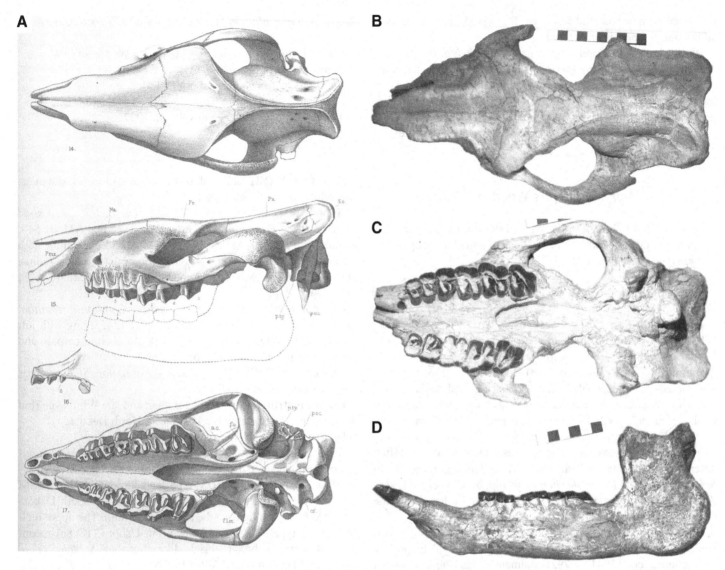

Figure 4.9. *Amphicaenopus platycephalus*. A. Osborn's (1898c) figure of type specimen. B–C. Actual condition of type skull (AMNH 542) in dorsal and palatal views. D. Referred left ramus (AMNH1444) in left lateral view. Scale bar = 2 cm increments.

Beyond this information, the only other new material that requires mention is a series of large limb elements (AMNH 548) from the Whitneyan of South Dakota. These were considered to be a large species of *Metamynodon planifrons* by Scott (1941, p. 865), who described them in detail, but fortunately did not name them as a new taxon. Because *Metamynodon* is unknown after the Orellan, it seems likely that this limb material might be referable to *Amphicaenopus*, which is only slightly larger than *Metamynodon planifrons*. Indeed, the ulna of AMNH 548 is nearly identical in size and morphology to an *Amphicaenopus* ulna, AMNH 12453. Thus, I refer this "Whitneyan *Metamynodon*" material to *Amphicaenopus*.

Subfamily Diceratheriinae Dollo, 1885
Subhyracodon (Brandt, 1878)

Rhinoceros Leidy, 1850 (in part) (*non* Linnaeus)
Aceratherium Leidy, 1851 (in part) (*non* Kaup)
Aceratherium (*Subhyracodon*) Brandt, 1878
Anchisodon Cope, 1879
Caenopus Cope, 1880 (in part)
Leptaceratherium Osborn, 1898
Subhyracodon Wood, 1927 (raised to generic rank)

Figures 4.10-4.14; Table 4.3

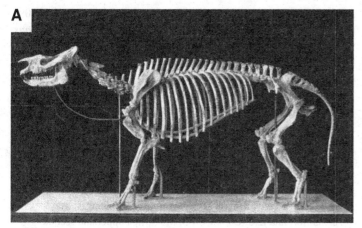

Figure 4.10. A. Composite skeleton of *Subhyracodon occidentalis*, formerly on display at the AMNH. B. Restoration of *S. occidentalis* by Charles R. Knight. (After Osborn, 1910).

Type species: *Subhyracodon occidentalis* (Leidy, 1850a)

Included species: Type and *S. mitis* (Cope, 1874a), *S. kewi* Stock, 1933.

Known distribution: Late Eocene to late Oligocene (earliest Chadronian to late Orellan, High Plains, Rocky Mountains, and Mississippi; latest Whitneyan to early Arikareean, California).

Diagnosis: Small to moderate-sized rhinocerotid with P2 completely molarized, P3–4 premolariform to nearly molarized (i.e., strong metaloph but protoloph still posteriorly deflected), heavy lingual cingula on upper cheek teeth; no nasal rugosities. Upper canines reduced or lost. Mc5 lost.

Discussion: As discussed in Chapter 3, the first rhinocerotids described from North America by Leidy (1850b) were referred to the extant genus *Rhinoceros*. They were then transferred to *Aceratherium* by Leidy (1851b). Brandt (1878) created the subgenus *Subhyracodon* for *Aceratherium mite, A. occidentale,* and *A. quadriplicatum*. This obscure document was written in Latin by a German scientist in a Russian monograph, and it was apparently unknown to Cope (1880a) when he erected the genus *Caenopus* for *Aceratherium mite*. Brandt's (1878) definition and description of *Subhyracodon* amounted to little more than a brief review of the information previously published by Leidy and Cope. However, because very few paleontologists read Latin these days, I provide a translation of the key sections below for benefit of future paleontologists:

B. Section or Subgenus *Subhyracodon* Nob.

Species 3. *Aceratherium mite* Cope

Crowns of premolars 2–3–4 wider than long. Mandibular symphysis greatly abbreviated and contracted. Species intermediate between *A. occidentale* and *H. nebrascensis,* about the size of a mule.

Species 4. *Aceratherium occidentale* Leidy.

Remains attributed to *A. occidentale*, owing their origin to some individuals, consist of an incomplete cranium lacking a mandible, equipped with well-preserved molars made of many

parts, almost complete lacking the nasal part, and part of the occiput. Besides this cranium, the greater part of the mandible and other fragments of the cranium are exposed, particularly numerous teeth, not only the molars, but also the incisors, the whole mandible until now based upon one fragment of the mandible, and an alveolus of a maxillary incisor. Some of the extremities of the bones (humerus, radius, ulna) were found, and fragments of the tibia and femur were together with it. The fourth anterior digit of the feet is not yet shown to be peculiar to the genuine *Aceratherium*.

The general appearance of the skull, being three-fourths the size of *Rhinoceros indicus*, with teeth on the whole even the number of the genuine *Rhinoceros*, recalls homologous parts of *A. incisivum*, particularly so in that it lacks a place for frontal insertion of the horn, but also in that it shows similarity with the cranium of *Oreodon culbertsonii*, except for the teeth. Even if the nasal part of the hitherto known skull lacked most of this part, something of the true appearance of the others described by Leidy could be interpreted to have had no nasal horn whatsoever based on the preserved remains.

With reference to Cope, p. 493, *A. occidentale* differs chiefly from *A. mite*: in greater size and elongation of the mandibular symphysis equipped with broad incisors.

From the figure of the crania of *A. incisivum* given by Kaup and of *A. occidentale* given by Leidy, you would conclude that its cranium had been straighter above, lower and narrower in the posterior part, and a little wider in the ventral part, equipped with a central crest, having less rounded orbits, lower zygomatic arch, and smaller mandibular incisors.

Remains have been found in the Miocene strata of the White River in the territory called "Badlands" and in the Dakota and Colorado territory.... (Brandt, 1878, pp. 30–32).

It is apparent that Brandt's (1878) concept of *Subhyracodon* included *Caenopus mitis*. There is nothing diagnostic of *Subhyracodon occidentalis* in his original description. Several authors (e.g., Wood, 1927; Scott, 1941) noted the inappropriateness of the name, because it falsely suggests that *Subhyracodon* is related to hyracodonts, not rhinocerotids. But as Wood (1927, pp. 54–55) has clearly shown, *Subhyracodon* is nevertheless the first valid supraspecific name for Leidy's (1850b) *Rhinoceros occidentalis*, and therefore it has priority.

In 1880, Cope erected the genus *Caenopus* for *Aceratherium mite*. This genus and species was based on an indeterminate jaw fragment (AMNH 6325), from the ?Chadronian of Colorado. The paratype palate (AMNH 6325 also) is usually taken to represent Cope's (1880b) concept of *Caenopus*. Cope (1880, p. 611) diagnosed *Caenopus* as follows: "Dentition I2/1, C0/1, M4/4 [meaning P4/4], M3/3. Digits 3-3. The typical species is *C. mitis* (*Aceratherium* Cope)." This definition is so broad that it could apply to any Eocene or Oligocene rhinocerotid except *Teletaceras, Trigonias,* or *Penetrigonias* (which have a fifth metacarpal). As a result, most late Eocene and Oligocene rhinocerotids described in the late nineteenth and early twentieth centuries were called *Caenopus*, and the name was widespread even after the name

Subhyracodon was rediscovered by Osborn and Matthew (1909) and also used in Cope and Matthew (1915). Eventually, *Caenopus* was restricted to just eight specimens by Wood and Wood (1937). Scott (1941, p. 776) provided a definition of *Caenopus* and *Subhyracodon* as follows:

> *Caenopus* Cope, 1879; small true rhinoceros: I2/2-1, C1-0/0, P4/3, M3/3; upper premolars more advanced than in *Subhyracodon*, or *Amphicaenopus*; cingulum of cheek-teeth much weaker than in former; order of change in upper premolars to molar pattern, P4, P2, P3. I1 and i2 not markedly elongated; manus tridactyl.
>
> *Subhyracodon* Brandt, 1878; medium size, manus tridactyl; dental formula I2/2, C0/0, P4/4-3, M3/3; first upper and second lower incisor enlarged tusks; P2 most advanced of the upper premolars, order of change P2 to P3 to P4; cingula of cheek teeth heavy; anterointernal cusp of upper molars set off from transverse crest; lower Brulé.

A

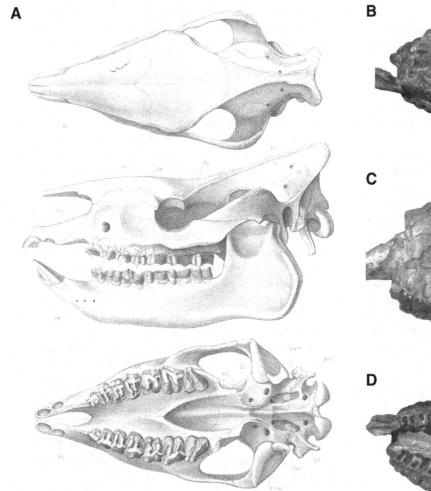

B

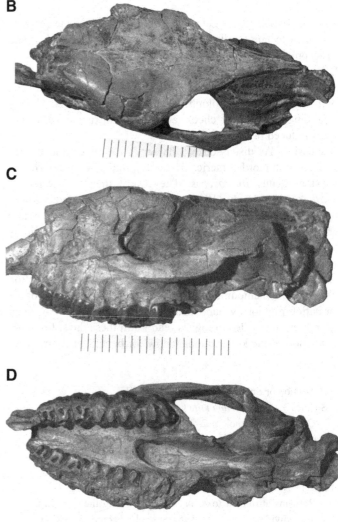

C

D

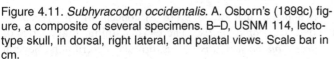

Figure 4.11. *Subhyracodon occidentalis*. A. Osborn's (1898c) figure, a composite of several specimens. B–D, USNM 114, lectotype skull, in dorsal, right lateral, and palatal views. Scale bar in cm.

This definition of *Caenopus* applies only to "*Caenopus*" *mitis*, but many authors continued to refer any small late Eocene or Oligocene rhinocerotid with weak lingula cingula to *Caenopus*. Besides the types specimen, these taxa included "*Caenopus*" *dakotensis* (see *Penetrigonias*), "*Caenopus*" *premitis* (see *Trigonias*), a specimen from the "Oligocene" of the Texas Gulf Coastal Plain originally referred to *Caenopus* cf. ?*premitis*, but referred to *Diceratherium annectens* (Albright, 1999b), and "*Caenopus*" *yoderensis* (see *Penetrigonias*). Nothing unites this heterogeneous assemblage except their small size.

Because the other species of "*Caenopus*" are referred to more appropriate taxa, the genus *Caenopus* is thus restricted to its type material (an indeterminate lower jaw and the paratype palate, AMNH 6325). The size of the paratype palate is intermediate between *S. occidentalis* and *Penetrigonias*, so size is not a valid character. The upper canine is variably present from specimen to specimen in *Subhyracodon*. The supposed weak lingual cingula are no weaker than many specimens of *Subhyracodon*. The supposed lack of an enlarged I1/2 (based on the badly broken alveoli of the type symphysis) is not diagnostic. Comparably sized female *S. occidentalis* have similar small incisors. This leaves the order of molarization of upper premolars (P4–P2–P3) as the only character of *Caenopus* not found in *Subhyracodon*.

The molarization of P2–3 in AMNH 6325 is comparable to the state in *Subhyracodon* "*trigonodus*" (= *S. occidentalis*). Only the aberrantly molarized P4 with its anterior spur (not a crochet) on the metaloph is unique to *Caenopus*. Of the many Chadronian specimens found in the last few decades, none shows this feature. I hesitate to place much importance on it, because upper premolar crests in primitive rhinocerotids are so notoriously variable (see Chapter 2). Therefore, I consider "*Caenopus*" a junior synonym of *Subhyracodon*. *Subhyracodon mitis* is treated as a valid species (see below), but there is no further justification for the genus *Caenopus*. That genus had become a wastebasket for specimens referable to *Hyracodon*, *Penetrigonias*, *Trigonias*, *Amphicaenopus*, *Subhyracodon*, *Menoceras*, and *Diceratherium*. It has long outlived its usefulness. Lucas *et al.* (1981) informally considered *Caenopus* to be a junior subjective synonym of *Subhyracodon*. This synonymy is formally adopted here. *Anchisodon* Cope, 1879 was shown by be indeterminate by Wood (1927, p. 68). Wood (1927, pp. 59–60) demonstrated that *Leptaceratherium* is a junior synonym of *Subhyracodon*, because the presence or absence of an upper canine is highly variable and not of generic significance.

Subhyracodon is the most common rhinocerotid of the late Eocene and early Oligocene, particularly in the rich Orellan deposits of the Lower Nodular Zone of the Scenic Member of the Brule Formation in the Big Badlands of South Dakota. It is readily distinguished from *Penetrigonias* by its larger size, larger I1, and greater degree of molarization. It is distinguished from contemporary Chadronian *Trigonias* by its smaller size (in primitive forms), lack of I3, and laterally broadened I1, distinctive skull shape, and absence of Mc5. It is distinguished from *Diceratherium* by its smaller size, lack of nasal rugosities in male skulls, and incompletely molarized P3–4.

Subhyracodon occidentalis (Leidy, 1850)

Rhinoceros occidentalis Leidy, 1850
Aceratherium occidentale Leidy, 1851
Rhinoceros hesperius Leidy, 1865 (in part)
Hyracodon quadriplicatus Cope, 1873
Aceratherium quadriplicatum Cope, 1875
Aceratherium (*Subhyracodon*) *occidentale* Brandt, 1878
Anchisodon quadriplicatus Cope, 1879
Aceratherium trigonodum Osborn and Wortman, 1894
Aceratherium (*Caenopus*) *mite* Osborn and Wortman, 1894 (in part)
Leptaceratherium trigonodum Osborn, 1898
Aceratherium copei Osborn, 1898 (in part)
Caenopus occidentalis Osborn, 1900
Caenopus occidentalis Hay, 1901
Aceratherium exiguum Lambe, 1908
Caenopus occidentalis Osborn and Matthew, 1909
Caenopus trigonodus allus Troxell, 1921
Caenopus tridactylus metalophus Troxell, 1921
Caenopus (*Leptaceratherium*) *trigonodus* Troxell, 1921
Subhyracodon trigonodus Wood, 1927
Subhyracodon occidentalis Wood, 1927
Subhyracodon metalophus Wood, 1927
Subhyracodon gidleyi Wood, 1927
Subhyracodon hesperius Wood, 1960

Figure 4.10, 4.11, 4.13, Table 4.3

Lectotype: USNM 114, a complete skull with well-preserved teeth (Fig. 4.11B–D). This specimen was designated by Osborn (1898c) as the "neotype" (actually a lectotype) because Leidy's original type specimens ("several molar teeth and fragments of maxillae") were never figured or adequately described, and are now apparently lost (Osborn, 1898c, pp. 151–152). The type specimen is from "the Missouri Territory," presumably from the early Orellan Scenic Member, Brule Formation, of the Big Badlands of South Dakota.

Hypodigm: There are thousands of specimens referable to *S. occidentalis* scattered about in virtually every natural history museum, university collection, and rock shop that has White River fossils. A partial listing would include:

Late Chadronian: From the "upper *Titanotherium* beds", Big Badlands, South Dakota.: AMNH 528, palate (type of "*A. trigonodum*"); AMNH 1131, skull (referred to "*A. trigonodum*"); AMNH 529, skull and jaws; from the Nuttall Rhino Quarry, Niobrara County, Wyoming: F:AM 112162, uncrushed skull and jaws, and much additional uncatalogued and unprepared material, including four rami, three skulls, numerous postcranials.

Early Orellan, Brule Formation, High Plains: From the Lower Nodular Zone, Scenic Member, Brule Formation, Big Badlands of South Dakota, AMNH collections: AMNH 38850, skull and lower jaw; 38851, skull; 12300, skull; 38895, skull and mandible; 12302, skull and mandible; 1135, mandible; 521, juvenile skull

Table 4.3. Cranial and dental measurements of *Subhyracodon* (in mm)

MEASUREMENT	*S. mitis*			*S. occidentalis*			*S. kewi*		
	Mean	SD	N	Mean	SD	N	Mean	SD	N
Nasals to occiput	404	3	2	458	30	4	—	—	—
Pmax to condyles	408	2	2	461	2	2	—	—	—
Width @ zygoma	198	—	1	228	33	17	—	—	—
Height of occiput	114	3	2	159	25	14	—	—	—
Width of occiput	—	—	—	133	18	16	—	—	—
I2 length	15	2	3	11	2	5	—	—	—
I2 width	10	2	3	8	1	4	—	—	—
Diastema length	36	2	3	38	1	4	—	—	—
P2-M3 length	145	1	2	170	10	17	162	—	1
M1-3 length	84	1	2	100	5	17	85	—	1
P2 length	18	1	4	22	2	12	22	2	3
P2 width	24	1	4	26	2	6	23	1	3
P3 length	20	1	4	23	2	11	25	3	3
P3 width	29	1	4	33	3	4	28	2	3
P4 length	21	1	4	24	2	11	27	2	3
P4 width	33	1	4	34	3	17	32	1	3
M1 length	27	1	6	33	5	17	34	3	3
M1 width	32	1	6	35	3	17	35	3	2
M2 length	29	1	4	37	4	17	41	3	3
M2 width	36	1	4	37	3	17	39	2	3
M3 length	32	2	3	35	2	8	35	1	2
M3 width	34	1	2	39	1	6	37	2	2
i2 length	9	1	3	16	1	2	—	—	—
i2 width	12	2	2	24	3	3	—	—	—
Diastema length	24	3	2	17	1	2	—	—	—
p2-m3 length	126	3	2	175	13	17	—	—	—
m1-3 length	73	6	3	102	8	17	—	—	—
p2 length	14	1	2	22	2	3	—	—	—
p2 width	8	—	1	14	1	2	—	—	—
p3 length	18	2	2	25	2	4	—	—	—
p3 width	13	2	2	15	1	2	—	—	—
p4 length	18	1	3	27	3	4	—	—	—
p4 width	14	2	3	19	1	2	—	—	—
m1 length	23	3	4	20	4	17	—	—	—
m1 width	18	2	3	23	2	17	—	—	—
m2 length	27	1	3	35	3	17	—	—	—
m2 width	18	2	3	23	2	17	—	—	—
m3 length	28	2	3	27	4	3	—	—	—
m3 width	17	2	3	21	1	2	—	—	—
Symphysis to angle	262	13	2	213	18	3	—	—	—
Ht. coronoid above angle	143	4	2	145	18	9	—	—	—
Length symphysis	46	4	2	81	21	3	—	—	—
Depth jaw below p2	39	1	2	45	5	3	—	—	—
Depth jaw below m2	47	1	2	48	7	3	—	—	—

(referred to "*A. trigonodum*"); 12297, juvenile skull; 9638, left maxilla with P2–M1; 38938, juvenile skull and ramus; 1134, right ramus; 1119, skull; 38897, palate; 38841, palate; 704, partial skull; 530, left ramus with p2–m3; 1107, left maxilla with P2–M3; 92653, right ramus with m1–3; 1160, maxilla with I2–3, P1–3; 22652, isolated teeth; 92647, right ramus with m2–3; 92623, maxilla with P3–4; 1149, right tibia; 1150, carpals and tarsals; 1117, radius and ulna; 12309, femur; 1139, mandible; 1103, carpals and tarsals; 1143, limb elements; 1146, articulated right carpus; 1148, right Mc3; 1254 b, cuneiform; 1132, mandible, skull fragments, most of skeleton; 1140, partial skeleton; 1128, mandible; 1125, juvenile skull; 1144, skull and partial skeleton; 12295, skull; 1145, humerus; 537, skull; 1106, right scapula; 1123, skull; 534, juvenile skull and mandible; 1489, skull and jaws. From the Brule Formation, Little Badlands, Stark County, North Dakota: F:AM 112165, left ramus with p2-m3; additional uncatalogued rami. From the White River Group (early Orellan), Seaman Hills, Niobrara County, Wyoming: F:AM 112161, partial skull, additional uncatalogued maxillae, rami, postcranials. From the Orellan Cedar Creek Member, Brule Formation, Logan and Weld Counties, Colorado: AMNH collections: 8839, skull; 8838, left ramus with m3; 8820, symphysis; 8842, skull; 8844, skull; 6638, left dP2–3 (type of *Anchisodon quadriplicatus*); 6328, skull; 6329, palate; 6339, right dP3–4; 8846, juvenile skull; 8847, scapula; 6330, right maxilla with P4–M2; 8845, partial skull; 8822, femur; 8840, maxillae; 6334, astragalus, magnum, Mt3; 6331, left ramus with p4–m3; 8841, scapula; 9235, unprepared skull.

Early Orellan, Cook Ranch l.f., Jefferson County, Montana: several specimens in the CM and University of Montana collections (Tabrum *et al.*, 1996, Table 10).

Late Orellan: From the Upper Nodules (upper Scenic Member), Big Badlands, South Dakota: AMHH 1108, skull; 1154, left Mc3; 1177, right ramus; 529a, maxillae. From Fitterer Ranch, Stark County, North Dakota: F:AM 112164, juvenile skull and jaws, additional uncatalogued postcranial material. From Frank Kostelecky Ranch, Stark County, North Dakota: F:AM 112163, left maxilla with P1–M1, additional uncatalogued material, including partial skull, several rami, miscellaneous postcranials.

Early Oligocene: MGS 1794, mandible, from the marine Byram Formation, Vicksburg Group, Hinds County, Mississippi (Manning, 1997).

Known distribution: Late Chadronian–late Orellan (late Eocene to late early Oligocene) of the Rocky Mountains and High Plains; early Oligocene of Mississippi.

Diagnosis: Largest species of *Subhyracodon*. P2 completely molariform. P3 and P4 usually have a trace of the posterior projection of the protoloph, but metaloph is strong and generally merged with protoloph. In some specimens (e.g., "*metalophus*"), P3–4 are almost completely molarized. Antecrochets may be present on M1–2.

Description: *Subhyracodon occidentalis* has been fully described before (Osborn, 1898c; Wood, 1927; Scott, 1941), so there is nothing further to add to these descriptions.

Discussion: Chapter 2 discusses the comparative molarization of the upper premolars of the "species" of *Subhyracodon*. The *trigonodus-occidentalis* morphologies are extremely similar, differing only slightly in size and in degree of development of the posterior projection of the protoloph. Likewise, the degree of molarization of P3–4 grades continuously from *copei* to *trigonodus* to *occidentalis* to *metalophus*. Such intergradation makes it impossible to separate these species. All these morphotypes occur together in Harvard Fossil Reserve Quarry, yet the eight skulls in this sample are nearly identical in size. This evidence is conclusive that they are all one species. *S. occidentalis* and *S. "trigonodus"* have identical stratigraphic ranges, and the range of *S. "metalophus"* (late Orellan) is entirely subsumed by that of *S. occidentalis*.

In earlier works, the small sample size gave the impression that the "species" of *Subhyracodon* formed a graded series, which succeeded each other in time (as stated by Wood, 1941; Scott, 1941, p. 817). But the larger sample sizes have obscured this morphocline. Instead the "species" of *S. occidentalis* are end members of nearly continuous variation within a single time frame. Both primitive and derived extremes of the morphocline persist throughout the entire late Chadronian to late Orellan.

Subhyracodon gidleyi (Wood, 1927) is based on a specimen (USNM 11337) that is identical in size and premolar morphology with *S. occidentalis*. Its only alleged distinguishing feature is the presence of strong antecrochets on M1–2. This feature is incipient on many specimens of *S. occidentalis*, so it is not distinctive. No other specimens with such strong antecrochets have turned up since 1927 in the hundreds of skulls of *Subhyracodon* that are known. It is therefore likely that *S. gidleyi* is not a good species based on a population with this tooth morphology, but based on an aberrant individual of *S. occidentalis*, as suggested by Scott (1941, p. 816).

Aceratherium exiguum (Lambe, 1908) from the Chadronian of Cypress Hills, Saskatchewan, is based on an edentulous symphysis of the right size and morphology for *S. occidentalis*, and is here considered to be a junior synonym.

Rhinoceros hesperius was named by Leidy in 1865 based on a lower jaw (MCZ 29118) found by J.D. Whitney in the "auriferous gravels" in Chili Gulch, Calaveras County, California (see Chapter 3). This specimen was described and extensively discussed by Wood (1960). Wood stated that its overall size compared with *S. tridactylus*, but that its teeth were the size of *S. occidentalis*. He also stated that it was larger than *Diceratherium annectens* and smaller than *D. armatum*. Because it did not compare well with any of the known species, Wood (1960) considered *Subhyracodon hesperius* to be a valid species.

In re-examining the specimen, I found that it is no larger than typical *S. occidentalis* from the Lower Nodules of the Big Badlands (e.g., AMNH 1135) in both its jaw and teeth. I see no comparison between it and *Diceratherium tridactylum* or any other species of *Diceratherium*. I therefore consider it a junior synonym of *S. occidentalis*.

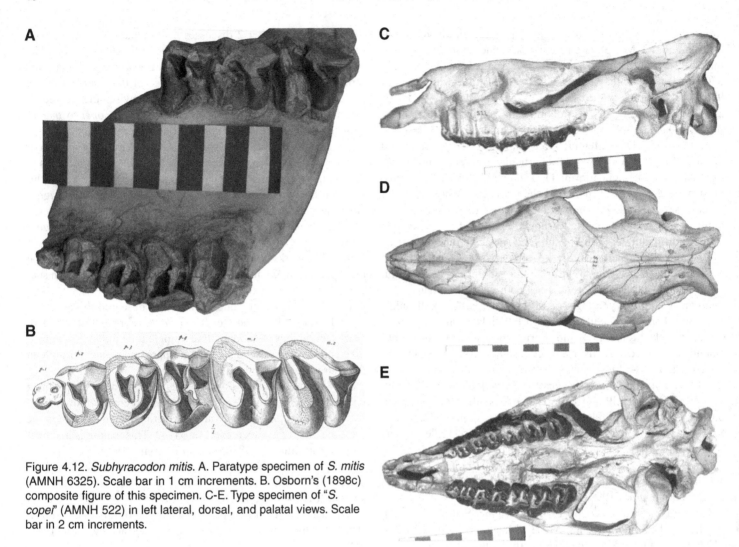

Figure 4.12. *Subhyracodon mitis.* A. Paratype specimen of *S. mitis* (AMNH 6325). Scale bar in 1 cm increments. B. Osborn's (1898c) composite figure of this specimen. C-E. Type specimen of "*S. copei*" (AMNH 522) in left lateral, dorsal, and palatal views. Scale bar in 2 cm increments.

Subhyracodon mitis (Cope, 1875)

Aceratherium mite Cope, 1875
Aceratherium (Subhyracodon) mite Brandt, 1878
Caenopus mitis Cope, 1880
Aceratherium (Caenopus) mite Osborn and Wortman, 1894
Aceratherium copei Osborn, 1898
Caenopus (= Subhyracodon) copei Osborn and Wortman, 1909
Caenopus copei Peterson, 1920
Caenopus copei Troxell, 1921 (in part)
Subhyracodon occidentalis Sinclair, 1924
Subhyracodon copei Wood, 1927
Subhyracodon copei Scott, 1941

Figure 4.12-4.13, Table 4.3

Holotype: AMNH 6325, a badly damaged and nearly edentulous mandible. The paratype (also AMNH 6325) consists of a palate and some postcranial fragments that were probably associ-ated. From the Chadronian Horsetail Creek Member, Brule Formation, Logan County, Colorado.

Hypodigm: *From the late Chadronian*: The type and AMNH 6236 (paratype material from Colorado); AMNH 14577, palate, from Government Slide, Beaver Divide, Natrona County, Wyoming; F:AM 111872, partial skull with right P3–M3, 20 feet above Ash G, Flagstaff Rim, Natrona County, Wyoming; F:AM 112160, palate, 185 feet below the 50-foot correlator ash, Ledge Creek, Natrona County, Wyoming; F:AM 111874, partial skull, from Douglas, Converse County, Wyoming; F:AM 111875, max-illla with right P4–M3, left P2–M3; F:AM 112161, partial skull, both from Spring Draw, Seaman Hills, Niobrara County, Wyoming; AMNH 9711, right dP4, incisors, other fragments, from Pipestone Springs, Jefferson County, Montana; AMNH 1268, fragmentary rami; 12299, right and left rami; 1157, left Mc4; 1014, patella; 1162, fragmentary foot bones; 12452, skull, all from the Chadron Formation, Shannon County, South Dakota.

From the early Orellan: AMNH 522 (a complete skull, the type of "*Aceratherium copei*"); AMNH 524, skull; F:AM 11873, skull

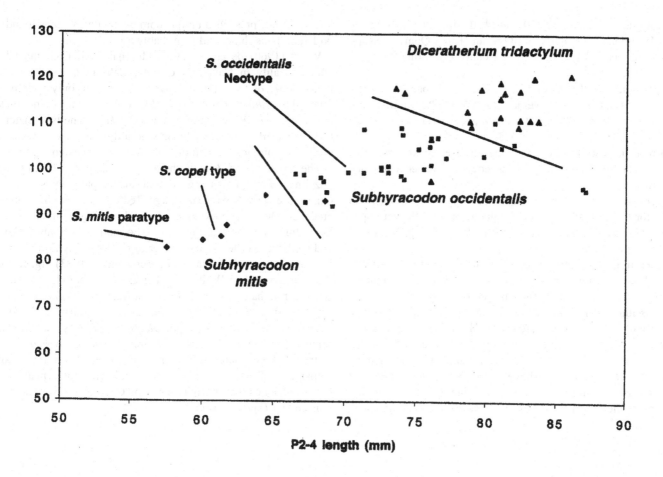

Figure 4.13. Graph of upper molar row length vs. upper premolar row length of *Subhyracodon mitis* (diamonds), *Subhyracodon occidentalis* (squares), and *Diceratherium tridactylum* (triangles). The three species show a trend in increasing size with only slight overlap of size ranges.

and right ramus; AMNH 38941, palate; AMNH 704, anterior portion of skull, all from the Orellan lower nodular beds of South Dakota.

Known distribution: Late Chadronian to early Orellan of the High Plains (South Dakota, Nebraska, Colorado, Wyoming) and Montana.

Diagnosis: Smaller species of *Subhyracodon*; P2 completely molariform. P3 and P4 with strong posterior extensions of protoloph; metaloph not connected to protoloph in P3 and only slightly so in P4.

Description The type material was described and figured by Osborn (1898c) and again by Scott (1941).

Discussion: *Subhyracodon mitis* is the common late Chadronian rhino. Although it appears to grade into *S. occidentalis*, in size and in the molarization of its upper premolars it can be consistently distinguished (Fig. 4.13). Both species are found sympatrically in the early Orellan lower Oreodon beds of the Big Badlands. The clear difference in size between the two species in these deposits strengthens the case for their separation into two species.

Wood (1927, p. 56) and Scott (1941) attempted to separate the type material of *S. mitis* from *S. copei*, based on the differences in size and molarization of the upper premolars. As discussed above, the paratype palate of *S. mitis* is one of the smallest specimens of this species known (Fig. 4.13), but within the currently known range of variation. In addition, the peculiarities of the teeth appear to be yet another manifestation of the highly variable crests on the upper premolars. No new specimens have appeared in over a century to corroborate the peculiarity of this specimen, so I regard it as an aberrant individual, not a distinct species.

Subhyracodon kewi Stock, 1933

"Probably *Rhinoceros hesperius*" Leidy, 1869, pp. 231
Subhyracodon kewi Stock, 1933
Subhyracodon occidentalis Lander, 1983

Figure 4.14, Table 4.3

Holotype: LACM (CIT) 1205, crushed partial skull, from the ?late Whitneyan or early Arikareean Kew Quarry, Sespe Formation, Las Posas Hills, Ventury County, California (Prothero *et al.*, 1996).

Hypodigm: LACM (CIT) 1221, 1222, 1225, paratype upper teeth, LACM (CIT) 1224, lower jaw; UCMP 37588, maxilla, from Kew Quarry. MCZ 9020, a left M3 and MCZ 9021, a partial left M2, from Chili Gulch, Calaveras County, California.

Known distribution: ?Late Whitneyan or early Arikareean of Ventura County, California; ?Oligocene of Calaveras County, California.

Diagnosis: Similar in size and morphology to *S. occidentalis* except for the accessory ribs on the lingual side of the ectolophs of P4, M2, and M3, lack of lingual cingula on the upper molars, and the incipient crochet on M3.

Description: Stock (1933) described and figured the material of *S. kewi* known to him, and Wood (1960) added additional materials. No new specimens have since been discovered.

Discussion: *Subhyracodon kewi* is almost identical to the type specimen of *S. "trigonodus"* (AMNH 528) in size and morphology, except for the accessory ribs on the lingual side of the ectolophs on P4, M2, and M3. This is not found in any other specimen of *Subhyracodon* I have seen. Some specimens of *Diceratherium* begin to show this feature, but their premolars are much more molarized than those of *S. kewi,* and they are much

larger. *S. kewi* presents a unique combination of features found in no High Plains rhino, and is probably a valid species.

Wood (1960) described two additional teeth from the Chili Gulch locality in Calaveras County, California, which closely match those of the type of *S. kewi*. They were only tentatively referred to "*Rhinoceros hesperius*" by Leidy, so they do not make the name *hesperius* a senior synonym of *kewi*. The type material of "*Rhinoceros hesperius*" belongs to a different taxon, which is probably a junior synonym of *S. occidentalis* (see above).

The presence of *S. kewi* in the early Arikareean of California is puzzling. Hunt (1998) argued that the amphicyonids of Kew Quarry are early Arikareean in age. Tedford *et al.* (1987) considered Kew Quarry approximately equivalent in age with the Gering Formation in Nebraska, the Sharps Formation in South Dakota, and the lower Turtle Cove Member of the John Day Formation in Oregon (all earliest Arikareean). But Lander (1983) argued that Kew Quarry is late Whitneyan, based on the oreodonts, rhinos, and other archaic groups. *S. kewi* is a form that might have been expected from the late Chadronian or early Orellan in the High Plains. By the early Arikareean, *Diceratherium* was the dominant rhinoceros in the High Plains and Oregon, and *Subhyracodon* (as restricted here) had been extinct for over 2 million years. Apparently, *S. kewi* is a primitive endemic that remained in California long after its High Plains and Oregon relatives had been replaced by *Diceratherium*.

A

B

C

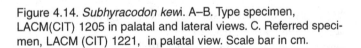

Figure 4.14. *Subhyracodon kewi*. A–B. Type specimen, LACM(CIT) 1205 in palatal and lateral views. C. Referred specimen, LACM (CIT) 1221, in palatal view. Scale bar in cm.

Diceratherium Marsh, 1875

Rhinoceros Marsh, 1870 (non Linneaus)
Diceratherium Marsh, 1875
Aceratherium Cope, 1879 (non Kaup)
Aceratherium Osborn, 1893 (in part)
Caenopus Hay, 1902 (in part)
Metacaenopus Cook, 1908
Epiaphelops Cook, 1912
Coenopus [sic] Gidley, 1924 (in Deussen, 1924)
Subhyracodon Wood, 1927 (in part)
Caenopus Wood and Wood, 1937 (in part)

Figures 4.13, 4.15–4.18, Table 4.4

Type species: *Diceratherium armatum* Marsh, 1875
Included species: Type species and *D. tridactylum* (Osborn, 1893); *D. annectens* (Marsh, 1873); *D. matutinum* (Marsh, 1870); *D. niobrarense* Peterson, 1906.
Known distribution: Early Whitneyan to ?late Hemingfordian, High Plains, Montana, Oregon, and New Jersey.
Diagnosis: Moderate to large rhinocerotids with paired ridge-like subterminal nasal horns or rugosities in males; hornless in females. P2–4 nearly completely molariform. Molars with strong cingula, relatively simple crown patterns, nearly always lacking crochets or cristae. Skull relatively long, with long basicranial axis and flat dorsal profile. Limbs elongate, but metapodials relatively short. Distinguished from *Subhyracodon* by the presence of nasal rugosities in male skulls, molarized upper premolars, and larger size. Distinguished from *Menoceras* by the dolichocephalic skull with long basicranial region and flat dorsal profile, subterminal ridge-like paired nasal horns (not the terminal spherical bosses seen in *Menoceras*), relatively simple teeth with no crochets or cristae, and larger size.
Discussion: The name *Diceratherium* has long been indiscriminately applied to all rhinoceroses with paired nasal horns. Naturally, this included the paired-horned *Menoceras* from Agate Springs Quarries in Nebraska. Because Agate *Menoceras* is by far the most common horned rhino in most American collections, it has become representative of "*Diceratherium*" in the minds of many paleontologists.

Troxell (1921d) first pointed out the differences between *Menoceras* and *Diceratherium*, but he was not taken seriously by later workers. Most of the other taxa Troxell created were demonstrably invalid, so their reaction is not surprising. Finally, Tanner (1969) pointed out just how different the two taxa are and brought *Menoceras* back from obscurity. Many museums in this country still have an Agate *Menoceras arikarense* mislabeled "*Diceratherium cooki*."

The distinction goes far beyond the differences in horn shape. *Diceratherium* s.s. is a primitive rhinoceros in most features and appears to be a sister-group of the *Subhyracodon* lineage. *Menoceras*, on the other hand, is a very derived rhino that has many features that unite it with higher rhinos. It does not have

North American sister-taxa, but instead appears to be an immigrant related to *Pleuroceros* from the late Oligocene and early Miocene of Europe, and *Menoceras zitteli* is known from the lower Miocene of Europe. Prothero *et al.* (1986) and Antoine (2002) demonstrated that the only advanced feature shared by *Diceratherium* and *Menoceras* is the paired nasal horns, and these are formed in entirely different ways in the two genera. This one feature is outweighed by the many synapomorphies uniting *Menoceras* with other higher rhinoceroses (Prothero *et al.*, 1986; Figure 4.1), a conclusion confirmed by Antoine (2002).

Diceratherium tridactylum (Osborn, 1893)

Aceratherium tridactylum Osborn, 1893
Diceratherium proavitum Hatcher, 1894
Caenopus tridactylus Hay, 1902
Aceratherium tridactylum Loomis, 1908
Caenopus tridactylus avus Troxell, 1921
Diceratherium tridactylum Osborn, 1923
Diceratherium avum Wood, 1927
Subhyracodon tridactylum [sic] Wood, 1927
Diceratherium tridactylus [sic] Scott, 1941
Diceratherium tridactylus avus [sic] Scott, 1941
Diceratherium tridactylus metalophus [sic] Scott, 1941

Holotype: AMNH 538, a complete articulated skeleton of a female individual from the *Protoceras* beds (late Whitneyan) of South Dakota.
Hypodigm: *Early Whitneyan Lower Poleslide Member, Brule Formation, Big Badlands, South Dakota*: AMNH 534, skull and mandible; AMNH 578, right ramus with dp1–3; AMNH 1123, skull, Mt4; AMNH 525, right forelimb; AMNH 1106, right scapula, pelves, sacrum, vertebrae; AMNH 526, forelimb; AMNH 537, skull, partial skeleton; AMNH 536, vertebrae, limb bones.
Middle to late Whitneyan Upper Poleslide Member, Brule Formation, Big Badlands, South Dakota: AMNH 1124, male skull and mandible; AMNH 541, male skull; AMNH 539, femur; AMNH 1127, mandible; AMNH 543, mandible; AMNH 564, scapula; AMNH 1112, right ramus with dp2–4; AMNH 1109, mandible; AMNH 1137, skull and rami; AMNH 1120, skull; AMNH 1121, female skull; AMNH 1126, male skull, mandible, partial skeleton; AMNH 1116, right Mt2, 3, left Mt3; AMNH 573a, carpus, metacarpals; AMNH 1152, partial left hindlimb; AMNH 544a, left cuneiform; AMNH 1122, skull; AMNH 1118, juvenile skull; additional uncatalogued skulls and jaws in the AMNH collections.
Middle Whitneyan (80 feet above Lower Whitney Ash), 66 Mountain, Goshen County, Wyoming: F:AM 112169, left maxilla with P1-M2, mandible.
Late Whitneyan, Harris Ranch, Fall River County, South Dakota: F:AM 112170, left maxilla with P3–M2, additional uncatalogued material.
Late Whitneyan, Killdeer Formation, Chalky Buttes, Slope County, North Dakota: AMNH 8092, radius, atlas, teeth, vertebral

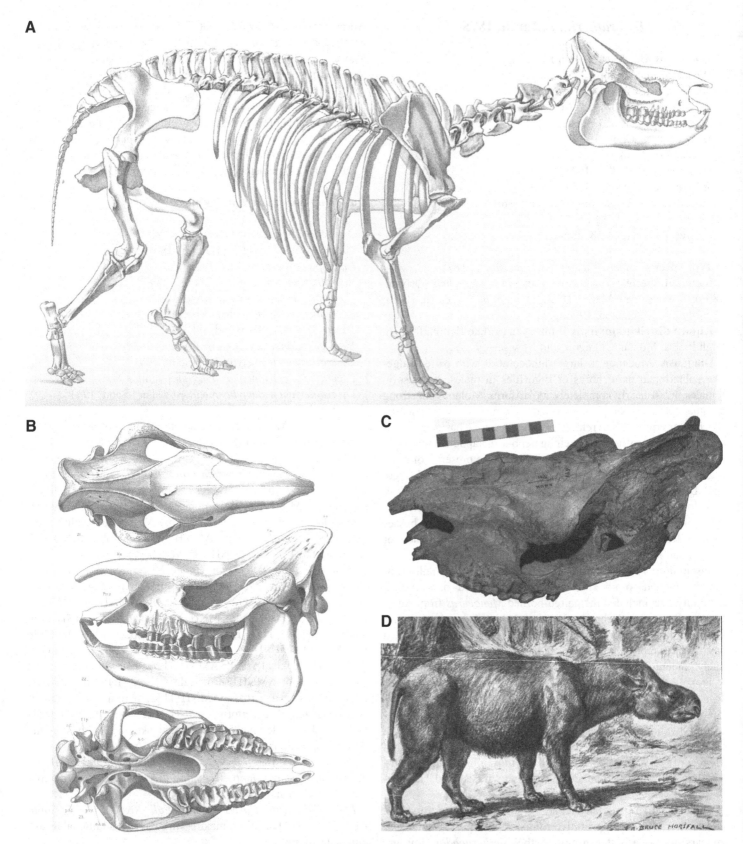

Figure 4.15. *Diceratherium tridactylum*. A. Type skeleton, AMNH 538, after Osborn (1898c). B. Type skull in dorsal and lateral views; palatal view after AMNH 541 (after Osborn, 1898c). C. Robust male skull (AMNH 541) with incipient horn ridges on nasals. Scale bar in 2 cm increments. D. Restoration by B. Horsfall (in Scott, 1913)

fragments; AMNH 8088, female skull; AMNH 8089, cranium; AMNH 8091, partial skull; AMNH 8090, left ramus with dp4–m2; AMNH 80228, right M2.

Vista Member, Logan County, Colorado: AMNH 8819, left ramus with m1–3; AMNH 8834, vertebrae; AMNH 8826, carpals; AMNH 8833, metapodials; AMNH 8837, skull; AMNH 8827, vertebrae.

Known distribution: Whitneyan of the High Plains (South Dakota, North Dakota, Wyoming, Nebraska, Colorado).

Diagnosis: Medium-sized species of *Diceratherium*; P2 molariform; P3–4 nearly or completely molariform. Rugosities extending anteroposteriorly on the anterolateral edge of nasals in male skulls; females have no noticeable rugosity.

Description: *D. tridactylum* was thoroughly described by Osborn (1898c) and Scott (1941). The type specimen is a complete skeleton in a panel mount, which was on display for a long time at the American Museum. Although many skulls and dentitions of *D. tridactylum* have been figured, the type skull has never had the jaws prepared away from the skull, because it was a display specimen. Thus, no one has ever seen the crowns of the type dentition. However, the variability of the upper premolars in *D. tridactylum* is great enough that the crown pattern is only of limited use in recognizing the species.

Discussion: *D. tridactylum* has usually been referred to *Subhyracodon* because it intergrades with *Subhyracodon occidentalis* in the White River Group. Hatcher (1894b), Osborn (1923), and Scott (1941) pointed out that the presence of strong nasal rugosities foreshadowed the nasal ridges of *D. armatum*. Some Whitneyan specimens (e.g., "*Diceratherium avus*") actually have fully developed paired nasal ridges (Fig. 4.15C). In addition, the dentition shows the derived characteristics of *Diceratherium*. I therefore follow Osborn (1923) and Scott (1941) in returning the species *tridactylum* to *Diceratherium*. It has been united with *Subhyracodon* in the past solely on the basis of primitive characters. Drawing the generic boundary between *S. occidentalis* and *D. tridactylum* is also more natural, since this definition unites all forms with paired nasal ridges or rugosities and with fully molarized premolars.

The supposed "intermediacy" of the "*Subhyracodon avus*" and "*S. metalophus*" subspecies (Wood, 1927; Scott, 1941) is not supported by the stratigraphic data and the morphologic variability of the collections now available. These "subspecies" are simply individual variants, and do not deserve taxonomic recognition.

Hatcher (1894b) named the species *Diceratherium proavitum*, based on a skull with slightly different premolars from the Whitneyan of South Dakota. This taxon was quickly transferred to *Aceratherium tridactylum* by Osborn and Wortman (1895), and then to *Caenopus tridactylus* by Troxell (1921d), and then to *Subhyracodon tridactylum* [*sic*] by Wood (1927, 1961) and finally to *Diceratherium tridactylus* [*sic*] by Scott (1941).

Diceratherium armatum **Marsh, 1875**

Diceratherium armatum Marsh, 1875
Aceratherium truquianum Cope, 1879
Diceratherium gregorii Peterson, 1920
Diceratherium lobatum Troxell, 1921
Metacoenopus gregorii Troxell, 1921

Figure 4.16, Table 4.4

Holotype: YPM 10003, adult male skull, from the lower John Day Formation, Oregon.

Hypodigm: *Early to middle Arikareean, Turtle Cove Member, John Day Formation*: YPM 12487 (type of "*Diceratherium lobatum*"); AMNH 7333, symphysis, right and left rami with m2–3 (type of "*Aceratherium truquianum*"); AMNH 7321, cranium; AMNH 7339, right maxilla with P4–M3; AMNH 7312, left dP4, left dP3; AMNH 7469, ulna; AMNH 7380A, humerus. UCMP 31930, palate; UCMP 76627, palate; UCMP 77432, skull; UCMP 582, palate; UCMP 31930, palate; UCMP 39241, ramus; UCMP 73432, palate; UCMP 75987, ramus; JODA 5794, male jaw.

Early Arikareean, Gering Formation, Schomp Ranch, 8.5 miles north of Mitchell, Sioux County, Nebraska,: F:AM 112172, skull.

Early Arikareean, Gering Formation, Little Muddy Creek, Niobrara County, Wyoming: F:AM 112173, palate.

Early Arikareean, Harris Ranch, ? Sharps Formation, Fall River County, South Dakota: F:AM 112181, left P2–M3.

Early Arikareean, Sharps Formation, Shannon County, South Dakota: SDSM 53484, left maxilla and ramus; SDSM 54399, P2–M1; SDSM 54165, M2; SDSM 64198 m1; SDSM 54188, right ramus with m1–2; SDSM 54144, palate; SDSM 53584, left P2–M3, right I2–P4, mandible.

Early to middle Arikareean, Cabbage Patch beds, Granite County, Montana: numerous specimens in the KU and University of Montana collections.

Middle Arikareean, Monroe Creek Formation, Muddy Creek, Niobrara County, Wyoming: F:AM 112174, skull, jaws; F:AM 112175, skull, ramal fragments.

Early late Arikareean, Harrison Formation, 77 Hill Quarry, Niobrara County, Wyoming: a large sample, including F:AM 112177, male skull, rami; F:AM 112178, limb elements.

Early late Arikareean, Harrison Formation, north of Keeline, Niobrara County, Wyoming: F:AM 112176, skull, mandible, atlas.

Early late Arikareean, Harrison Formation, North Ridge locality, Niobrara County, Wyoming: F:AM 112175, skull and ramal fragments, additional uncatalogued material, limb elements.

Early late Arikareean, ?Harrison Formation, Niobrara River, near mouth of Bear Creek, Cherry County, Nebraska: F:AM 112179, radius.

Early late Arikareean, Turtle Butte Formation, Tripp County, South Dakota: F:AM 42954, right radius; F:AM 42958, proximal right Mc3.

Arikareean, Madison Valley Formation, near Belgrade, Montana: USNM 11682, anterior portion of male skull (Wood, 1933).

Late Arikareean Toledo Bend l.f., Newton County, Texas: Numerous specimens documented by Albright (1999b).

Known distribution: Early to late (but not latest) Arikareean, High Plains, Montana, Oregon, and Texas.

Diagnosis: Largest species of *Diceratherium*. P2–4 completely molariform. Molars simple, without any cristae or crochets, except for faint precursors of crochets on M2–3. Prominent rugose lateral ridges on nasals of male skulls; female skulls with no ridges or rugosities.

Description: The cranial material of *D. armatum* was thoroughly described by Peterson (1920). Additional well-preserved skulls from the Arikareean of Wyoming (Fig. 4.16E-J) show the widely flaring nasal ridges even better than the type material. Virtually all the previously undescribed postcranial skeleton of *D. armatum* is now known from large samples at 77 Hill Quarry. It will be described in detail below (Chapter 5). The most striking feature of the postcranial skeleton is the relatively long limbs. *D. armatum* and especially *D. niobrarense* were clearly long-limbed, running rhinos, much more so than any other North American Oligocene or Miocene rhinoceros.

Discussion: *D. armatum* is the best-known species of *Diceratherium*. Although it was originally described from the John Day Formation of Oregon, it is also very abundant until the late Arikareean in the High Plains. During most of this time, it was sympatric with its smaller relative, *D. annectens*. The two species are similar in all morphologic features except size.

Diceratherium gregorii was described by Peterson (1920) based on a female skull from the "Lower Rosebud beds" of South Dakota. According to Macdonald (1963, 1970), specimens referable to both *D. gregorii* and *D. armatum* occur from near the base of the Sharps Formation to the top of the Arikaree Group in South Dakota. Thus, the temporal ranges of *D. gregorii* and *D. armatum* are coextensive. In morphology, the only feature that distinguishes *D. gregorii* is size. *D. gregorii* is usually on the small end of the *D. armatum* size range (Table 4.4), and in some dimensions it is intermediate between *D. armatum* and *D. annectens*. However, none of the other diagnostic features given by Peterson (1920, pp. 421–423) is valid. The type skull has been crushed dorsoventrally, but otherwise it is indistinguishable from the normal range of variation seen in *D. armatum*. It is slightly less robust and rugose, but this is probably because it is a female skull. There seems to be no valid reason for retaining *D. gregorii*.

As Peterson (1920, p. 412–413) has shown, Cope's (1879) "*Aceratherium truquianum*" is probably a based on a symphysis of *D. armatum*. Green (1958) pointed out that Troxell's (1921d)

Diceratherium lobatum is also a *D. armatum* with slightly aberrant premolars. P3–4 in the type of *D. lobatum* (YPM 12487) have cristae, a weak crochet, and a "mure" between the protoloph and metaloph. The type of *D. lobatum* is well within the range of size variation of *D. armatum*, and comes from the same Turtle Cove Member of the John Day Formation that yields the type material of *D. armatum*. Because upper premolars are notoriously variable in these rhinos (see Chapter 2), this single aberrant individual does not merit taxonomic recognition. In the same paper, Troxell (1921d) referred *D. gregorii* to Cook's misguided genus, *Metacaenopus*, although he considered the species to be of doubtful validity.

Albright (1999b) has since documented specimens of *D. armatum* from the late Arikareean Toledo Bend l.f., Newton County, Texas, extending the range of this species beyond the Pacific Northwest, Rocky Mountains, and High Plains.

Diceratherium annectens (Marsh, 1873)

Rhinoceros annectens Marsh, 1873
Diceratherium nanum Marsh, 1875
Diceratherium annectens Peterson, 1920
Diceratherium cuspidatum Troxell, 1921
Coenopus [*sic*] sp. Gidley, 1924 (*in* Deussen, 1924)
Caenopus premitis Wood and Wood, 1937

Figure 4.17, Table 4.4

Holotype: YPM 10001, left P2–4 and I1, John Day beds, Oregon (stratigraphic level unknown).

Hypodigm: *Early to middle Arikareean, lower John Day beds (Turtle Cove Member), Oregon*: YPM 10004 (type of "*D. nanum*"); AMNH 7324, a complete skull; YPM 00000 (type of "*D. cuspidatum*"); AMNH 7312, fragmentary radius, metapodial; AMNH 7346, right ramus with p2–m3, right p2; AMNH 7317, mandible; AMNH 7325, palate; AMNH 7342, right maxilla with dP2–3; AMNH 7364, fragmentary right ramus; AMNH 7330, left and right ramus; AMNH 7744, right ramus with p4; AMNH 7343, symphysis, right ramus with p2–m3, left m3, right maxilla with p2–m3; AMNH 7314, partial skull; AMNH 7464, cranium; AMNH 7332, juvenile skull; AMNH 7345, juvenile palate; AMNH 7335, left ramus; AMNH 7351, right ramus with dp2–3; AMNH 7319, mandible; AMNH 7318, mandible, humerus; AMNH 7337, tooth fragments; UCMP 92, left maxilla; UCMP 2098, left and right maxillae; UCMP 39213, right maxilla; UCMP 77457, maxilla; UCMP 2095, maxilla; UCMP 31931, maxilla; UCMP 39175, female skull; UCMP 39176, partial skull; UCMP 77432, skull; UCMP 31931, maxilla; UCMP 39175, male skull;

Figure 4.16. (opposite page). *Diceratherium armatum*. A–D. Type skull (YPM 10003) in palatal, dorsal, right lateral, and posterior views. E–F. Large robust male skull, F:AM 112176, in palatal and right lateral views. G–H. Lower jaw of F:AM 112176, showing male i2 tusks. I–J. Comparison of a male skull of *D. armatum* (F:AM 112176) with a male skull of *Menoceras arikarense* (F:AM 86224) from Agate Springs Quarries, showing the contrast in skull shape and proportions and the subterminal nasal flanges of *Diceratherium* contrasted with the terminal nasal bosses of *Menoceras*. Scale bar in 1 cm increments.

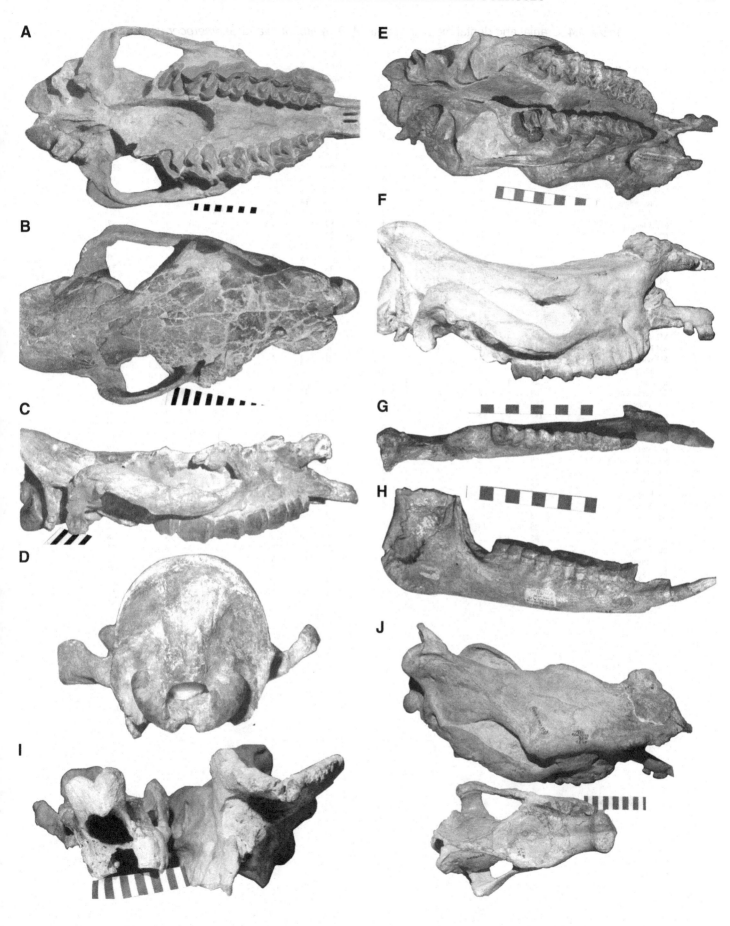

Table 4.4. Cranial and dental measurements of *Diceratherium* and *Skinneroceras* (in mm)

MEASUREMENT	D. tridactylum			D. armatum			D. annectens			D. niobrarense			S. manningi
	Mean	SD	N	Mean	SD	N	Mean	SD	N	Mean	SD	N	F:AM 111853
Nasals to occiput	499	9	3	560	27	4	416	39	3	512	69	2	414
Pmax to condyles	480	24	7	524	49	4	416	25	4	503	48	3	403
Width @ zygoma	265	24	8	272	18	4	225	10	3	275	62	2	156
Height of occiput	161	19	8	188	—	1	119	2	2	173	17	3	114
Width of occiput	92	12	6	165	2	3	127	5	3	160	39	2	92
I1-M3 length	205	20	8	356	18	2	270	37	3	321	53	2	228
I2 length	12	1	4	33	9	2	—	—	—	27	4	2	18
I2 width	13	4	5	16	5	2	—	—	—	9	1	2	9
Diastema length	51	5	5	69	6	3	68	14	3	53	2	2	44
P2-M3 length	176	9	13	224	13	4	177	15	4	223	31	4	151
M1-3 length	103	7	13	132	10	4	82	47	4	129	16	4	97
P2 length	25	2	10	28	3	4	22	2	4	28	4	4	18
P2 width	32	2	9	37	9	4	31	4	4	38	3	4	27
P3 length	29	4	11	30	3	4	26	3	4	33	4	4	20
P3 width	40	3	8	45	5	4	38	5	4	43	5	4	31
P4 length	32	3	10	32	1	4	27	2	4	33	4	4	29
P4 width	44	5	7	48	8	4	42	5	4	49	5	4	33
M1 length	34	5	13	39	2	4	33	5	4	42	5	4	24
M1 width	43	4	13	51	6	4	44	4	4	51	6	4	36
M2 length	37	3	13	57	3	4	39	3	4	43	5	4	32
M2 width	45	4	13	56	5	4	44	5	4	53	6	4	38
M3 length	37	3	3	46	2	4	36	2	4	43	6	4	40
M3 width	40	3	3	51	2	4	41	2	4	51	5	4	38
i2 length	21	2	2	25	5	5	16	5	3	22	1	3	13
i2 width	29	1	2	22	4	5	20	4	3	26	7	3	18
Diastema length	60	9	2	54	4	4	45	11	3	44	7	3	45
p2-m3 length	189	16	13	218	15	5	170	14	3	233	10	3	147
m1-3 length	113	6	13	134	6	5	104	6	5	140	2	4	90
p2 length	27	2	2	26	1	3	18	2	2	26	4	3	19
p2 width	17	1	2	17	1	3	12	1	2	18	1	3	12
p3 length	32	1	2	30	2	4	23	1	4	34	3	4	19
p3 width	19	1	2	23	2	4	18	1	4	22	2	4	14
p4 length	32	1	2	33	4	5	28	2	5	38	2	4	20
p4 width	24	2	2	27	2	5	22	1	5	27	2	4	18
m1 length	31	3	13	41	2	5	30	5	5	43	4	4	23
m1 width	24	2	13	29	2	5	23	2	5	29	2	4	19
m2 length	38	3	13	45	2	5	35	2	5	45	3	4	32
m2 width	25	2	13	30	2	5	24	2	5	30	2	4	22
m3 length	45	1	2	58	4	5	38	4	5	51	2	4	38
m3 width	29	1	2	30	2	5	23	2	5	29	3	4	21
Symphysis to angle	437	—	1	437	13	5	349	17	5	454	38	2	323
Ht. coronoid above angle	261	—	1	224	15	4	177	6	5	238	5	2	150
Length symphysis	103	—	1	95	25	5	60	10	5	91	5	2	61
Depth jaw below p2	70	—	1	62	7	5	50	6	5	63	3	2	52
Depth jaw below m2	74	—	1	72	5	5	62	7	5	77	2	2	61

UCMP 105, jaw; UCMP 92, maxilla; UCMP 1780, teeth; UCMP 1373, teeth; UCMP 39213, teeth; UCMP 2098, teeth; UCMP 77456, jaw; UCMP 74589, jaw; UCMP 76105, jaw; UCMP 462, jaw; UCMP 824, juvenile jaw; UCMP 2103, jaw; JODA 1036, mandible; JODA 1205, female mandible; JODA 832, juvenile maxilla; JODA 3713, right M1–3;

Early Arikareean, Harris Ranch, ?Sharps Formation, Fall River County, South Dakota: F:AM 112184, right ramus with m3.

Early Arikareean, top of Cedar Pass, Sharps Formation, Jackson County, South Dakota: F:AM 112171, male skull.

Early Arikareean, Gering Formation, 3.5 miles east of Tremain, Goshen County, Wyoming: F:AM 112182, juvenile skull.

Early Arikareean, Gering Formation, Schomp Ranch, 8.5 miles north of Mitchell, Sioux County, Nebraska: F:AM 112185, skull.

Arikareean, Deep River beds, Fort Logan Formation, Meagher County, Montana: F:AM 112183, juvenile skull.

Middle Arikareean, Monroe Creek Formation, 3.5 miles west of Muddy Creek bridge, Niobrara County, Wyoming: F:AM 112186, articulated skull, mandible.

Middle Arikareean, Monroe Creek Formation, Fanning Ranch, Scottsbluff County, Nebraska: F:AM 99792, skull and jaws.

Middle Arikareean, Turtle Butte Formation, Tripp County, South Dakota: F:AM 42967, proximal radius.

Early late Arikareean, 10 miles north of Bannack, Beaverhead County, Montana: F:AM 112212, left ramus with m1–3.

Early late Arikareean, Harrison Formation, north of Keeline, Niobrara County, Wyoming: F:AM 112198, female mandible; F:AM 112199, right ramus.

Early late Arikareean, North Ridge, 0.5 mile west of Highway 85, Harrison Formation, Niobrara County, Wyoming: F:AM 112195, male skull; F:AM 112196, back of skull; F:AM 112197, left ramus with I2, P2–M3.

Early late Arikareean, 77 Hill Quarry, Harrison Formation, Niobrara County, Wyoming: F:AM 112194, male skull; F:AM 112200, skull; F:AM 112201, skull; F:AM 112202, skull; F:AM 112203, skull; F:AM 112204, skull; F:AM 112205, female skull;

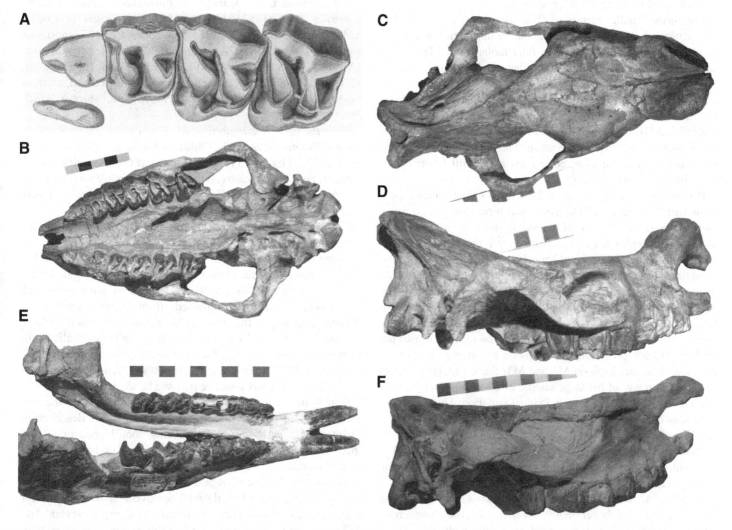

Figure 4.17. *Diceratherium annectens*. A. Type specimen (YPM 10001), after Peterson (1920). B–D. Referred male skull (F:AM 7524) showing characterstic horn bosses. E. Mandible (F:AM 112187) showing typical male i2 tusks. F. Female skull (F:AM 112171) showing reduced nasal ridges. Scale bar in 2 cm increments.

F:AM 112206, skull; F:AM 112207, right maxilla with P3–M3; F:AM 112208, right maxilla with P4–M3; F:AM 112209, posterior half of skull; F:AM 112210, posterior cranium; F:AM 112211, left maxilla with P3–M3; F:AM 112187, adult male skull; F:AM 112188, post cranials.

Early late Arikareean, ?Harrison Formation, mouth of Bear Creek, Cherry County, Nebraska: F:AM 99791, male skull, mandible.

Latest Arikareean, Morava Ranch Quarry, Anderson Ranch Formation, Box Butte County, Nebraska: F:AM 112213, astragalus.

From the late Arikareean Derrick Farm locality, Cedar Run l.f., Washington County, Texas: USNM 6573, left maxilla with dP1, P2–4 (see Wood and Wood, 1937; Prothero and Manning, 1987; Albright, 1999b).

From the late Arikareean Toledo Bend l.f., Newton County, Texas: Numerous specimens documented by Albright (1999b).

Known distribution: Early to late Arikareean, Oregon, Montana, and the High Plains, and Texas.

Diagnosis: Smaller species of *Diceratherium* (P1–M3 length = 174–205 mm). As in *D. armatum*, simple upper molars with no cristae or crochets. All premolars fully molariform. Male skulls have prominent rugose lateral nasal ridges, as *D. armatum*; female skulls have smooth nasals.

Description: The type of *D. annectens* consists only of P2–4, but Peterson (1920, pp. 418–421) described a referred specimen (AMNH 7324) that shows most of the features of the skull, and establishes the validity of the taxon. The postcranial skeleton is now known from the large collections at 77 Hill Quarry. It is described in Chapter 5.

Discussion: The John Day and 77 Hill Quarry samples both show that two species of *Diceratherium* were sympatric through most of the Arikareean. This difference in size cannot be attributed to sexual dimorphism, because skulls with both male and female nasal morphology are known from both size clusters. The first valid name to be applied to the smaller John Day dicerathere is *D. annectens* (Marsh, 1873). As Peterson (1920, p. 417) demonstrated, *D. nanum* Marsh, 1875, is in the same size range, and is therefore a junior synonym. Troxell (1921d) named *D. cuspidatum* for a small John Day dicerathere (the size of *D. annectens*), which has slightly aberrant premolars and cusps on the mures between the protoloph and metaloph on M1 and M3. Since this feature does not appear on any of the large sample of *Diceratherium* from 77 Hill Quarry, it is probably an individual variation. Thus, *D. cuspidatum* is here considered to be an aberrant *D. annectens*, not a valid species.

Prothero and Manning (1987) suggested that the long-misunderstood rhino maxilla from the Cedar Run l.f. of Washington County, Texas (Wood and Wood, 1937) is not *"Caenopus"* but a specimen of *M. arikarense*. Albright (1999b) has since shown that the specimen is a better match for *Diceratherium annectens*. Albright (1999b) also documented numerous additional specimens from the late Arikareean Toledo Bend l.f., Newton County, Texas.

Diceratherium niobrarense Peterson, 1906

Diceratherium niobrarensis [*sic*] Peterson, 1906
Aceratherium egregius Cook, 1908
Diceratherium petersoni Loomis, 1908
Aceratherium egrerius [*sic*] Loomis, 1908
Metacaenopus egregius Cook, 1909
Epiaphelops virgasectus Cook, 1912
Diceratherium niobrarense Peterson, 1920
Diceratherium egrerius [*sic*] Peterson, 1920
Metacaenopus niobrarensis Troxell, 1921
Diceratherium egrerius [*sic*] Wood, 1927

Figure 4.18, Table 4.4

Holotype: CM 1271, young male skull, from latest Arikareean Quarry A, Anderson Ranch Formation, Agate Springs, Nebraska.

Hypodigm: *From the latest Arikareean Anderson Ranch (= upper Harrison, = Marsland) Formation, Sioux County, Nebraska*: AMNH 83591, type of *"Aceratherium egregius"*; AMNH 82679, 83495, CM 1274, foot bones; CM 1272, posterior portion of skull; CM 1907A, juvenile right femur, left femur, tibia; CM 1898, left ramus; CM 1910, partial skeleton (ramus, atlas, humerus, scapula, tibiae, miscellaneous ribs and podials), all from Agate Springs Quarry A; F:AM 129669, skull, Morava Ranch Quarry.

From the early Hemingfordian Runningwater Formation, Sioux County, Nebraska: AMNH 98081, referred to *"Diceratherium egrerius"* [*sic*] by Wood (1927), Bridgeport Quarries, Nebraska;

From the ?late Arikareean or ?late Hemingfordian Mollie Gulch locality, Railroad Canyon, Lemhi County, Idaho: F:AM 111848, anterior portion of skull.

Known distribution: Latest Arikareean to early Hemingfordian, High Plains; ?late Hemingfordian, Idaho.

Diagnosis: Medium-sized *Diceratherium* (length of P1–M3 = 185–255 mm) with raised bosses in the center of elongate lateral nasal ridges on male skulls. Weak crochets on some molars. I2 occasionally lost. Distinguished from earlier species of *Diceratherium* by the distinct boss on the nasal ridges. Distinguished from *Menoceras* by its more dolichocephalic skull, subterminal nasal bosses which are developed on lateral ridges, presence of I2, and weaker crochets. Unlike *Menoceras*, its skull profile is only slightly concave (not saddle shaped).

Description: Peterson (1920, pp. 424–431) described the cranial material of *D. niobrarense*. The postcranials are described in Chapter 5. *D. niobrarense* is the most long-legged of all the species of *Diceratherium*, and therefore was probably also the most cursorial.

Discussion: The larger Agate Springs rhino (originally known only from Quarry A) has always been overshadowed by the far more abundant *Menoceras arikarense*. As a result, several late Arikareean and early Hemingfordian rhinos were compared only to *Menoceras* and later rhinos, but not to *Diceratherium*. The consequence of this practice was that several invalid taxa were creat-

A

B

C

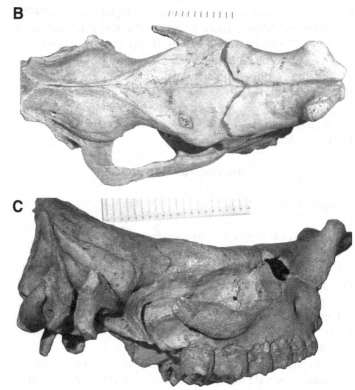

Figure 4.17. *Diceratherium niobarense.* A-C. Type skull (CM 1271) in palatal, dorsal, and right lateral views. Note that the nasal ridges in this male skull are beginning to develop small bosses, but they are subterminal extensions of the lateral nasal ridge, not terminal knobs as in *Menoceras* (see Fig. 4.15I-J, and Fig. 4.19). Scale in 1 cm increments.

ed, because they differed noticeably from *Aphelops* or "*Caenopus*," but no comparisons were made with *Diceratherium*.

Loomis (1908) named *Diceratherium petersoni* (from Agate Quarry A) for a left M2–3 larger than the type of *D. niobrarense*. The type of *D. niobrarense* is the smallest known individual of this species. Several larger individuals of *D. niobrarense* are now known that are the size of *D. petersoni*. This makes *D. petersoni* a junior synonym of *D. niobrarense*.

Cook (1908, 1909) created the new genus and species *Metacaenopus egregius* (misspelled "*egrerius*" by Loomis, 1908, Peterson, 1920, and Wood, 1927) based on a skull and jaws of a hornless rhino from the "lower Harrison beds," 4 miles west of Agate Springs Quarry. Its faunal association (Cook, 1909) clearly indicates an Agate-equivalent age for this locality, suggesting that it comes from the upper Harrison Formation of Peterson (1920). Cook compared this specimen to "*Caenopus*" but not to *Diceratherium* or *Menoceras*, even though he must have been aware that the female skulls of *Diceratherium* and *Menoceras* lacked horns. In most respects, particularly in size, and in the dolichocephalic skull with narrow zygoma and lambdoid, "*Metacaenopus egregius*" is a good match for many female specimens of *D. niobrarense*. The only difference is the apparent loss of I2, which resembles the condition in *Menoceras*. However, the overall skull shape seems to be a much more reliable indicator than the loss of this single tooth in one individual, a character that has proven to be highly variable.

Cook (1912a, 1912b) erected another invalid genus and species (in fact, none of his rhino taxa have proven to be valid) for a

mandible from the Runningwater Formation, 18 miles east of Agate. It was associated with a typically Runningwater fauna. Cook compared it only to "*Caenopus*" and *Aphelops*, but not to *Diceratherium*, and thus named it a new genus and species, *Epiaphelops virgasectus*. It is a perfectly normal *D. niobrarense* mandible with no features that might suggest any other affinities. *Epiaphelops virgasectus* is here synonymized with *D. niobrarense*.

Tanner (1969) considered *D. niobrarense* to be a junior synonym of *Menoceras arikarense*. His only justification for this synonymy was the supposed "frontal convexity" of the types of *D. niobrarense* and *M. arikarense*, and their occurrence in the same quarry. To clarify the latter point first, Quarry A at Agate is very different from the main quarries, and yields a different fauna. Secondly, Peterson (1920) clearly demonstrated that *D. niobrarense* is a very different animal from *Menoceras*. *D. niobrarense* has a much larger, more dolichocephalic skull, with much simpler upper teeth with weak crochets on the molars. It has the subterminal nasal bosses developed on a *Diceratherium*-style lateral nasal ridge, not the terminal spherical nasal knobs of *Menoceras*. Because Tanner (1969) was the first to clarify the long-obscured distinction between *Diceratherium* and *Menoceras*, it is surprising that he confused the two. *D. niobrarense* is a valid species, not a junior synonym of *M. arikarense*.

The front portion of a male skull (F:AM 111848) was recovered by Charles Falkenbach from the Railroad Canyon area, Lemhi County, Idaho. It clearly is referable to *Diceratherium niobrarense*. Although fossils of Arikareean, Hemingfordian, and

Barstovian age are known from Railroad Canyon (Nichols, 1998), Barnosky and Nichols (pers. comm.) believe that the specimen came from the Mollie Gulch locality. Barnosky and Nichols (pers. comm..) regard this locality as late Hemingfordian in age. If so, this is the youngest known occurrence of *Diceratherium*. However, Tedford *et al.* (2004, fig. 6.2I) regard this locality as latest Arikareean, consistent with its appearance in the High Plains, and requiring no range extension of *Diceratherium* into the late Hemingfordian.

Skinneroceras new genus

Figure 4.19, Table 4.4

Type species: *Skinneroceras manningi* new species.
Included species: Type species only
Etymology: In memory of Morris Skinner (1906–1989), one of the greatest paleontologists and fossil collectors of the twentieth century, who found the type specimen.
Known distribution: "Brown siltstone member" of the White River Formation (Tedford *et al.*, 1996), Nebraska (earliest Arikareean).
Diagnosis: Small rhino, smaller than any known species of *Diceratherium* and most species of *Subhyracodon*, but with a similar elongated skull and flat dorsal skull profile (unlike that of *Menoceras*). Its most distinctive feature is the well developed parasagittal temporal crests. Strong antecrochets on molars.
Description: F:AM 111843 is smaller than the smallest individual of *D. tridactylum* or *D. annectens*, and close to *Subhyracodon mitis* in size, even though it is fully adult with well-worn teeth, which are too advanced to be referable to *S. mitis*. F:AM 111843 has been laterally crushed and distorted, but otherwise it is complete and relatively well preserved. The most prominent features of the skull are the strong parasagittal temporal crests, which meet at the occiput and connect with heavy supraorbital rugosities. These crests are faint in most skulls of *Subhyracodon* or *Diceratherium*, but in no specimen are they as strongly developed as in *Skinneroceras*. The dorsal skull profile is slightly concave, as in *Diceratherium* rather than saddle-shaped, as in higher rhinos. The complete nasals are thin and slender, with no sign of rugosities for horn bases. The nasal incision reaches the level of anterior P3 (typical also of *Diceratherium*). The premaxilla is long and slender, with well-worn I1 and no trace of an alveolus for I2. There is a distinct infraorbital foramen above P3. The zygomatic arches are slender and only slightly flared laterally, although this is difficult to assess due to crushing. The lambdoid crests divide at the union with the sagittal crests. The occiput was high and narrow, as is also typical of *Diceratherium* (taking into account the lateral compression and distortion).

In ventral view, the crushing is more evident. The teeth are laterally crushed, so that the M3s are rotated. They were so worn in life that almost no crown pattern remains. It is clear, however, that P2–3 were completely molariform, as is also typical of *Diceratherium*. The surprising feature is the strength of the ante-

crochets on M2–3, a feature seen only in higher rhinos. The dP1 was present on both sides.

The basicranium is difficult to interpret due to the crushing. The postglenoid processes both appear to be broken, but were originally quite slender with an anteriorly oriented articulation for the mandibular condyle. The paroccipital processes were long and narrow, and the short mastoid process was fused to the postglenoid process, resulting in a close external auditory meatus. This condition is typical of most *Diceratherium*.

The associated mandible (F:AM 111844) is also laterally crushed. Its teeth are extremely worn, so almost no crown pattern remains. Another less worn ramus shows a typical rhinocerotid lower molar crown pattern, with strong labial and lingual cingula. Small button-like i1 and a tusk-like i2 are also present. The i2 has the broad-based shape that is characteristic of female rhinocerotids (see Chapter 2). This corroborates the suggestion (based on the lack of horn rugosities) that this individual was a female. The posterior diastema is quite long, and the symphysis procumbent, in typical *Diceratherium* fashion. There is a distinct "chin" due to a slight inflection of the ventral surface of the symphysis. Mental foramina are present beneath the diastema and beneath p2, although the exact number is difficult to determine due to the breakage. The anterior edge of the coronoid process is almost perpendicular to the tooth row, and tapers to a slender tip. The posterior margin of the coronoid process is also nearly vertical. The condyles are horizontal and nearly flat, with a strong oblique post-condylar crest, as in all higher rhinocerotids. The angle of the jaw has a thickened posterior margin, with some sculpturing of the interior surface.

Associated with F:AM 111843 is an axis and atlas. The atlas has unusually broad articulations for the occipital condyle. The transverse processes are unusually short and stubby, with a much shallower alar notch than is typical for rhinocerotids. The axis is broken so that the odontoid process and anterior portion are still inserted into the atlas. The neural arch of the axis is complete, with a relatively low dorsal crest.

Discussion: In most features, *Skinneroceras manningi* is simply a primitive diceratheriine rhinocerotid. Of all the rhino genera, it is most similar to *Diceratherium* in overall skull morphology, but there are no unique derived characters that unite it with *Diceratherium*. Because the type skull was apparently that of a female (judging from the i2 tusks), it is impossible to tell whether the males of this taxon had horns like *Diceratherium*. The strength of the parasagittal temporal crests is a unique feature that justifies its placement in a distinct genus. The loss of I2 is known in some advanced *Diceratherium*, although it is more common in higher rhinos. The strong antecrochets on M2–3 are known only in teleoceratines and other advanced rhinos, which have features not seen in *Skinneroceras manningi*. The small size is unique among post-Whitneyan rhinos—no North American rhino after *Penetrigonias dakotensis* is ever as small as *Skinneroceras manningi*. It might be considered an example of dwarfing if the sister-group relationships were more clear-cut.

Based on all these distinctions, I erect a new genus for this material, because it has such a large number of apparently autapomorphic characters. However, it is still known only from one skull and a few jaws. It shows no unique *Diceratherium* synapomorphies, and its skull shape rules out *Penetrigonias, Trigonias, Subhyracodon, Menoceras,* and any more advanced rhino genera. It was apparently very restricted in time and space, because it is still known from a single locality. It was probably a short-lived local offshoot of the widespread *D. tridactylum–armatum–annectens* diversification that was occurring in the early Arikareean.

Skinneroceras manningi new species

Holotype: F:AM 111843, skull with jaws and a few cervical vertebrae. From the base of Roundhouse Rock, Morrill County, Nebraska.

Hypodigm: F:AM 111844, a left ramus with p3-m3, from the type locality.

Etymology: In honor of Earl Manning, who transformed the systematics of North American rhinos, and who first recognized the distinctiveness of this specimen.

Known distribution: Same as for genus.

Diagnosis: Same as for genus.

Description: Same as for genus.

Subfamily Menoceratinae Prothero, Manning, and Hanson, 1986

Menoceras (Troxell, 1921)

Diceratherium Barbour, 1906
Diceratherium Peterson, 1906
Aceratherium Loomis, 1908 (in part)
Diceratherium Loomis, 1908 (in part)
Diceratherium Cook, 1912
Diceratherium Peterson, 1920 (in part)
Diceratherium (*Menoceras*) Troxell, 1921
Diceratherium (*Menoceras*) Wood, 1964
Moschoedestes Stevens, 1969
Menoceras Tanner, 1969
Aphelops Dalquest and Mooser, 1974

Type species: *Menoceras arikarense* (Barbour, 1906)
Included species: Type and *M. barbouri* (Wood, 1964).

Known distribution: Latest Arikareean, Nebraska, Wyoming, Texas, and possibly New Jersey and Florida; early to middle Hemingfordian, High Plains, New Mexico, Texas, and Florida.

Diagnosis: Small to medium-sized rhinos with paired terminal bosses on the tips of the nasals in male skulls. Strong crochets on the upper molars. I2 lost. Basicranium short relative to palatal length. Upper molar lingual cingula weak or absent.

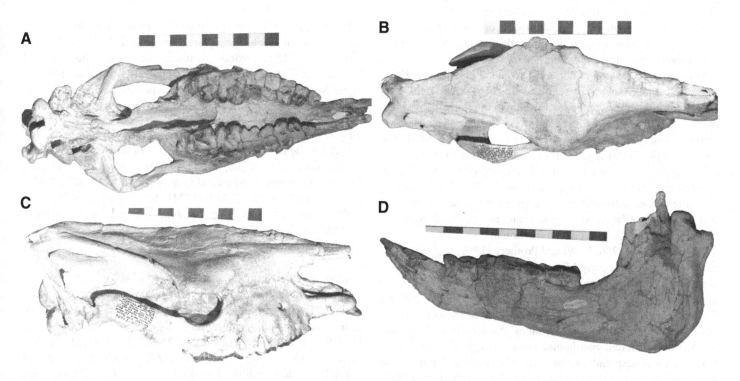

Figure 4.19. *Skinneroceras manningi* n.gen., n.sp. A-C. Type skull (F:AM 111843) in palatal, dorsal, and right lateral views. D. Referred left ramus (F:AM 111844). Scale bar in 2 cm increments.

Description: The cranial material of this genus was well described by Peterson (1920). The postcranial materials are described in Chapter 5.

Discussion: With the discovery of the rich fossil beds at Agate Springs, Nebraska, in 1904 (Peterson, 1906a, 1906b; Barbour, 1906), the classic Arikareean fauna of Wood *et al.* (1941) became known to science. The most common animal in this fauna was a small rhinoceros with paired knobs on the nasals of male skulls, which presumably supported small paired keratinous horns. It was described by Peterson (1906c, 1920) as *Diceratherium cooki*, and large samples of this rhino became part of the collections of museums all over the world. As a result, many reconstructions and museum displays were labeled "*Diceratherium cooki*," and this has spread into the popular books as well.

Unfortunately, much of what has been written about the little Agate rhino is wrong or misleading. There are probably more misconceptions about this fossil than there are for any other extinct North American mammal. As discussed below, the following common ideas about the small Agate Springs rhino are incorrect:

1) Its name. Although nearly every museum label and book in print calls it "*Diceratherium cooki*," the valid name is *Menoceras arikarense*. As discussed in Chapter 3, Troxell (1921d) first applied the name *Menoceras* to the common Agate Springs rhino as a subgenus of *Diceratherium*, and this was followed by later workers (including Wood, 1964). Tanner (1969) finally clarified the distinction between the two genera and raised *Menoceras* to generic rank.

2) Its relationships. *Menoceras* was originally placed in *Diceratherium*, then considered a subgenus of *Diceratherium*, based on the possession of paired horns. As discussed above under *Diceratherium*, *Menoceras* is not only a distinct genus, but it is not even closely related to *Diceratherium* s.s.

3) Its origins. Several authors (e.g., Wood, 1927; Tanner, 1969) suggested that *Menoceras* evolved in North America from *Diceratherium* or *Subhyracodon*. As shown in Figure 4.1 and discussed by Prothero *et al.* (1986, 1989), its closest relatives are actually late Oligocene European taxa. Antoine (2002) and Antoine *et al.* (2003) argue that the menoceratines are the sister-group of the Eurasian elasmotheres. Heissig (1999) and Agusti and Anton (2002) even refer *Menoceras zitteli* (Schlosser, 1902) from the earliest Miocene of Germany to the genus. *Menoceras* was apparently a latest Arikareean immigrant to North America, and persists until the middle Hemingfordian.

Prothero *et al.* (1986, 1989) and Prothero (1998) suggested that the late Oligocene to early Miocene European rhinoceros *Pleuroceros pleuroceros* (Duvernoy, 1853) might be related to *Menoceras*. However, Antoine *et al.* (2003) have shown that true *Pleuroceros pleuroceros* is a much more derived rhinoceros, possibly sister-taxon to the rhinocerotines. As several authors have pointed out, different non-homologous versions of paired nasal horns were widespread in the early Miocene of Europe, also occurring in the primitive teleoceratins *Diaceratherium asphaltense* (Depéret and Douxami, 1902) and *Prosantorhinus douvillei* (Antoine, pers. comm.). This further corroborates the idea that

Menoceras and *Diceratherium* evolved paired nasal horns independently, and there is no monophyletic group "Diceratheriina" that unites them (*contra* McKenna and Bell, 1997, p. 482).

4) Because of its great abundance at Agate, *Menoceras* is often considered "typical" of the Arikareean in most dioramas, and in most conceptions of that land mammal age. In fact, true *Diceratherium* is the only North American rhino through about 9 million years of the Arikareean, and *Menoceras* shows up only at the very end of that land mammal age. *Menoceras* is thus a very late addition, not the "typical" Arikareean rhino.

Menoceras arikarense (Barbour, 1906)

Diceratherium arikarense Barbour, June 15, 1906
Diceratherium cooki Peterson, August 31, 1906
Aceratherium stigeri Loomis, 1908
Diceratherium schiffi Loomis, 1908
Diceratherium aberrans Loomis, 1908
Diceratherium loomisi Cook, 1912
Diceratherium cooki Peterson, 1920
Diceratherium (Menoceras) cooki Troxell, 1921
Menoceras arikarense Tanner, 1969
Menoceras cooki Tedford and Hunter, 1984

Figure 4.20, 4.21, Table 4.5

Holotype: UNSM 62008, a skull with associated palate from latest Arikareean Agate Springs Quarry, Anderson Ranch Formation, Sioux County, Nebraska.

Hypodigm: An enormous amount of material of this rhino is known from the Agate Springs quarries, especially in the AMNH, CM, and UNSM collections, so it will not be listed here. Except for specimens from Quarry A, all rhinoceros material from the Agate Springs quarries belong to this species. In addition to the Agate specimens, the following are also referred to *M. arikarense*:

From the latest Arikareean Anderson Ranch Formation of Hunt (2002) (= "upper Harrison Formation" of Peterson, 1909, 1920) (= Marsland Formation), Sioux County, Nebraska: Galusha's *Stenomylus* Quarry (= Marsland Quarry): F:AM 112243, left M1; Morava Ranch Quarry: F:AM 112241, femur, tibia, tarsals.

From the latest Arikareean Anderson Ranch Formation, Niobrara County, Wyoming: southeast of Van Tassel: F:AM 116085, left maxilla; Royal Valley: F:AM 112217, skull and mandible; F:AM 112218, skull, mandible, postcranials; F:AM 116086, articulated tibia, tarsus.

From the latest Arikareean Anderson Ranch Formation, Goshen County, Wyoming: southeast of Rawhide Buttes: F:AM 112226, male skull, mandible, forelimb, tibia; Spoon Buttes: F:AM 112219, partial skull, mandible; F:AM 112220, femur, tibia/fibula, tarsus; 16 mile district: F:AM 112221, skull, ramus; F:AM 112222, female mandible; F:AM 112223, humerus; 20 mile district: F:AM 112224, articulated skull, jaws, tibia, astragalus; F:AM 112225, juvenile skull; Sand Gulch: F:AM 112233, humerus, radius/ulna, two tibiae; F:AM 112235, right ramus, maxilla; F:AM 112234, mandible, partial skeleton; F:AM 112236, skull, vertebrae, pelvis, limbs; F:AM 112237, skull, mandible, partial skeleton; Jay Em section: F:AM 112227, partial right ramus; F:AM 112228, male skull, symphysis; F:AM 112229, articulated female skull and jaws; F:AM 112230, juvenile left ramus; F:AM 112231, crushed skull, mandible; F:AM 112232, skull, mandible; 2 mile district: F:AM 112238, left ramus.

From the latest Arikareean Anderson Ranch Formation, Platte County, Wyoming: Roll Quarry: F:AM 112244, male skull, mandible, partial skeleton; F:AM 112245, male skull, mandible; F:AM 112246, articulated skull, mandible; F:AM 112247, quarry block with several individuals, disarticulated; F:AM 112248, skull; F:AM 112249, articulated skull, jaws; F:AM 112250, male skull; F:AM 112251, male skull; F:AM 112252, partial skeleton; F:AM 112253, partial male skull; F:AM 112254, male skull;

F:AM 112255, female skull, mandible; F:AM 112256, female skull; F:AM 112257, female rostrum, symphysis; F:AM 112258, partial female skull; F:AM 112259, skull; F:AM 112260, partial skull, symphysis; F:AM 112261, palate; F:AM 112262, mandible, femur; F:AM 112263, mandible; F:AM 112264, left ramus; F:AM 112265, left ramus; F:AM 112266, left ramus; F:AM 112267, right ramus; F:AM 112268, left ramus; F:AM 112269, right ramus; F:AM 116301, juvenile skull; F:AM 116302, two juvenile skulls, rami; F:AM 116303, juvenile skull; F:AM 116304, juvenile skull; F:AM 116305, juvenile skull; F:AM 116306, juvenile mandible; F:AM 116307, juvenile mandible; F:AM 116308, juvenile mandible; F:AM 116309, partial juvenile skeleton; F:AM 116310, seven humeri; F:AM 116311, radius, atlas; F:AM 116312, ulna; F:AM 116313, radius/ulna; F:AM 116314, femur, tibia/fibula, tarsals; F:AM 116315, hindlimb elements; F:AM 116316, four

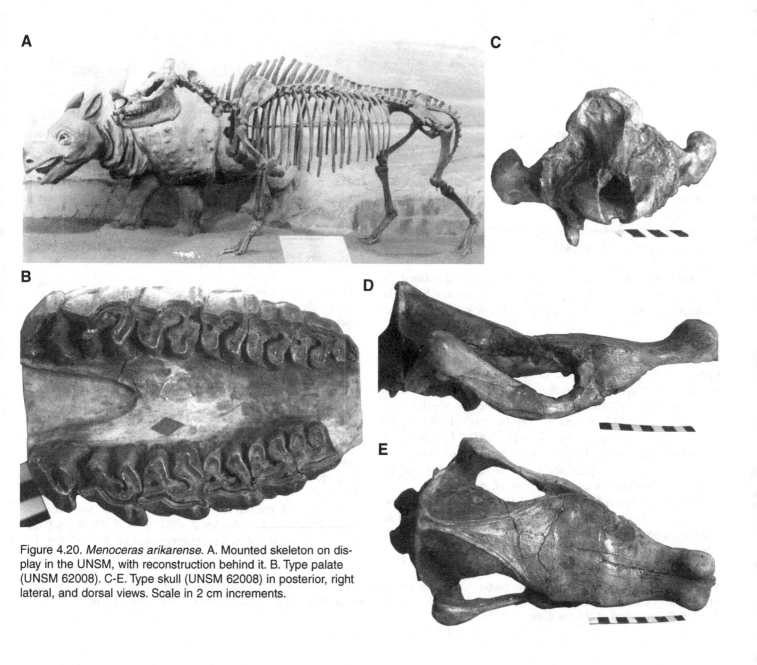

Figure 4.20. *Menoceras arikarense.* A. Mounted skeleton on display in the UNSM, with reconstruction behind it. B. Type palate (UNSM 62008). C-E. Type skull (UNSM 62008) in posterior, right lateral, and dorsal views. Scale in 2 cm increments.

Figure 4.21.
Reconstruction of
Menoceras arikarense by
B. Horsfall (in Scott, 1913).

femora; F:AM 116317, three tibiae/fibulae; F:AM 116318, articulated hindfoot; F:AM 116319, tibia, tarsals; F:AM 116320, two tibiae/fibulae, tarsals.

Known distribution: Latest Arikareean (early Miocene), Nebraska, Wyoming, and possibly late Arikareean of New Jersey and Florida.

Diagnosis: Small *Menoceras* (M1–3 length = 90–105 mm) with relatively gracile skull and postcrania. Teeth less hypsodont than those of *M. barbouri*.

Description: *M. arikarense* was thoroughly described by Peterson (1920), so no further description is necessary here. The postcranial anatomy of this species is further described in Chapter 5.

Discussion: Tanner (1969) demonstrated that the species *arikarense* (Barbour, June 15, 1906) has almost two months' priority over Peterson's (August 15, 1906) better-known species name, *cooki*. It appears that *arikarense* is the valid species name for all Agate Springs *Menoceras* following the rule of priority.

Peterson (1920) demonstrated that Loomis' (1908) species *stigeri, schiffi, aberrans*, and Cook's (1912) "*D.*" *loomisi* are all variants of typical Agate *M. arikarense*. Since that initial period of typological oversplitting, no new names have been proposed for the large, homogeneous sample of Agate *M. arikarense*.

Tedford and Hunter (1984) reported *M.* "*cooki*" (= *arikarense*) from the Farmingdale l.f., Monmouth County, New Jersey, and the Martin-Anthony l.f., Marion County, Florida. It is uncertain on what material these identifications were based, but if they are cor-

rect, then *M. arikarense* occurs in the late Arikareean of the Atlantic Coastal Plain. According to Albright (1998) and Tedford *et al.* (2004), these occurrences are late (but not latest) Arikareean ("Ar3" in their terminology), so that it appears that *Menoceras* reached the Gulf Coast several million years before it appears in the High Plains.

Menoceras barbouri (Wood, 1964)

Diceratherium (*Menoceras*) *barbouri* Wood, 1964
Diceratherium niobrarensis [*sic*] "geologic variety" Stecher, Schultz, and Tanner, 1962
Menoceras marslandensis Tanner, 1969
Moschoedestes delahoensis Stevens (*in* Stevens *et al.*, 1969)
Menoceras falkenbachi Tanner, 1972
Aphelops sp. Dalquest and Mooser, 1974
Menoceras barbouri Prothero and Manning, 1987

Figure 4.22, 4.23, Table 4.5

Holotype: MCZ 4452, palate, basicranium, and occiput, from the early Hemingfordian Thomas Farm l.f., Gilchrist County, Florida.

Hypodigm: *From the early Hemingfordian Thomas Farm l.f., Gilchrist County, Florida*: MCZ 4061, 7441–7466, 9328–9329, teeth and isolated bones representing most of the skeleton; UF

Figure 4.22. (opposite page) *Menoceras barbouri*. A. Reconstructed skeletons of male and female individuals from Bridgeport Quarries on display at the UNSM. B. Close-up of mounted male skull. C. Type palate (MCZ 4452). D-E. Type skull of "*M. falkenbachi*," UNSM 1241. F-G. Type skull of "*M. marslandensis*," UNSM 62008.

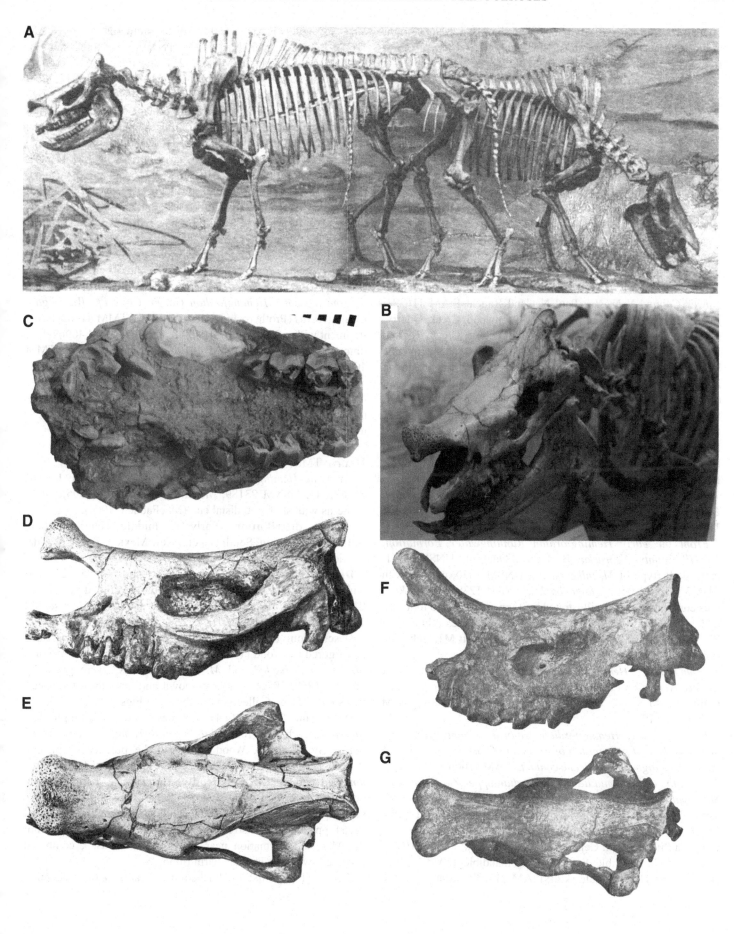

200456, right lunar; UF 199036, manubrium; UF 199067, right radius; UF 199125, right radius; UF 19857, vertebra; UF 203786, right radius; UF 204154, left Mc3.

From the early Hemingfordian Martin Canyon Formation, Weld County, Colorado: Clay Quarry: F:AM 116069, left I1.

From the early Hemingfordian, Batesland Formation, Bennett County, South Dakota: Flint Hill Quarry: F:AM 116071, distal tibia.

From the early Hemingfordian Runningwater Formation, Box Butte County, Nebraska: From UNSM locality Bx-28: UNSM 62003, male skull (type of *M. marslandensis*); Dry Creek Prospect B: F:AM 116065, juvenile left ramus; F:AM 116066, rear left zygomatic; F:AM 116067, female tusk (I2); 0.25 miles west of *Stenomylus* Quarry, Morava Ranch: F:AM 116068, occiput; Runningwater Quarry: F:AM 9969, calcaneum, astragalus; F:AM, 99695, partial ulna.

From the early Hemingfordian Runningwater Formation, Dawes County, Nebraska: Paul Neeland Ranch: F:AM 116321, calcaneum, partial right M2, M3; Cottonwood Creek Quarry: F:AM 116083A–B, calcaneum, patella; Pebble Creek Region: F:AM 116321, calcaneum, right M2–3; F:AM 116081, juvenile left maxilla; F:AM 116082, ulna; Potter Quarry, Sand Canyon Region: F:AM 116080, juvenile mandible; 1.25 miles northwest of Marsland: F:AM 116078, skull; southeast of Marsland: F:AM 82849, skull and mandible; F:AM 116079, radius/ulna; "B" Quarry: F:AM 116075, partial skull, ramus, teeth; F:AM 116084, left ramus; Dunlap Camel Quarry: F:AM 116073, partial skull; F:AM 116074, ulna; F:AM 116076, right ramus; F:AM 116077, left juvenile ramus.

From the early Hemingfordian Runningwater Formation, Cherry County, Nebraska: 2 miles west of Pole Creek: F:AM 116072, right unciform.

From the early Hemingfordian Runningwater Formation, Morrill County, Nebraska: Bridgeport Quarries: UNSM 1241, male skull (type of *M. falkenbachi*); UNSM 6050, maxilla with P1-M3 (paratype of *M. falkenbachi*); UNSM 1238, female skull, plus complete skeletons illustrated by Tanner (1972) and mounted in the UNSM; AMNH 98081, two right dp4s, left m2, right P4, left P3, left P4, 2 left i2s (male and female), 3 I1s, left M1, right M2, right M1.

From the early Hemingfordian Runningwater Formation, Sioux County, Nebraska: East of Coyote Spring: F:AM 82845, proximal radius; F:AM 82832, calcaneum; F:AM 82854, left dp4; F:AM 82806, right P2.

From the early Hemingfordian, Niobrara County, Wyoming: northeast side of pass on top of west end of east Seaman Hills, 5 feet above Oligocene–Miocene contact: F:AM 116064, left ramus.

From the early to middle Hemingfordian upper Zia Formation, Sandoval County, New Mexico: Ceja Prospect: F:AM 116052, astragalus; north fork Canyada Moquino near top of red cliffs: F:AM 116051, distal humerus, pisiform; Straight Cliff Prospect, Canyada Pilares, north Ceja del Rio, Puerco Area: F:AM 116033, juvenile skull, mandible; F:AM 116034, mandible; F:AM 116035, skull; F:AM 116036, male skull; F:AM 116037, scapula, radius,

ulna, astragali; F:AM 116038, distal humerus, tibia; F:AM 116039, thoracic vertebra; F:AM 116040A–C, three humeri; F:AM 116041, ulna; F:AM 116042, radius/ulna; F:AM 116043, radius/ulna; F:AM 116044, Mc2, Mc3, Mc4; F:AM 116045, Mc3, Mc4, Mc5; F:AM 116046, three femora; F:AM 116047A–B, two tibiae; F:AM 116048, patella; F:AM 116049, femur, tibia; F:AM 116054, mandible; F:AM 116063, limbs (female skull, mandible, scapula, tibiae, femur, astragalus); Jemez Area, cliff in Arroyo Pueblo, 12–15 feet above Jeep Quarry horizon: F:AM 116062, skull; Blick Quarry (Blick Hill): F:AM 116050, proximal ulna; F:AM 116053, calcaneum; Jeep Quarry: F:AM 116055, left ramus; F:AM 116056, right juvenile ramus; F:AM 116057A–C, three juvenile rami; F:AM 116058, radius; F:AM 116059, radius/ulna, humerus; F:AM 116060, tibia; F:AM 116061, skull.

From the early Hemingfordian Castolon l.f., Delaho Formation, Brewster County, Texas: TMM 40964-2, mandible and upper teeth (type of "*Moschoedestes delahoensis*" Stevens et al., 1969).

From the early Hemingfordian Garvin Gully l.f., Washington County, Texas (Prothero and Manning, 1987): TMM 41662-1, jaw fragment with m2–3; TMM 31048-48, left dP4; TMM 40067-124, immature right femur; TMM 40106-4, lumbar vertebra; TMM 40067-190, fragmentary skull with poor dentition; TMM 40067-71, tooth fragment; TMM 40067-178, broken metacarpals; TMM 40067-189, tooth fragments; TMM 31048-54, endentulous partial right ramus; TMM 31048-3, right lower m2; TMM 31048-39, tooth fragment.

From the early Hemingfordian Zoyatal l.f., Aguascalientes, Mexico: TMM 41536-1, right M1 or M2.

From the Hemingfordian Gaillard Cut l.f., Panama: USNM 23187, I1; USNM 23189, M3 and p2; USNM 171005, partial humerus with shaft and distal end (MacFadden, 2004).

Known distribution: Early to middle Hemingfordian, Nebraska, Wyoming, South Dakota, New Mexico, Texas, Florida, Mexico, and Panama.

Diagnosis: Large *Menoceras* (M1–3 length = 105–125 mm) with proportionately larger and more robust skull and postcrania. Cheek teeth significantly more hypsodont than *M. arikarense*. M2 unusually long anteroposteriorly.

Description: The type material of *M. barbouri* was described and figured by Wood (1964). Additional descriptions and figures of *M. "falkenbachi"* and *M. "marslandensis"* were given by Tanner (1969, 1972). The best-known and least distorted specimens in the Frick Collection are shown in Figs. 4.22 and 4.23.

Discussion: Wood (1964) first described a Hemingfordian *Menoceras*, which he referred to *Diceratherium (Menoceras) barbouri*. By doing so, Wood continued the practice of placing Troxell's (1921d) genus *Menoceras* as a subgenus of *Diceratherium*, and perpetuated its obscurity. The Thomas Farm material described by Wood (1964) was badly crushed and poorly preserved, so the complete skull morphology was unknown. Wood (1964, pp. 378–379) also mentioned that some isolated teeth from the Marsland Formation near Bridgeport, Nebraska, compared favorably with the Thomas Farm rhino.

Tanner (1969) first established the clear generic distinction

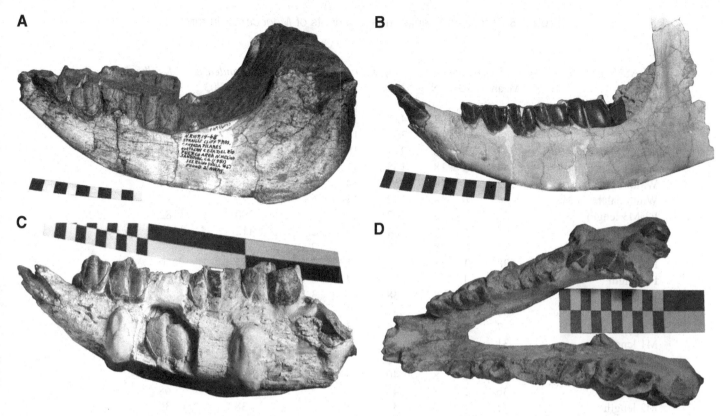

Figure 4.23. Lower jaws of *Menoceras barbouri*. A. F:AM 116033 from the Zia Formation, New Mexico. B. F:AM 116064 from the Seaman Hills, Wyoming. Both show very hypsodont dentitions. C–D. Type specimen of "*Moschoedestes delahoensis*" Stevens, 1969 (TMM 49694-2) in lateral (side of jaw prepared to show hypsodonty) and occlusal views. Note that the size and hypsodonty are a close match for jaws of *Menoceras barbouri*. Scale bars in 1 cm increments, except (A), which is in 2 cm increments.

between *Menoceras* and *Diceratherium*, although he considered them to be closely related. In the same paper, he created a new species, *M. marslandensis* [*sic*], based on a nearly complete skull, UNSM 62003, from the upper part of the Marsland Formation. Although Tanner (1969, p. 401) briefly mentioned Wood's species, he made no direct comparison with the new Marsland species. According to Tanner (pers. comm.), the editor removed his comparisons with *M. barbouri* from the manuscript. Tanner compared the Marsland form only with *M. arikarense*, and naturally found that it was much larger. *M. "marslandensis,"* however, is nearly identical in size with *M. barbouri* (Table 4.5). Nearly all of the skull characters cited by Tanner (1969) are also found in *M. barbouri*, with two exceptions. UNSM 62003 has unusually long nasal horn bosses, and a slight swelling in the frontals that suggests a possible frontal horn. Both of these features appear to be unique to this specimen, because they do not occur in many of the numerous uncrushed skulls from the early Hemingfordian (e.g., F:AM 116035, F:AM 116036 from Straight Cliff Prospect, upper Zia Formation, New Mexico; UNSM 62004, UNSM 1240, or F:AM 116078 from UNSM locality Bx-28, the type locality in the Marsland Formation). If this morphology appeared in more than one specimen in such a large sample, there might be justification for retaining the name *marslandensis*. However, the unusu-

al features of UNSM 62003 appear to be due to individual variation. Based on size and most other features, *M. "marslandensis"* is here considered a junior synonym of *M. barbouri*.

Tanner (1972) erected another species, *Menoceras falkenbachi*, based on specimens of rhinos from the lower Marsland Bridgeport Quarries, Morrill County, Nebraska (UNSM localities Mo-113, Mo-114, Mo-115, Mo-116, and Mo-118). This time, he made no comparisons with *M. barbouri*. His justification for the new species was that it was intermediate in size, morphology, and stratigraphic position between *M. arikarense* and *M. "marslandensis."* Actually, the type specimen of *M. falkenbachi* matches other early Hemingfordian *Menoceras* (including *M. "marslandensis"*) very closely (Fig. 4.22, Table. 4.5). The skull dimensions of the type specimen of *M. "marslandensis"* are slightly larger than those of the type of *M. falkenbachi*, but this is due to the fact that the former is based on a single skull of a robust male (see above). In most measurements, *M. falkenbachi* and *M. "marslandensis"* both fall within the same range of variation that includes *M. barbouri* (Table 4.5). *M. falkenbachi* is clearly a junior synonym of *M. barbouri*.

Stevens (*in* Stevens *et al.*, 1969) described a partial mandible and associated upper and lower dentitions from the Castolon l.f., Brewster County, Texas. She erected a new genus and species,

Table 4.5. Cranial and dental measurements of *Menoceras* (in mm)

MEASUREMENT	*M. arikarense*			*M. barbouri*			*M. marslandensis*	*M. falkenbachi*
	Mean	SD	N	Mean	SD	N	UNSM 62003	UNSM 1241
Nasals to occiput	350	1	2	427	33	9	474	444
P2 to occiput	323	14	18	391	16	9	402	388
Width @ zygoma	198	25	18	243	18	9	250	245
Height of occiput	144	13	18	158	11	7	140	154
Width of occiput	122	11	18	152	4	8	145	154
Width palate @ M2	116	10	18	142	13	6	—	—
P2-M3 length	155	7	8	207	—	1	200	182
M1-3 length	96	5	18	114	9	2	112	105
P2 length	22	2	8	28	1	2	28	—
P2 width	28	1	8	30	1	2	35	40
P3 length	23	1	8	28	4	3	28	42
P3 width	34	2	8	38	1	3	41	43
P4 length	31	2	8	38	2	3	30	28
P4 width	37	2	8	46	3	3	46	40
M1 length	31	2	8	40	6	3	32	34
M1 width	40	2	8	48	3	3	52	41
M2 length	33	4	8	40	3	3	40	39
M2 width	36	2	8	48	1	3	52	48
M3 length	31	2	8	34	3	3	38	30
M3 width	36	2	8	43	1	2	47	38
p2-m3 length	157	4	8	199	1	2	—	—
m1-3 length	92	4	10	118	2	3	—	—
p2 length	19	1	7	22	4	3	—	21
p2 width	14	1	7	16	1	2	—	17
p3 length	23	1	8	28	6	4	—	25
p3 width	17	1	8	20	2	4	—	17
p4 length	27	1	8	31	2	3	—	29
p4 width	19	1	8	22	1	4	—	19
m1 length	27	2	8	36	5	4	—	32
m1 width	20	1	8	22	3	4	—	22
m2 length	33	2	8	43	3	4	—	36
m2 width	21	1	8	24	1	3	—	24
m3 length	36	2	8	43	1	3	—	23
m3 width	20	1	8	24	1	3	—	20

Moschoedestes delahoensis, for these specimens, because they matched no taxon known to her. She compared the mandible with *Menoceras* "*cooki*" and some other rhinos, and noted that it was larger and much more hypsodont than Agate *Menoceras*.

Comparison of the Castolon rhino with early Hemingfordian rhinos from New Mexico and Wyoming (Fig. 4.23) shows clearly that it is very similar in size and hypsodonty to *M. barbouri*. There are no features of *Moschoedestes delahoensis* that are not found within the range of variation of this larger sample of *M. barbouri* from the Frick Collection. Therefore, I consider *Moschoedestes delahoensis* to be a junior synonym of *Menoceras barbouri*.

This identification suggests that the age of the Castolon l.f. is younger than the latest Arikareean age argued by Stevens (1977). The Castolon l.f. contains a puzzling mixture of late Arikareean and early Hemingfordian elements. As Stevens (1977, p. 62) pointed out, the hedgehog *Brachyerix hibbardi* and the camel *Michenia australis* have their closest relatives in the early Hemingfordian. According to Rich (1981), *Brachyerix* first appears in the early Hemingfordian, not latest Arikareean. The oreodont *Merychyus* cf. *M. calaminthus* and the rabbit *Archaeolagus* cf. *A. acaricolus* are also known from the Tick Canyon l.f. of southern California, which may be latest Arikareean or early Hemingfordian. Stevens (1977, p. 39) indicated that the oreodont *?Phenacocoelus leptoscelos* is more specialized in the direction of the later form *Ustatochoerus* than the late Arikareean species *P. typus*. The Hemingfordian camel *Aguascalientia wilsoni* has its closest relatives in the early Hemingfordian Thomas Farm l.f. of Floriday, Garvin Gully l.f. of Texas, and Zoyatal l.f. of Mexico.

On the other hand, the Castolon stenomyline is smaller and more primitive than the late Arikareean *Stenomylus*. *Nanotragulus ordinatus*, *Nothocyon* cf. *N. annectens*, and *Gregorymys riograndensis* are also more similar to forms from the late Arikareean.

If Stevens (1977) is correct in considering the Castolon l.f. to be late Arikareean, then *M. barbouri* and the other early Hemingfordian forms listed above must have evolved in isolation in the Big Bend region of Texas about a million years before they appear in the Texas Gulf Coastal Plain, Florida, New Mexico, or the High Plains. It seems much more parsimonious to suggest that the Castolon l.f. is earliest Hemingfordian (younger than Agate Springs Quarry) but equivalent to or slightly older than the Marsland fauna. This makes the advanced members of the fauna less anomalous, and implies that the few taxa with definite Arikareean aspect are probably relicts.

Dalquest and Mooser (1974, p. 3) referred an isolated M1 or M2 from the Zoyatal l.f. of Aguascalientes Province, Mexico, to *Aphelops* sp. They based this identification on a comparison with a rhino skull from the Barstovian Cold Spring l.f. of Texas, TMM 31219-227, which they refered to *Aphelops*. Prothero and Manning (1987) have shown that this skull belongs to a dwarf species of *Peraceras*, *P. hessei*. Based on a single tooth, it is difficult to clearly identify TMM 41536-1 to genus. Dalquest and Mooser (1974) apparently chose to refer the specimen to *Aphelops* because they believed the fauna to be Barstovian in age, and the

dwarf *Peraceras* is the only Barstovian rhino that would have had such a small tooth. Stevens (1977) argued that the presence of a floridatraguline and other elements in the Zoyatal l.f. suggest a mid-Hemingfordian age. If this is so, then the Zoyatal rhino tooth could also belong to *Menoceras*. Its dimensions (anteroposterior length = 37.6 mm; width = 40.0 mm) are within the range of variation of *M. barbouri*, too small for *Diceratherium* or *Floridaceras* (Table 4.6), and too large for *M. arikarense*. Based on size, this specimen matches *M. barbouri* better than any other Hemingfordian rhino, and is tentatively referred to that taxon.

Whitmore and Stewart (1965) mentioned specimens from the Hemingfordian Cucaracha Formation of Panama, which they referred to *Diceratherium*. MacFadden (2004) has referred some of this material to *Menoceras barbouri*; the rest are referred to *Floridaceras whitei*.

Subfamily Aceratheriinae (Dollo, 1885)

Discussion: The taxon Aceratheriinae has long been used as a taxonomic wastebasket for all hornless rhinoceroses. Primitively, all rhinocerotids are hornless, so this character no longer has any bearing on diagnosing monophyletic groups. In addition, some species of of the aceratheriine *Peraceras* (see below) may have had horns on males, and the Eurasian *Hoploaceratherium tetradacylum* and *Aceratherium* (*Alicornops*) *simorrense* apparently had horns in both sexes (K. Heissig, pers. comm.). Matthew (1931) first recognized that the North American Aceratheriinae s.s. is a good monophyletic group , characterized by several synapomorphies: the premaxilla is greatly reduced; the nasal incision is retracted so that it lies above anterior P4; the medial flange of the lower tusk is reduced. Primitively, the aceratherines retain the narrow skull and long-limbed skeletal proportions of other rhinocerotids, and in this respect they are readily distinguished from the broad-skulled, short-legged teleoceratines, such as *Teleoceras*. A few derived species of *Chilotherium* and *Peraceras* superficially mimic the feature of *Teleoceras*, but they also retain the derived aceratheriine features that clearly indicate their affinities.

Floridaceras Wood, 1964

Figure 4.24, Table 4.6

Type and only species: *Floridaceras whitei* Wood, 1964
Known distribution: Early Hemingfordian of Florida, Nebraska, Oregon, and Panama.
Diagnosis: Large aceratheriine rhino with long limbs, and probably retaining I1 and a large premaxilla. Nasals retracted to the level of anterior P3. Mc5 large and functional.
Discussion: As discussed by Prothero *et al.* (1986), the affinities of *Floridaceras* are still unclear. It is known only from partial specimens, with the only known skull badly crushed, so most of the anatomy of the skull and skeleton are unknown. It is an unusu-

ally large rhinocerotid for the early Hemingfordian, easily distinguished from the much smaller contemporary *Menoceras barbouri*, which also occurs in the Thomas Farm local fauna. In most features (e.g., the brachydont teeth, narrow occiput, sagittal dorsal crest, lack of nasal horns, unreduced p2, and long limbs) it was a very primitive rhinocerotid. It apparently lacked some of the synapomorphies of the Aceratheriinae, such as the nasal incision retraction (it is at the level of anterior P3 in *Floridaceras*) and the loss of I1 with concomitant reduction of the premaxilla (the paratype has a thegosis facet on i2 for an I1). But a few features suggest that it is a very primitive aceratheriine that has yet to develop all the aceratheriine synapomorphies. Like other aceratheriines, *Floridaceras* has a long post-i2 diastema and a reduced medial flange on the i2.

Wood (1964) was particularly impressed by the retention of the Mc5 in *Floridaceras*. It is interesting that among the higher rhinocerotids, only the aceratheriines have an unreduced Mc5. These are now known not only from *Floridaceras* (Wood, 1964), but also from *Aphelops*, *Peraceras*, and *Aceratherium* s.s. This could be interpreted in two ways: either the Mc5 is reduced independently in *Subhyracodon*, *Diceratherium*, *Menoceras*, the teleoceratines, and the rhinocerotines, or the aceratheriines have secondarily enlarged the Mc5 (Prothero *et al.*, 1986). Although the latter scenario seems more parsimonious, Mc5 reduction to form a tridactyl manus is also very easily done in many lineages of mammals. Thus, the presence of Mc5 in *Floridaceras* is not a clear synapomorphy uniting it with the aceratheriines, but it is highly suggestive of such relationships.

Floridaceras whitei Wood, 1964

Holotype: MCZ 4046, a crushed skull with left P2–M3 and right P3–M3, from the early Hemingfordian Thomas Farm l.f., Gilchrist County, Florida;

Hypodigm: *From the early Hemingfordian Thomas Farm l.f., Gilchrist County, Florida*: MCZ 4435, left ramus (paratype); MCZ 4047–4053, 7467–7556, partial skulls and mandibles, most parts of the skeleton; over 60 specimens in the Florida Museum of Natural History, Gainesville, Florida.
From the early Hemingfordian J.L. Ray Ranch, Dawes County, Nebraska: F:AM 115199, partial skeleton.
From the early Hemingfordian Warm Springs l.f., Jefferson County, Oregon: LACM 2707, a proximal left Mc4; JDNM 150A, left M1.
From the Hemingfordian Gaillard Cut l.f., Panama: USNM 18360, mandible; USNM 23178, partial mandible; USNM 23189 (part), p2; USNM 23190, p4 (MacFadden, 2004).

Diagnosis: Same as for genus.

Description: Wood (1964) described most of the skeleton of *Floridaceras whitei* based on the Thomas Farm material.

Discussion: In addition to these materials, there is also a large partial skeleton from the J.L. Ray Ranch (F:AM 115199) of Dawes County, Nebraska, which consists of a left scapula, articulated humerus, radius, and ulna, and associated femur, tibia, fibula, and tarsals. No cranial remains are found with this specimen, so its systematic affinities are difficult to determine. In addition, the specimen is very badly crushed and poorly preserved, so it is difficult to describe it in detail. In overall size and proportions,

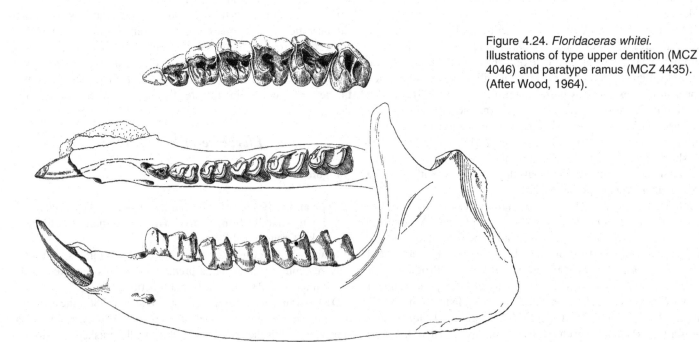

Figure 4.24. *Floridaceras whitei*. Illustrations of type upper dentition (MCZ 4046) and paratype ramus (MCZ 4435). (After Wood, 1964).

Table 4.6. Cranial and dental measurements of *Floridaceras* and *Galushaceras* (in mm)

MEASUREMENT	*F. whitei*			*G. levellorum*		
	Mean	SD	N	Mean	SD	N
P2 to occiput	—	—	—	419	—	1
Lambdoid crest to nasals	—	—	—	492	—	1
Length of nasals (to notch)	—	—	—	165	—	1
Width @ zygoma	325	—	1	274	—	1
Height of occiput	—	—	—	150	—	1
Width of occiput	—	—	—	—	—	—
P2-M3 length	212	9	2	192	—	1
P2-4 length	102	11	2	88	—	1
M1-3 length	140	1	2	109	—	1
dP1 length	23	—	1	22	—	1
dP1 width	19	—	1	16	—	1
P2 length	26	—	1	26	—	1
P2 width	38	—	1	32	—	1
P3 length	35	1	2	31	—	1
P3 width	50	1	2	43	—	1
P4 length	39	1	4	34	—	1
P4 width	58	2	5	46	—	1
M1 length	44	1	3	35	—	1
M1 width	60	4	2	45	—	1
M2 length	47	1	3	48	—	1
M2 width	62	1	2	42	—	1
M3 length	50	1	2	34	—	1
M3 width	57	—	2	40	—	1
Diastema length	96	—	1	—	—	—
p2-m3 length	245	—	1	—	—	—
p2-4 length	106	—	1	—	—	—
m1-3 length	148	—	1	—	—	—
p2 length	27	—	1	—	—	—
p2 width	16	—	1	—	—	—
p3 length	36	—	1	—	—	—
p3 width	23	—	1	—	—	—
p4 length	38	—	1	—	—	—
p4 width	27	—	1	—	—	—
m1 length	43	—	1	—	—	—
m1 width	28	—	1	—	—	—
m2 length	47	—	1	—	—	—
m2 width	28	—	1	—	—	—
m3 length	58	—	1	—	—	—
m3 width	28	—	1	—	—	—
Symphysis to angle	553	—	1	—	—	—
Ht. coronoid above angle	280	—	1	—	—	—
Depth jaw below p2	80	—	1	—	—	—
Depth jaw below m2	90	—	1	—	—	—

however, it is clearly not *Menoceras, Aphelops, Peraceras*, or *Teleoceras*, or any other known Miocene rhino. The proximal elements are comparable in size to a large *Aphelops*, but the distal elements are proportionately much longer. Instead, the limb lengths match the proportions of *F. whitei* as reported by Wood (1964, tables 3 and 4). This interpretation is corroborated by the provenance of F:AM 115199. It was collected from the J.L. Ray Ranch (S1/2 NE1/4 Sec. 4, T31N R47W, Dawes County, Nebraska). The geology of this locality has never been published, but associated with this rhino is a jaw which Beryl Taylor identified as *Sinclairomeryx*, a dromomerycid which is characteristic of the Hemingfordian (Frick, 1937). *Sinclairomeryx* is known from many localities, including the late Hemingfordian Sheep Creek Formation in Sioux County, Nebraska, the middle Hemingfordian Box Butte Formation in Dawes County, Nebraska, and the early Hemingfordian Runningwater Formation in Sioux County, Nebraska (Frick, 1937). Thus, the likeliest possibility is that the J.L. Ray Ranch locality is a Runningwater equivalent and early Hemingfordian in age, although the details of the geology has yet to be mapped and published properly.

Dingus (1990) referred a proximal fragment of a left Mc4 (LACM 2707) from the early Hemingfordia Warm Springs l.f., Jefferson County, Oregon, to *Floridaceras whitei*. It is slightly smaller than the *Floridaceras* Mc4 from Thomas Farm (MCZ 7495), but otherwise very similar in morphology, especially in having a groove for a large Mc5. It is clearly far too large and robust for any species of *Menoceras* or other Hemingfordian rhino.

It appears that *Floridaceras*, although rare, was widely distributed in North America, ranging from Florida to Nebraska to Oregon, and even to Panama (Whitmore and Stewart, 1965; MacFadden, 2004). However, many other early Hemingfordian sites with large collections (e.g., the Martin Canyon Formation of Colorado, the Batesland Formation of South Dakota, the Garvin Gully fauna of the Texas Gulf Coastal Plain, and the Zia Formation of New Mexico) still do not yield *Floridaceras*, but so far produce only *Menoceras* or *Teleoceras*.

Galushaceras new genus

Figure 4.25, Table 4.6

Type species: *Galushaceras levellorum* new species.
Included species: Type species only.
Etymology: In honor of Ted Galusha, who discovered and initially recognized the distinctiveness of this specimen.
Known distribution: Early late Hemingfordian, Dry Creek Prospect A, Box Butte Formation, Box Butte County, Nebraska (Galusha, 1975b).
Diagnosis: Aceratheriine rhino showing highly retracted nasal incision and reduced premaxilla without I1. Differs from primitive *Aphelops* and other aceratheriines in its extreme retraction of nasal incison (to the level of posterior P4 or anterior M1). Differs from

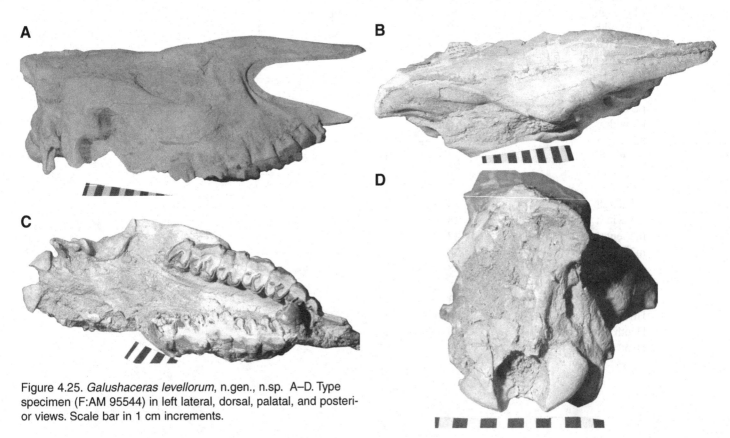

Figure 4.25. *Galushaceras levellorum*, n.gen., n.sp. A–D. Type specimen (F:AM 95544) in left lateral, dorsal, palatal, and posterior views. Scale bar in 1 cm increments.

all other known rhinos in its uniquely flat dorsal skull profile. Much smaller than any other North American Miocene rhinoceros except *Menoceras*.

Description: F:AM 95544 is a nearly complete skull which is badly damaged on the left side, and slightly crushed laterally. However, the rest of the specimen is unusually well preserved, with a good basicranium and unbroken nasals.

In overall shape, the skull is very distinctive, and unlike any other North American rhino. The dorsal skull profile is very flat, with only a slight dorsal flexure of the lambdoid crest and a slight swelling above the frontals. This is a highly derived autapomorphic condition, because all other known rhinos from North America show a strong dorsal flexure of the occiput, and *Aphelops* and other primitive aceratheriines have dorsally convex frontals. The single sagittal crest is faint and narrow, as in *Aphelops*. The occiput is very high and narrow, the primitive condition seen in *Aphelops*. The nasals have a faint rugosity at the tip, but this is so slight that it is not clear whether it indicates the presence of a nasal horn or not. The nasals appear extremely narrow and slender because the nasal incision is unusually large, both dorsoventrally and anteroposteriorly. The incision reached the level of anterior M1, a condition seen only in the most derived Hemphillian species of *Aphelops*. There is a large teardrop-shaped infraorbital foramen on the posteroventral margin of the nasal incision. There is a moderately developed supraorbital ridge, and the usual preorbital rugosity seen in most rhinos. The zygomatic arches are slender, as in *Aphelops*.

The basicranium is moderately well preserved, although crushing has obscured most of the foramina. The postglenoid process is strong and laterally compressed, as in *Aphelops*. The mastoid and paroccipital processes are much more slender than in *Aphelops* or *Peraceras*. The paroccipital processes in particular are very pointed and lack the typical triangular cross-section. There is a shallow medial pocket between the mastoid and paroccipital processes, as seeen in primitive *Peraceras*. The mastoid and glenoid processes do not close beneath the auditory canal. The occipital condyles are broad, larger and more flaring than in *Peraceras profectum* or *Aphelops megalodus*. The deep cleft between the occipital condyles and the paraoccipital process contains a wide hypoglossal foramen. There is a sharp angle between the basioccipital ridge and the basisphenoid ridge, as seen in most rhinos.

The palate, although laterally crushed, has the shelf of the internal nares that terminates at the level of anterior M2. The dentition is well worn, and the left cheek teeth show some damage, so that the left P3 is rotated medially. The teeth are brachydont and smaller (Table 4.6) than any other Hemingfordian rhino from North America except *Menoceras*. A small antecrochet is visible on M2 and small crochets are present on most of the premolars and molars. In all other respects, the dentition shows the condition primitive for aceratheriines and some *Menoceras*, with fully molarized upper premolars. The right P2 is less worn than the rest of the teeth, so the protoloph is still unconnected to the ectoloph. A dP1 is also retained on the right side.

The anterior part of the premaxilla is broken off, but judging from the slenderness of the anterior maxillary process, it must have been very reduced and could not have supported an I1 chisel. This is one of the most distinctive aceratheriine features of the skull.

Discussion: Galusha (1975b, pp. 59–60) briefly discussed this specimen and recognized that it was distinct from any other rhinoceros then known. He refrained from naming it, though, and listed it as "?new caenopine genus." Comparisons of F:AM 95544 with other rhinos in the Frick Collection clearly shows that it cannot be referred to any existing genus from North America, as Galusha realized. The size and the skull and dental morphology are reminiscent of a small primitive *Aphelops megalodus* or *Peraceras profectum*, but the nasal retraction is unusually precocious and reminiscent of a Hemphillian *Aphelops*. The flat dorsal skull profile and several other features discussed above are unique to this taxon, and require that it be placed in a new genus and species.

Galushaceras levellorum new species

Holotype: F:AM 95544, a nearly complete skull with left side crushed (Fig. 4.25).

Hypodigm: Type specimen only.

Etymology: In honor of the LeVelle family, also long-time denizens of the Great Plains.

Known distribution: Type locality only.

Diagnosis: Same as for genus.

Description: Same as for genus.

Aphelops Cope, 1874

Aceratherium Cope, 1873 (in part)
Peraceras Cope, 1884 (in part)
Rhinoceros Leidy, 1890 (in part)
Teleoceras Matthew, 1899 (in part)

Type species: *A. megalodus* (Cope, 1873)

Included species: *A. malacorhinus* Cope, 1878, and *A. mutilus* (Matthew, 1924).

Known distribution: Late Hemingfordian to late Hemphillian, North America.

Diagnosis: Aceratheriine rhino with the derived features of a dorsally arched nasofrontal profile, and long diastema between i2 and the first premolar. It retains a large number of primitive aceratheriine features, including the greatly reduced premaxilla and the loss of I1; broad, unfused, hornless nasals with only slightly downturned lateral edges; nasal incision retracted to the level above anterior P4; brachydont teeth without cement (becoming more hypsodont in *A. mutilus*); upper molars with weak antecrochets; primitive lambdoid crests with skull triangular in posterior view; narrow zygomatic arches; the i2 tusk subcircular in cross-section; and primitive rhinocerotid skeletal proportions.

Description: The type skull of *A. megalodus* was figured and described by Cope and Matthew (1915), and other materials were described by Matthew (1931, 1932), Prothero and Manning (1987), and in the sections below.

Discussion: Cope (1873a) first referred *A. megalodus* to the genus *Aceratherium* which was then a wastebasket for all hornless fossil rhinoceroses. Cope (1874a) erected the genus *Aphelops* when he determined that the North American form differed from *Aceratherium* in lacking Mc5. (In fact, some *Aphelops* do have a functional Mc5, but it is rarely preserved.) Cope mistakenly thought that *Aphelops* had I1/i1, and so restored it in his drawings. However, the type specimen of *A. megalodus* clearly has a vestigial premaxilla that lacks I1. This can be corroborated by the lack of a thegosis facet for I1 on the lower tusk (i2).

Once the genus *Aphelops* had been proposed, it too became a wastebasket genus for nearly all North American Miocene rhinos, including "*Aphelops*" (now *Teleoceras*) *meridianus* (Leidy, 1865), "*Aphelops*" *jemezanus* Cope, 1875 (now considered indeterminate), "*Aphelops*" (now *Teleoceras*) *fossiger* Cope, 1878, "*Aphelops*" *planiceps* Osborn, 1904 (here referred to *Teleoceras medicornutum*), and "*Aphelops*" *brachyodus* Osborn, 1904 (here referred to *Peraceras profectum*). It is ironic that the more distinctive and derived North American rhinos (*Teleoceras, Peraceras*) were long confused with the more primitive form simply because it was the first available name. This is the opposite of the common occurrence of the most derived form being recognized and named first.

Matthew (1918, 1932) was the first to diagnose the genus *Aphelops* correctly. Since that time, several additional species have been referred to *Aphelops*. As discussed below, most of these are based on inadequate comparisons or are distinguished by insignificant features that are clearly due to individual variation. The large quarry samples available in the Frick Collection now make it possible to assess this variation and eliminate several invalid taxa.

Next to *Teleoceras, Aphelops* is the most wide-ranging and successful rhinoceros in the Miocene of North America. It occurs with *Teleoceras* in several quarries, usually in lesser numbers, although at certain times (e.g., the early Clarendonian) it is very scarce. In the late Hemphillian, this balance between *Aphelops* and *Teleoceras* seems to change. Several late Hemphillian quarries (e.g., Edson Quarry in Kansas, Hill and Coffee Ranch Quarries in Texas) contain nearly all *A. mutilus* and only scraps of *Teleoceras*. But the nearby late Hemphillian localities of Guymon, Oklahoma, contain only *Teleoceras*. Although there is probably some ecological separation between the browsing, long-limbed *Aphelops* and the grazing, short-limbed *Teleoceras*, it is peculiar that this separation appears only in the late Hemphillian. Whatever its significance, this separation led Matthew (1932) and others to suggest that only *Aphelops* survived into the late Hemphillian. However, at least two different species of *Teleoceras* are also known from this time.

Aphelops megalodus (Cope, 1873)

Aceratherium megalodum Cope, 1873a
Aphelops megalodus Cope, 1874a

Figures 4.26, 4.27, Table 4.7

Holotype: AMNH 8291, skull and mandible, from the ?early Barstovian Pawnee Creek Formation, Colorado (Galbreath, 1953).

Hypodigm: *?Late Hemingfordian, Pebble Creek Prospect, ?Sheep Creek Formation, Dawes County, Nebraska*: F:AM 114713, radius.

Late Hemingfordian, Sheep Creek Formation, Sioux County, Nebraska: Hilltop Quarry: F:AM 114602, left ramus; F:AM 114624, pelvis; F:AM 114621, tibia; F:AM 114623C,E, two patellae; F:AM 114619C, Mt3; F:AM 114617L, Mc4; F:AM 114616C,E,G, 3 Mc2s; F:AM 114615A, Mt4; F:AM 114614A–C, 3 Mc3s; F:AM 114613D, Mc4; F:AM 114612C, humerus. Ravine Quarry: F:AM 114616D, Mt2. Ashbrook Quarry: F:AM 114623D, patella; F:AM 114617F–G, 2 Mc4s; F:AM 114611C, ulna; F:AM 114610B, radius. Long Quarry: F:AM 114615B–D, 3 Mt4s; F:AM 114614F, Mc3; F:AM 114613B, Mc4; F:AM 114610D, radii. Thistle Quarry: F:AM 114612E, humerus; F:AM 114610C, radius. Buck Quarry: F:AM 114610E, radius. *Pliohippus* Draw: F:AM 114603, left ramus; F:AM 114604, right ramus; F:AM 114605, left ramus; F:AM 114606, right ramus; F:AM 114617B, C, H, I, 4 Mc4s; F:AM 114614 D–E, 2 Mc3s; F:AM 114613E, Mc4; F:AM 114612B, humerus; F:AM 114611B,D, two ulnae; F:AM 114610A, radius. Thompson Quarry, Stonehouse Draw: F:AM 114620A–B, two astragali; F:AM 114607, atlas; F:AM 114623A, patella; F:AM 114619B,D, 2 Mt3s; F:AM 114618A–E, five calcanea; F:AM 114617D,E,J,K, 4 Mc4s; F:AM 114616A–B, 2 Mt2s; F:AM 114615 E-F, 2 Mt4s; F:AM 114613C, Mc4; F:AM 114611A, ulna. Greenside Quarry, Ranch House Draw: F:AM 114608, cervical vertebra; F:AM 114623B, patella; F:AM 114619A, Mt3; F:AM 114617A, Mc4; F:AM 114616F, Mt2; F:AM 114613A, Mc4; F:AM 114612A,D, two humeri. Ranch House Draw: F:AM 114622, femur. Sheep Creek Channel, Antelope Draw: F:AM 114609, axis; Rhino Quarry: F:AM 114302, skull.

Latest Hemingfordian, Massacre Lake l.f., Washoe County, Nevada: UCR 15633, mandible; UCR 15682, mandible; UCMP 61848, m1, plus numerous other specimens described by Morea (1981, p. 152).

Early Barstovian, Pawnee Creek Formation, Weld County, Colorado: No. 1 Canyon: F:AM 114732, humerus; No. 4 Canyon: F:AM 114731, ulna; Pawnee Quarry: F:AM 114723, humerus; F:AM 114724A–B, two femora. Miocene Section 8: F:AM 114726, radius; F:AM 114725A–D, four partial rami; F:AM 114727A–B, two tibiae; Big Springs Quarry: F:AM 114728, radius; F:AM 114729, radius/ulna; F:AM 114730, scapula.

Early Barstovian, Olcott Formation, Sioux County, Nebraska:

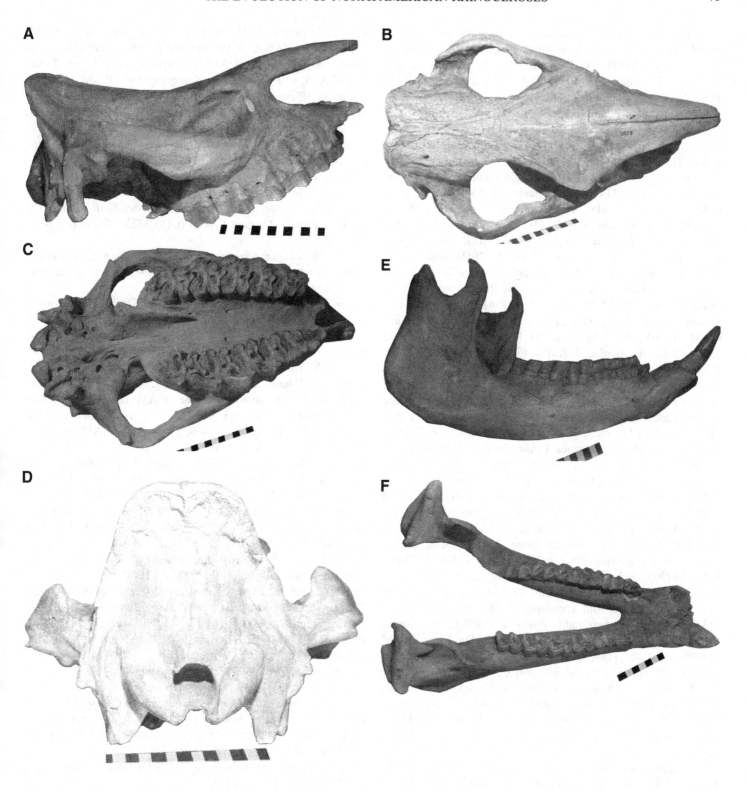

Figure 4.26. *Aphelops megalodus*. Type skull and jaws (AMNH 8291) in (A) lateral, (B) dorsal, (C) palatal, (D) posterior views of skull, and (E-F) right lateral and occlusal view of jaw. Scale bar in 1 cm increments.

Version Quarry: F:AM 114689A, calcaneum; Boulder Quarry: F:AM 115204A–B, two ulnae; Humbug Quarry, Ranch House Draw: F:AM 114301, skull; F:AM 114672, mandible; F:AM 114688, tibia/fibula; F:AM 114680A–D, four femora; F:AM 114677A–C, three humeri; F:AM 115213A–B, two radii; F:AM 115212A–B, two ulnae; F:AM 115207, ulna; *Synthetoceras* Quarry: F:AM 114308, partial skull; F:AM 114681, femur. Echo Quarry, Antelope Draw: F:AM 114657, maxilla with left M1–3; F:AM 114660, right ramus; F:AM 114667, mandible; F:AM 114668, right ramus; F:AM 114669, right ramus; F:AM 114670, left ramus; F:AM 114671, right ramus; F:AM 114673, atlas; F:AM 114686, fibula; F:AM 115200, humerus; F:AM 115206A–C, three radii; F:AM 114687A–C, three tibiae; F:AM 114685A–C, E, four tibiae; F:AM 114679B,D,E, three femora; F:AM 115208 A–B, two ulnae; F:AM 115205B, ulna; F:AM 115204C, ulna; Campsite Echo Quarry, Antelope Draw: F:AM 114647A–C, skull and jaws; F:AM 114648A–B, palate; F:AM 114687C, tibia; F:AM 114685D, tibia; F:AM 114679A,C, two femora; Jenkins Quarry, Sinclair Draw: F:AM 114656, left ramus; F:AM 114691A–B, 2 Mt3s; F:AM 114690, astragalus; F:AM 114683A–C,three fibulae; F:AM 114682A–F, six tibiae; F:AM 115214A–D, four femora; F:AM 115201A–B, two radii. Channel Horizon C, East Sinclair Draw: F:AM 115065, left ramus; F:AM 115066, mandible; F:AM 115203 A,B,D, three radii; New Surface Quarry, East Sinclair Draw: F:AM 114689D, calcaneum; F:AM 114684A, tibia; Quarry 2 (East Surface Quarry), East Sinclair Draw: F:AM 114662, left ramus; F:AM 114663, left ramus; F:AM 114664, right ramus; F:AM 114665, left ramus, F:AM 114666, left ramus; F:AM 114676, humerus; F:AM 115210A–F, six radii; F:AM 114689C, calcaneum; F:AM 114684C-D, two tibiae; F:AM 114678A,C, two femora; F:AM 115208A,C, two ulnae; F:AM 115204D, ulna; F:AM114661, mandible; F:AM 115209, radius; East Sand Quarry, West Sinclair Draw: F:AM 115202, radius; West Sand Quarry, West Sinclair Draw: F:AM 114683A, fibula; F:AM 114305, back of skull; F:AM 114649, nasal; West Surface Quarry, West Sinclair Draw: F:AM 114651, left ramus; F:AM 114654, right ramus with symphysis; F:AM 114675, humerus; F:AM 115211A–B, two radii; F:AM 114684B,E, two tibiae; F:AM 114678B, femur; Quarry 7, West Sinclair Draw: F:AM 114306, skull and jaws; Quarry 8, West Sinclair Draw: F:AM 114655, left ramus; Channel Horizon C, West Sinclair Draw: F:AM 114650A–B, mandible; F:AM 114652, right ramus (juvenile); F:AM 114653, left ramus (juvenile); F:AM 114658, maxilla with left P4-M2; F:AM 114659A–B, left and right rami; F:AM 114674, scapula; F:AM 114689B, calcaneum; F:AM 114684F, tibia.

Early Barstovian, Olcott Formation equivalent, Dawes County, Nebraska: Observation Quarry: F:AM 114303, skull; F:AM 114304, skull; F:AM114625, maxilla with right P4–M3; F:AM 114626, maxilla with left P4–M3; F:AM 114627, maxilla with right P2–P4; F:AM 114628, left ramus; F:AM 114629A–E, five rami (juvenile); F:AM 114630, left ramus; F:AM 114631, right ramus; F:AM 114632, right ramus; F:AM 114633, right ramus; F:AM 114634, left ramus; F:AM 114635, humerus; F:AM

114636, humerus; F:AM 114637A–D, four radii; F:AM 114638, ulna; F:AM 114639A–B, two femora; F:AM 114640A–B, calcaneum and astragalus; F:AM 114641A–E, five tibiae; F:AM 114642A–B, two fibulae; Hay Springs Quarry: F:AM 114643D, astragalus; Pioneer Quarry: F:AM 114643A–C, three astragali; Survey Quarry: F:AM 114644, right ramus; F:AM 114645, radius; F:AM 114646, tibia/fibula.

?Barstovian, Barstow Formation, San Bernardino County, California: Sunset Wash: F:AM 114720, left and right forelimbs.

Early Barstovian, Virgin Valley beds, Humboldt County, Nevada: UCMP 11607, left ramus; UCMP 35269, male mandible.

Early Barstovian, Battle Creek Formation, Skull Springs l.f., Malheur County, Oregon: LACM (CIT) 32256, footbones, Mc2, astragalus.

Early Barstovian, Barstow Formation, San Bernardino County, California: Lower Green Hills: F:AM 114719, left femur; Rak Camp: F:AM 114750, partial skeleton.

Early Barstovian, Tesuque Formation, Skull Ridge Member, Santa Fe County, New Mexico: Tesuque Locality: F:AM 114692, mandible. Skull Ridge: F:AM 114693, calcaneum; F:AM114694, ramus (juvenile).

Early Barstovian, Burkeville Fauna, Texas: Trinity River Pit 1, San Jacinto County, Texas. F:AM 108892, 2 Mc5; F:AM 108906, 4 fibulae; F:AM 108905, 3 tibiae; F:AM 108904, 3 patellae; F:AM 108903, 4 patellae; F:AM 108902, 3 partial femora; F:AM 108901, 2 femora; F:AM 108900, partial ulna; F:AM 108899, 2 radii; F:AM 108898, 4 humeri; F:AM 108927, 3 Mc4; F:AM 108926, 3 Mc3; F:AM 108925, 4 Mc2; F:AM 108924, entocuneiform; F:AM 108923, 2 ectocuneiforms; F:AM 108922, 2 naviculars; F:AM 108921, pisiform; F:AM 108920, trapezoid; F:AM 108919, 3 unciforms; F:AM 108918, 3 unciforms: F:AM 108917, cuneiform; F:AM 108916, 2 lunars; F:AM 108915, 4 scaphoids; F:AM 108914, 3 calcanea; F:AM 108913, 3 calcanea; F:AM 108912, 3 calcanea; F:AM 108911, 3 astragali; F:AM 108910, 4 astragali; F:AM 108909, 2 M1s; F:AM 108908, left M3; F:AM 108931, 2 Mt4; F:AM 108930, 3 Mt3; F:AM 108929, 2 Mt2. Woodville localities, Tyler County, Texas: F:AM 108932, proximal right scapula; F:AM 108943, left Mc2; F:AM 108942, right Mc2; F:AM 108941, right Mt3; F:AM 108940, left Mt4; F:AM 108938, 3 isolated worn upper molars; F:AM 108936, left calcaneum; F:AM 108935, right scaphoid; F:AM 108934, right calcaneum; F:AM 108933, proximal ulna.

Late Barstovian, Temblor Formation, North Coalinga l.f., Kern County, California: LACM (CIT) 115945, right tibia; LACM (CIT) 1634, left M2; LACM (CIT) 1744, right M3; LACM (CIT) 52191, left M3; LACM (CIT) 52192, right M3.

Late Barstovian, Roosevelt l.f., Klickitat County, Washington: LACM (CIT) 33572, left M3.

Late Barstovian, Cold Spring Fauna, Texas: 2 miles northeast of Cold Spring, San Jacinto County, Texas: F:AM 108956, juvenile mandible; 3 miles north of Cold Spring, San Jacinto County, Texas: F:AM 108955, right femur; McMurray Pits, San Jacinto County, Texas: F:AM 108959, patella; F:AM 108958, 3 tibiae; F:AM 108957, 3 femora; F:AM 108954, right Mt2; F:AM

108953, left Mc2; F:AM 108952, right Mc3; F:AM 108950, 3 calcanea; F:AM 108951, astragalus; F:AM 108949, cuboid; F:AM 108948, 3 fibulae; F:AM 108947, 2 ulnae; Near Swartaut, Polk County, Texas: F:AM 108939, right M1; F:AM 108937, right i2.

Early Late Barstovian, Pawnee Creek Formation, Weld County, Colorado: Mastodon and Horse Quarry: F:AM 114733, mandible (juvenile); F:AM 114734, skull (juvenile); F:AM 114735, left maxilla with M1–3; F:AM 114736, left ramus; F:AM 114737, right ramus; F:AM 114738, left ramus; F:AM 114739, right ramus; F:AM 114740, scapula; F:AM 114741, atlas; F:AM 114742A–E, five humeri; F:AM 114743A–B, two radii; F:AM 114744A–C, three ulnae; F:AM 114745A–B, 2 Mc3s; F:AM 114746A–C, three femora; F:AM 114747A–C, three astragali; F:AM 114748A–D, four calcanea; F:AM 114749A–D, four tibiae.

Early late Barstovian, Esmerelda Formation, Esmerelda County, Nevada: Stewart Valley l.f.: UCMP 31495, proximal femur; Stewart Spring l.f., UCMP 58628, astragalus; Cedar Mountain l.f., UMCP 58171, ramus.

Late Barstovian, Barstow Formation, San Bernardino County, California: New Year Quarry: F:AM 114715A–D, femur, tibia, metatarsals, calcaneum; Hidden Hollow Quarry: F:AM 114716, calcaneum; Leader Quarry: F:AM 114717, calcaneum; *Hemicyon* Strata: F:AM 114718, left M3.

?Late Barstovian, Elko County, Nevada: Antelope Creek: F:AM 114312, partial skull and jaw.

Late Barstovian, Juntura Formation, Malheur County, Oregon: UO F-10263, mandible; F-10528, mandible; F-6695, mandibles (Shotwell, 1963).

Late Barstovian, Wood Mountain Formation, Saskatchewan: numerous teeth in the SMNH and ROM collections listed by Storer (1975, p. 51).

Late Barstovian, Valentine Formation, Cornell Dam Member, Brown County, Nebraska: Norden Bridge Quarry: F:AM 114698, supraoccipital; F:AM 114699, right P2.

Late Barstovian, Valentine Formation, Devil's Gulch Member, Brown County, Nebraska: Devil's Gulch Horse Quarry: F:AM 114700, right P4–M2.

Late Barstovian, Valentine Formation, Devil's Gulch Member, Cherry County, Nebraska: Ripple Quarry: F:AM 114701, left P2–3; Nenzel Quarry: F:AM 114702, Mt4; Sawyer Quarry: F:AM 114703A–B, two humeri; F:AM 114704A–B, two fibulae; F:AM 114705, partial right ramus with m3; F:AM 114706, right male i2 tusk; F:AM 114707, radius.

Late Barstovian, Valentine Formation, Burge Member, Cherry County, Nebraska: Burge Quarry: F:AM 114708, skull (juvenile); F:AM 114710, tibia/fibula; F:AM 114711A–B, two ulnae; F:AM 114752, radius; Burge Channel Sands: F:AM 114709, tibia; Rattlesnake Gulch: F:AM 114753, humerus.

Late Barstovian, Valentine Formation, Burge Member, Brown County, Nebraska: Lucht Quarry: F:AM 114751, humerus.

Late Barstovian, Valentine Formation, Burge Member, Todd County, South Dakota: Bailey Ranch: F:AM 114714, femur.

Barstovian/Clarendonian, Tesuque Formation, Pojoaque Member, Santa Fe County, New Mexico: Pojoaque Bluffs: F:AM 114695, femur; F:AM 114696, femur; F:AM 114697, ulna; South Pojoaque Butte: F:AM 114309, skull and jaws.

Early Clarendonian, Dove Spring Formation, Ricardo Group, Kern County, California: UCMP 27107, left ramus; UCMP 270110, mandible; UCMP 27109, right ramus with symphysis.

Early Clarendonian, Valentine Formation, Burge Member, Sheridan County, Nebraska: Paleo Channel Quarry: F:AM 114712, femur.

Early Clarendonian, Tesuque Formation, ChamA–El-Rito Member, Rio Arriba County, New Mexico: Ojo Caliente Locality: F:AM 114806, left and right rami.

Early Clarendonian, ?Lower Snake Creek Formation, Laucomer Member, Sioux County, Nebraska: Antelope Draw Talus: F:AM 114810, astragalus.

Clarendonian, Ash Hollow Formation, Todd County, South Dakota: 3 miles west of St. Francis, low on roadcut: F:AM 114763, right Mc2.

Clarendonian, Tesuque Formation, Rio Arriba County, New Mexico: Chimayo Wash No. 3: F:AM 114808A–B, left and right maxillae with P2–M3.

Clarendonian, Ogallala Formation, Lipscomb County, Texas: Cole Highway Pit: F:AM 114820, astragalus; Schwab Farm: F:AM 114821, maxilla with M1–2; F:AM 114822, left ramus.

Clarendonian, Churchill County, Nevada: Nightingale Road Locality: F:AM 114812, partial distal tibia, distal Mc3.

Clarendonian, Lyon County, Nevada: Jersey Valley: F:AM 114811, femur, radius, calcaneum, metatarsals.

?Middle Clarendonian, Tesuque Formation, Chama–El-Rito Member, Rio Arriba County, New Mexico: Conical Hill Quarry: F:AM 114311, mandible.

?Middle Clarendonian, ?Ash Hollow Formation, Cap Rock Member, Todd County, South Dakota: Hollow Horn Bear Quarry: F:AM 114759, mandible; F:AM 114760, tibia.

Middle Clarendonian, ?Ash Hollow Formation, Cap Rock Member, Mellette County, South Dakota: George Thin Elk Gravel Pit: F:AM 114755, M3.

Middle Clarendonian, Ash Hollow Formation, Cap Rock Member, Keya Paha County, Nebraska: Horton Ranch: F:AM 114754, left ramus (juvenile); 2.5 miles west of Burton Basal Cook Ranch: F:AM 114758, left M1–2.

Middle Clarendonian, Ash Hollow Formation, Cap Rock Member, Cherry County, Nebraska: 3 miles above Garner Bridge: F:AM 114756, Mc3.

Middle Clarendonian, Ash Hollow Formation, Cap Rock Member, Brown County, Nebraska: East Clayton Quarry: F:AM 114757, right maxilla with P2–4.

Middle to late Clarendonian, Ash Hollow Formation, Cherry County, Nebraska: Crane Bridge: F:AM 114762, left ramus with p3-m2.

Late Clarendonian, Tesuque Formation, Pojoaque Member, Santa Fe County, New Mexico: Santa Cruz Area: F:AM 114807, mandible.

Late Clarendonian, Tesuque Formation, Pojoaque Member, Rio Arriba County, New Mexico Santa Clara District: F:AM114310,

Table 4.7. Cranial and dental measurements of the species of *Aphelops* (in mm)

MEASUREMENT	*A. megalodus*			*A. malacorhinus*			*A. mutilus*			*A. "kimballensis"*
	Mean	SD	N	Mean	SD	N	Mean	SD	N	UNSM 5788
P2 to occiput	433	50	4	548	32	2	604	28	4	672
Lambdoid crest to nasals	484	34	3	494	13	2	545	—	1	710
Width @ zygoma	316	15	4	335	42	2	375	19	4	394
Width of occiput	137	15	3	130	—	1	142	13	3	255
Height of occiput	184	10	3	195	—	1	240	15	3	282
P2-M3 length	220	17	5	277	16	3	292	16	4	—
P2-4 length	94	5	5	123	13	3	140	6	4	—
M1-3 length	122	5	4	155	10	4	157	6	4	167
P2 length	28	3	5	35	9	3	44	2	4	53
P2 width	38	2	5	45	8	3	58	2	4	67
P3 length	33	1	5	33	1	4	46	4	4	57
P3 width	46	3	4	59	8	4	69	4	5	75
P4 length	35	2	5	47	3	4	50	3	5	63
P4 width	49	6	5	65	7	5	68	3	5	70
M1 length	39	3	4	52	3	4	49	3	5	65
M1 width	52	2	5	65	4	4	67	6	5	74
M2 length	43	3	5	53	2	4	55	2	5	68
M2 width	53	2	5	65	4	4	69	3	5	65
M3 length	52	2	5	50	5	4	57	2	5	65
M3 width	49	2	4	60	5	4	62	2	5	64
p2-m3 length	214	7	6	255	42	2	321	8	5	280
p2-4 length	96	4	6	117	2	2	148	6	5	82
m1-3 length	125	4	6	158	9	4	179	4	5	—
p2 length	30	2	6	33	4	2	40	2	5	—
p2 width	19	2	4	—	—	—	31	3	5	—
p3 length	33	2	6	39	1	3	52	2	5	36
p3 width	34	2	6	33	3	3	38	3	4	24
p4 length	35	2	6	44	4	4	55	2	5	48
p4 width	38	1	6	39	2	4	42	3	4	36
m1 length	38	2	6	46	3	4	56	5	6	52
m1 width	30	2	6	36	3	4	40	2	5	35
m2 length	44	1	6	54	5	4	60	2	6	61
m2 width	29	2	6	36	4	4	38	2	5	35
m3 length	46	2	6	57	6	4	62	2	6	65
m3 width	28	1	5	33	4	4	36	3	5	32

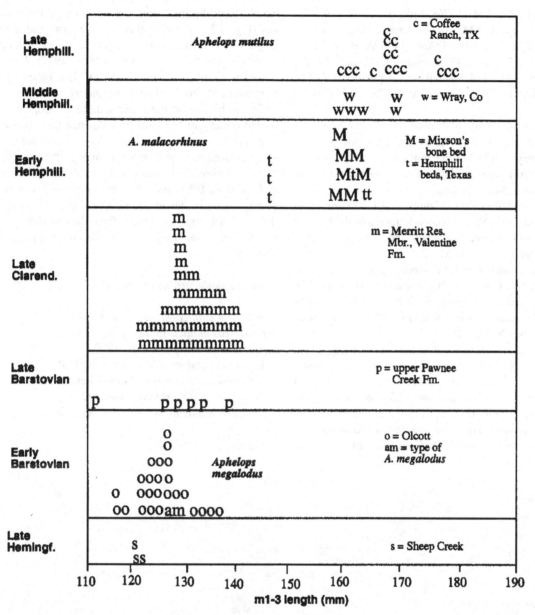

Figure 4.27. Plot of m1-3 dimensions of *Aphelops* specimens through time.

skull, left and right rami; F:AM 114805A–B, femur, pelvis, ulna, tibia, foot bones; F:AM 114809, metacarpals II and III.

Late Clarendonian, Ash Hollow Formation, Merritt Reservoir Member, Cherry County, Nebraska: Xmas Quarry: F:AM 114313, skull; F:AM 114314, skull; F:AM 114315, skull; F:AM 114316, skull and rami; F:AM 114317, skull and mandible; F:AM 114318, jaw; F:AM 114319, jaw; F:AM 114791A–B, palate; F:AM 114792, left ramus; F:AM 114793, right ramus with p4–m3; F:AM 114795, right ramus with m1–3; F:AM 114796, left ramus with p4–m3; F:AM 114797, scapula; F:AM 114798, scapula; F:AM 114800, humerus; F:AM 114801A–B, two ulnae; F:AM 114802, radius; F:AM 114803A–C, three femora; Between Xmas Quarry and Midway Quarry: F:AM 114799, humerus; *Leptarctus* Quarry: F:AM 114320, skull; F:AM 114321, skull; F:AM 114322,

mandibles; F:AM 114323, skull; F:AM 114324, skull; F:AM 114786, left maxilla with P2-M3; F:AM 114787, right ramus; F:AM 114788, right ramus; F:AM 114789A–B, right and left rami (juvenile); F:AM 114790, tibia; Balanced Rock Quarry: F:AM 114761, humerus; Hans Johnson Quarry: F:AM 114328, skull; F:AM 114766, right ramus; F:AM 114780, mandible; Morton Ranch: F:AM 114771, ulna; Norden Bridge Quarry: F:AM 114804, left ramus (juvenile); Cap Rock: F:AM 114781, femur; Kat Quarry: F:AM 114770, radius (juvenile); F:AM 114767, humerus; Trailside Kat Quarry: F:AM 114773, Mt3; East Kat Quarry: F:AM 114778, mandible; F:AM 114779, right ramus (juvenile); West Line Kat Quarry: F:AM 114327, skull; F:AM 114775A–B, left and right rami; F:AM 114776, palate with left P4–M2; F:AM 114777, tibia; F:AM 114794, mandible;

Quarterline Kat Quarry: F:AM 114768, right ramus with p4–m3; Connection Kat Quarry: F:AM 114325, crushed skull; F:AM 114326, skull; F:AM 114329, skull; F:AM 114769A–B, two radii; *Machairodus* Quarry: F:AM 114772, magnum; F:AM 114774, Mt3; Wade Quarry: F:AM 114782, left ramus (juvenile); F:AM 114783, femur; F:AM 114784, tibia; F:AM 114785 right maxilla with P1–4.

Late Clarendonian, Ash Hollow Formation, Merritt Reservoir Member, Brown County, Nebraska: Emry Quarry: F:AM 114764, right M3. Pratt Quarry: F:AM 114765, humerus.

Known distribution: Late Hemingfordian to late Clarendonian (early to late Miocene), North America.

Diagnosis: Small *Aphelops* (M1–3 length = 110–140 mm) with relatively brachydont teeth. Nasal incision retracted to the level of posterior P4. Weak to moderate crochets in the upper teeth; antecrochets weak or absent except on M1–2.

Description: The cranial material of *Aphelops megalodus* was described and figured by Cope (1873), Cope and Matthew (1915), and Osborn (1904). Additional figures are included here for reference (Fig. 4.26). Measurements are given in Table 4.7. The postcranial skeleton is described in Chapter 5.

Discussion: *Aphelops megalodus*, originally known only from the Barstovian of Colorado, has since proven to be a very widespread and long-lived taxon. At over 9 million years in duration (late Hemingfordian to late Clarendonian, 17.5 to 8.5 Ma), it was exceeded only by the 11-million-year duration as *Diceratherium armatum* and *D. annectens*. There are large quarry samples from the late Hemingfordian Sheep Creek Formation of Nebraska,

which fall at the small end of the size range of typical *A. megalodus* (Fig. 4.27), but are otherwise identical. Large samples of *A. megalodus* are found in the early Barstovian Olcott Formation, Sioux County, Nebraska (Skinner *et al.*, 1977). A similar range in variation in size is seen in the smaller sample from the late Barstovian Mastodon and Horse Quarry, Weld County, Colorado, which is the closest topotypic sample for Cope's type specimen (found on an isolated bluff near Pawnee Buttes). Relatively little *Aphelops* is known from the early Clarendonian, but the large samples from the late Clarendonian Merritt Reservoir Member of the Ash Hollow Formation (particularly Xmas Quarry—Skinner and Johnson, 1984) show the same mean size and range of variation as the early Barstovian sample (Fig. 4.27). Thus, the species *A. megalodus* appears to range from the late Hemingfordian to the late Clarendonian with little significant change in size or morphology.

Aphelops malacorhinus Cope, 1878

Aphelops malacorhinus Cope, 1878
Peraceras malacorhinus Cope, 1880
Rhinoceros longipes Leidy, 1890
Teleoceras malacorhinus Matthew, 1899
Aphelops malachorhinus [*sic*] Sellards, 1914
Peraceras malacorhinus Lane, 1927
Aphelops malachorhinus [*sic*] Cook, 1930

Figure 4.28, Table 4.7

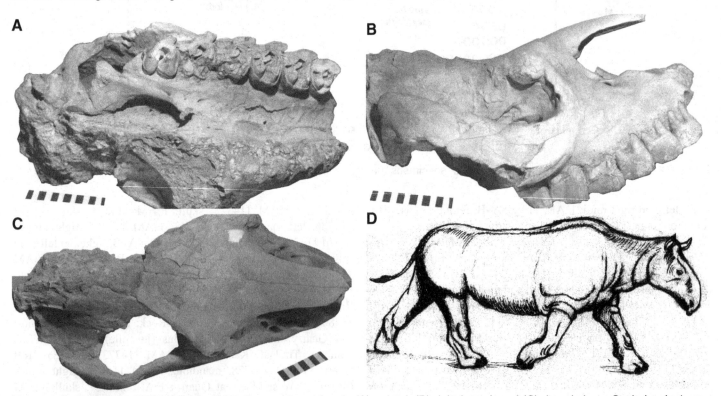

Figure 4.28. Type skull (AMNH 8381) of *Aphelops malacorhinus* in (A) palatal, (B) right lateral, and (C) dorsal views. Scale bar in 1 cm increments. (D) Reconstruction by Ben Nafus.

Holotype: AMNH 8381, skull with damaged left side, from the early Hemphillian, Sappho Creek, Decatur County, Kansas (Hitchcock County, Nebraska, *fide* Cope and Matthew, 1915).

Hypodigm: *Ogallala Formation, Ellis County, Oklahoma*: Port of Entry Pit: F:AM 114332, skull; F:AM 114334, skull; F:AM 114823, right ramus with m1–3; F:AM 114824, right ramus with p3–m3; F:AM 114825, left ramus with m2–3; F:AM 114826, right ramus with p2–m3.

Ogallala Formation, Lipscomb County, Texas: Box T Ranch, West Draw: F:AM 114833B, tibia. Box T Ranch, West Draw, Sandy Clay: F:AM 114833C, tibia; F:AM 114838, patella; F:AM 114837, unciform; F:AM 114836A–B, two calcanea; F:AM 114835, astragalus; F:AM 114834, radius; F:AM 114832, ulna; F:AM 114831A–B, two humeri; F:AM 114830, right ramus with m1–3; F:AM 114829, left and right rami (juvenile); Box T Ranch, West Valley: F:AM 114833A, tibia; F:AM 114828, humerus; F:AM 114827, atlas; Box T Ranch, Pit 1: F:AM 114842A–D, four astragali; F:AM 114844, tibia/fibula; F:AM 114843, tibia; F:AM 114841A–B, humeri; Box T Ranch, Bridge Creek: F:AM 114840, left ramus; F:AM 114839, mandible; F:AM 114333, palate.

Ogallala Formation, Clark County, Kansas: Rhino Quarry: F:AM 114330, skull; F:AM 114331, skull; F:AM 114814, femur; F:AM 114815, humerus; F:AM 114816, right ramus with m1–3; F:AM 114817, left symphysis with M2, I2; Jack Swazy Quarry: F:AM 114818, scapula; F:AM 114819, astragalus, Mc3.

Ogallala Formation, Yuma County, Colorado: Wray Area, Locality B: F:AM 114354, ramus; F:AM 114357, humerus; F:AM 114359, ramus; F:AM 114361, femur; F:AM 114363, humerus; F:AM 114371, skull; Wray Area, Locality C: F:AM 114338, jaw; F:AM 114339, broken skull; F:AM 114340, ramus; F:AM 114342, maxilla; Wray Area, East Fork Haskel Canyon: F:AM 114372, skull.

Humboldt County, Nevada: Lime Quarry: F:AM 114337, partial skull.

Thousand Creek Formation, Humboldt County, Nevada: UCMP 70244, Mc3; UCMP 70410, metapodials.

Coal Valley Formation, Smiths Valley l.f., Lyon County, Nevada: LACM (CIT) 17283, Mc2.

Alachua Clays, Alachua County, Florida: Love Bone Bed: Over 700 catalogued specimens in the Florida Museum of Natural History (Webb *et al.*, 1981).

Known distribution: Early Hemphillian, Florida, Texas, Kansas, Nevada, and Oklahoma

Diagnosis: Large species of *Aphelops* (M1–3 length = 140–168 mm). Nasal incision retracted to level of anterior M1. Teeth subhypsodont. Strong crochets on all premolars and molars. Cristae occasionally present. Anterior maxilla deeper than in *A. megalodus*. Upper molar lingual cingula better developed. Elevation of occiput usually higher than in *A. megalodus*.

Description: Osborn (1904, p. 311), Leidy and Lucas (1896), and others have described cranial material referable to *A. malacorhinus*. Additional illustrations and measurements are given in Fig. 4.28 and Table 4.7. The postcranial skeleton is described in Chapter 5.

Discussion: *Aphelops malacorhinus* was first recognized by Cope (1878) and since has proven to be a stable and highly distinctive early Hemphillian taxon. It is intermediate in size and morphology between *A. megalodus* of the late Clarendonian and *A. mutilus* of the middle and late Hemphillian.

Leidy (1890) described a second species of "*Rhinoceros*" from the early Hemphillian Mixson's Bone Bed of Florida as *Rhinoceros longipes*. The type material included a lower i2 tusk (misidentified as a canine) and Mc2 and Mc4. Lucas (Leidy and Lucas, 1896, pp. 45–48) compared this material with *A. malacorhinus*, and decided they were synonymous. This synonymy was followed by Trouessart (1904) and Sellards (1914), but Osborn (1904, p. 314) considered *longipes* as provisionally valid until it could be shown that *malacorhinus* was as long-limbed. Such a comparison is now possible. Samples of early Hemphillian *A. malacorhinus* Mc2 and Mc4 from the Texas Panhandle match those from Mixson's Bone Bed (both the types, and all the abundant referred material in the Frick Collection) very closely. In size (Fig. 4.27), the Mixson *Aphelops* is within the range of the collection from Box T Ranch area, Lipscomb County, Texas (Schultz, 1977). The available skulls and dentitions are also indistinguishable from the normal range of variation seen in early Hemphillian *A. malacorhinus*. Thus, there seems to be no further reason to separate Florida *Aphelops* from the related specimens on the High Plains.

Aphelops mutilus (Matthew, 1924)

Aphelops malacorhinus mutilus Matthew, 1924
Aphelops malacorhinus longinaris Cook, 1930
Aphelops mutilus Matthew, 1932
Aphelops kimballensis Tanner, 1967
Aphelops longinaris Tanner, 1967
Aphelops kimballensis Dalquest, 1983

Figures 4.27, 4.29, Table 4.7

Holotype: AMNH 17584, complete uncrushed skull, middle Hemphillian *Aphelops* Draw locality, Johnson Member, Snake Creek Formation, Sioux County, Nebraska (Skinner *et al.*, 1977, p. 356).

Hypodigm: *Middle Hemphillian, Ogallala Formation, Yuma County, Colorado*: Spencer Canyon: F:AM 114849, tibia; F:AM 114850, left ramus; Wray Area, Locality B: F:AM 114851, left ramus; F:AM 114852A–B, two I2 's; F:AM 114853, right ramus with p4–m3; Wray Area, Locality C: F:AM 114854A–B, left and right rami; F:AM 114855A–B, two male I2s; F:AM 114856A–C, three female I2s; F:AM 114888, left ramus; F:AM 114889, right ramus with m2–3; F:AM 114890, left ramus.

Middle Hemphillian, Ogallala Formation, Lipscomb County, Texas: Box T Ranch, Ranch West Draw: F:AM 114335, skull; F:AM 114336, skull; Box T Ranch, First Valley, West Pit 1: F:AM 114845, femur; F:AM 114846, tibia; F:AM 114847A–B, two ulnae; F:AM 114848, scapula.

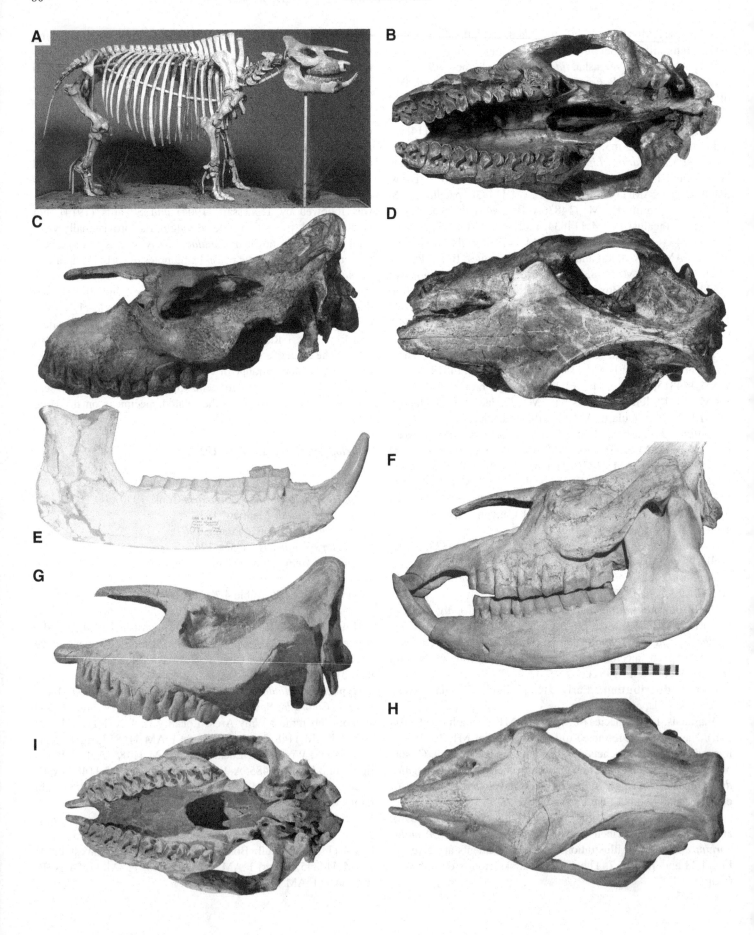

Late Hemphillian, Ogallala Formation, Wallace County, Kansas: Rhino Hill Quarry: F:AM 114368, skull and mandible.

Late Hemphillian, Ogallala Formation, Sherman County, Kansas: Edson Quarry: Hundreds of specimens in the KU collections, plus F:AM 114370, skull; F:AM 114376, skull; F:AM 114895, humerus; F:AM 114896A–D, four radii; F:AM 114897A–B, two ulnae; F:AM 114899, atlas; F:AM 114900, atlas; F:AM 114979A–B, two femora; F:AM 114980, femur; F:AM 114981, articulated foot; F:AM 114898, radius, ulna, carpals, metacarpals; F:AM 114982, femur; F:AM 114983, two tibiae.

Late Hemphillian, Ogallala Formation, Armstrong County, Texas: Hubbard Place: F:AM 114985, humerus; Goodnight Area, J.T. McGehee Place: F:AM 114341, ramus; F:AM 114343, ramus; F:AM 114986, ramus; F:AM 114987A–B, two radii; Goodnight Area, Christian Pit 2: F:AM 114984, radius; Goodnight Area, Hill Quarry: F:AM 114344, ramus; F:AM 114345, ramus; F:AM 114346, ramus; F:AM 114356, broken skull, F:AM 114360, skull; F:AM 114362, ramus; F:AM 114364, posterior skull; F:AM 114365, upper dentition; F:AM 114988, right ramus; F:AM 114989, left ramus; F:AM 114990, symphysis; F:AM 114991, symphysis; F:AM 114992, left ramus; F:AM 114993, left ramus; F:AM 114994, left ramus (juvenile); F:AM 114995, right maxilla; F:AM 114997 A–B, radius and ulna; Goodnight Area, Hill Quarry, Horace Baker Place: F:AM 114996B, humerus; F:AM 114998C, tibia; Hill Pit: F:AM 114999, femur.

Late Hemphillian, Hemphill beds, Hemphill County, Texas: Coffee Ranch: F:AM 114879A, radius; F:AM 114881B, femur; F:AM 114878B, humerus; F:AM 114347, skull; F:AM 114348, broken rami; F:AM 114349, ramus; F:AM 114350, ramus; F:AM 114351, skull; F:AM 114352, ramus; F:AM 114353, mandible; F:AM 114355, ramus; F:AM 114358, ramus; F:AM 114366, skull; F:AM 114367, skull; F:AM 114373, skull; F:AM 114374, skull; F:AM 114375, skull; F:AM 114862, left ramus; F:AM 114863, left ramus; F:AM 114865, right ramus; F:AM 114876, left ramus; 114880, femur; F:AM 114883A–F, six tibiae; F:AM 114893, left ramus; F:AM 114879B, radius; F:AM 114887, left maxilla; F:AM 114881A, femur; F:AM 114860, left ramus; F:AM 114861, right ramus; F:AM 114868, right ramus; F:AM 114869, right ramus; F:AM 114870, right ramus; F:AM 114871, left ramus; F:AM 114872, right ramus; F:AM 114874, right ramus; F:AM 114875, female left ramus; F:AM 114877, left ramus; F:AM 114891, left ramus; F:AM 114894 left ramus; F:AM 114878A, humerus; F:AM 114864, left ramus; F:AM 114866, right ramus; F:AM 114867, left ramus; F:AM 114873, left ramus; F:AM 114882A–B, two femora; F:AM 114885, right maxilla; F:AM 114892, right ramus; F:AM 114886, left maxilla.

Late Hemphillian, Snake Creek Formation, Johnson Member, Sioux County, Nebraska: Merychippus Draw: F:AM 114858, cal-caneum; *Aphelops* Draw: F:AM 114857, left ramus; *Aphelops* Draw, Hipparion Channel: F:AM 114859, humerus.

Late Hemphillian, Chamita Formation, Rio Arriba County, New Mexico: Opposite Lyden Quarry: F:AM 114369, mandible; San Juan: F:AM 114884, ramal fragment (juvenile).

Late Hemphillian, Chalk Buttes Formation, 2 miles southeast of Harper, Malheur County, Oregon: LACM (CIT) 88831, astragalus.

Known distribution: Middle to late Hemphillian, Nebraska, Kansas, Colorado, Oklahoma, and Texas.

Diagnosis: Largest species of *Aphelops* (M1–3 length = 155–185 mm). Nasal incision retracted to the level of anterior M2. Teeth hypsodont, with strong crochets. Cristae usually well developed, frequently connecting with ectoloph crochets to form medifossettes in the upper molars. Maxilla very deep and robust. Elevation of occiput higher than in *A. malacorhinus*.

Description: Matthew (1932) gave a thorough description of the cranial and postcranial material of *A. mutilus* from the Coffee Ranch l.f. of Texas. This is the only previous description of *Aphelops* postcranials in print. This description is amplified in Chapter 5.

Discussion: Matthew (1924) originally considered *A. mutilus* to be a subspecies of *A. malacorhinus*, because they overlap in size (Fig. 4.27). However, in 1932 Matthew raised the subspecies *mutilus* to species rank, probably because more complete material from the late Hemphillian (particularly from Coffee Ranch, Texas) made the distinction between the two species more obvious. Although the size overlap still exists, I feel that the many distinctive features (increased robustness, larger size, deeper narial incision, more complicated tooth pattern) warrant the recognition of the middle to late Hemphillian form as a distinct species. Enough material is now available from enough localities to establish that these differences are real and consistent.

Cook (1930) described a new subspecies, *A. malacorhinus longinaris*, from the middle Hemphillian Wray locality, Yuma County, Colorado. He made no attempt to compare it to *A. mutilus*, and all the characters he cited are either those of *A. mutilus* (larger size, more elevated occiput) or due to individual variation (slightly reduced premolars, long nasals). This last character, in particular, was considered so significant that it was the basis for the subspecific name *longinaris*. In fact, the nasal length is a highly variable feature in rhino populations. For example, among the dozens of skulls of *A. megalodus* from the early Barstovian Olcott Formation of Nebraska, one specimen (F:AM 114301, completely within the size range of the population) has unusually long nasals compared to all the others. The length of the nasals does not seem to be significant other than as an individual variation, and certainly does not merit a new subspecies based on a single aberrant

Figure 4.29.(opposite) *Aphelops mutilus*. A. Skeleton on display at the Panhandle-Plains Historical Museum (by permission of G. Schultz and J. Indeck). B–D. Type skull (AMNH 17584) in palatal, left lateral, and dorsal views. E. Referred jaw (F:AM 114353) with large male tusks. F. Skull and jaws (DMNH 1299) from the Wray rhino locality. G–I. Type specimen (UNSM 5788) of *Aphelops "kimballensis"* Tanner, 1967, in lateral, dorsal, and palatal views. Scale bar in cm.

skull. In every other feature, including size, "*A. malacorhinus longinaris*" falls within the normal range of variation of *A. mutilus* from Wray.

Tanner (1967) erected a new species, *A. kimballensis*, for a male skull (UNSM 5788), mandible (UNSM 5789), and undescribed skeletal elements from UNSM locality Ft-40. This locality, in the "Kimball Formation" of Schultz and Stout (1961), was once thought to be post-Hemphillian in age. Schultz and Stout (1961) erected a "Kimballian" land mammal age to reflect this. The "Kimballian" has since been demonstrated to be equivalent to the early Hemphillian (Breyer, 1981; Tedford *et al.*, 1987), so specimens from UNSM locality Ft-40 are not as young as Tanner, Schultz, Stout, and others had long assumed.

It appears that these erroneous age assignments influenced Tanner's decision to name a new species for what he thought was the "last species of the genus *Aphelops*." Tanner (1967) compared *A. kimballensis* only to the type skull of *A. mutilus* from the Snake Creek Formation of Nebraska. Larger samples of middle to late Hemphillian *A. mutilus* from Wray, Coffee Ranch, Edson, and Hill Quarries show that all of the diagnostic features cited by Tanner fall within the range of variation of *A. mutilus*. Several of the Coffee Ranch and Edson *Aphelops* match *A. kimballensis* very well in size, elevation of occiput, hypsodonty, retraction of narial notch, and development of cristae. The skull length of UNSM 5788 given by Tanner (1967, table I) is 672 mm from P1 to condyle, which is very similar to the dimensions (P2 to condyle = 668 mm) of F:AM 114371 from the middle Hemphillian of Wray. Tooth dimensions, which are more reliable in estimating size, since they are less easily distorted, show that *A. kimballensis* falls well within the range of variation of *A. mutilus*. Indeed, the M1–3 length *of A. kimballensis* (given by Tanner, 1967, table I, as 167 mm) is identical with the same dimension in the type of *A. mutilus*, although Tanner (1967, table I) failed to mention this.

Thus, in size and all other characters listed by Tanner (1967), *A. kimballensis* is based on a large, robust male specimen of *A. mutilus* from the middle Hemphillian, not the post-Hemphillian ("Kimballian"). The lineage suggested by Tanner (1967, 1975) of *Aphelops malacorhinus* (Clarendonian) to *mutilus* (lower Hemphillian) to *kimballensis* (upper Hemphillian-Kimballian) is reduced to only two species, *A. malacorhinus* (early Hemphillian) and *A. mutilus* (middle to late Hemphillian).

Dalquest (1983) described a large new sample of *Aphelops* from the Coffee Ranch Quarry, Hemphill County, Texas, which was part of the original type area of the Hemphillian land mammal age in the definition of Wood *et al.* (1941). He referred these specimens to *Aphelops kimballensis* Tanner, 1967, even though he admitted that they were not as large as Tanner's type material of this species. Surprisingly, he made few comparisons to *A. mutilus*, even though Matthew (1932) had originally described the Coffee Ranch rhinos as referable to that species. The measurements of the Coffee Ranch specimens are entirely within the range of *A. mutilus*, and since *A. kimballensis* Tanner, 1967, is an invalid species, it is clear that the Coffee Ranch *Aphelops* are *A. mutilus*.

Peraceras Cope, 1880

Peraceras Cope, 1880
Aceratherium Matthew, 1899 (in part)
Aphelops Matthew, 1901 (in part)
Aphelops (*Peraceras*) Osborn, 1904 (in part)
Aphelops (*?Diceratherium*) Osborn, 1904
Aceratherium Osborn, 1904 (in part)
Caenopus Osborn, 1904 (in part)
Aphelops Douglass, 1903
Aphelops Douglass, 1908
Aphelops Hesse, 1943
Diceratherium Quinn, 1955
Diceratherium Tanner, 1969
Diceratherium Tanner, 1977

Type species: *Peraceras superciliosus* [*sic*] Cope, 1880
Included species: *P. profectum (*Matthew, 1899), and *P. hessei* Prothero and Manning, 1987
Known distribution: Late Hemingfordian to late Clarendonian, western North America.
Diagnosis: Rhinocerotids with a brachycephalic skull, procumbent lambdoid crest and occiput, shortened nasals, flat dorsal skull profile, upturned symphysis in females, short diastema between i2 and the premolars, and lingual cingula on most cheek teeth.
Description: The cranial and dental material of *Peraceras* was described by Cope (1880a), Matthew (1899), and Prothero and Manning (1987). Postcranials were described by Prothero and Manning (1987) and are described further in Chapter 5.
Discussion: Cope (1880a) first described the genus *Peraceras* based on a complete skull (AMNH 8380) from "the Loup Fork of the Republican River Valley, Nebraska." The precise age and location of this specimen may never be determined, but recent research (F.W. Johnson and R.H. Tedford, pers. comm.) suggests that it came from an early Clarendonian Burge Formation equivalent, on Driftwood Creek, Hitchcock County, Nebraska.

Cope's (1880a) original definition of *Peraceras* emphasized the very derived features of this skull, including the extreme brachycephaly, the anteriorly inclined occiput, and the broad zygomatic arches. He also included hornlessness and extreme shortening of the nasals in his definition, not realizing that the nasals were simply broken. Osborn (1904) followed Cope's (1880a) diagnosis, and referred additional material to this species. Unlike Cope, Osborn clearly recognized that the abbreviation of the nasals was due to breakage.

Matthew (1918) described a new species, *Peraceras troxelli*, which also happened to have broken nasals. His definition of this taxon was essentially like that of Cope and Osborn, except that the hornless, short nasals take on greater significance. Matthew's (1932) final posthumous discussion of the three genera of Miocene rhinos is little changed from his concepts of 1918.

Peraceras has been virtually ignored since that time. Stock and Furlong (1926) and Dalquest and Hughes (1966) described rami that they tentatively referred to *Peraceras* (because the lower jaw of *Peraceras* was unknown at the time). "*Aceratherium*" *profectum* was frequently mentioned, but not associated with *Peraceras*, because the type specimen was so poor. Small rhinos from Texas were referred to *Aphelops* or *Diceratherium* by Hesse (1943), Quinn (1955), and Patton (1969), but none considered the possibility of affinities with *Peraceras*. Tanner (1969) first considered "*Diceratherium jamberi*" to be referable to *Peraceras*, but later (1977) placed this species in *Diceratherium*. In short, the genus *Peraceras* was in great confusion in the literature prior to 1987.

This confusion can now be clarified with the enormous sample of rhinos in the Frick Collection. Much of the problem outlined above is due to the fact that the original definition of *Peraceras* was only applicable to its largest and most derived species, *Peraceras superciliosum*. Complete skulls, jaws, and associated skeletal material referable to *Peraceras* in the Frick Collection forced Prothero and Manning (1987) to broaden the definition of the genus to include more primitive forms. *Peraceras* was apparently a very successful genus in the Barstovian and Clarendonian. The most primitive species, *P. profectum*, was especially common in the west and southwest, outnumbering *Aphelops* and *Teleoceras* in New Mexico, and reported also from California and Nevada. The largest, most derived species, *P. superciliosum,* apparently tried to mimic *Teleoceras* in its morphology, and may have competed directly with *Teleoceras* as a large-bodied grazer. *P. superciliosum* is mostly known from Montana, South Dakota, Nebraska, and New Mexico, but not the southern plains (Kansas, Oklahoma, Texas), which were dominated by *Teleoceras*. The dwarf species, *P. hessei,* occurs primarily as a Barstovian endemic in the unusual faunas of the Texas Gulf Coastal Plain (Prothero and Manning, 1987); it is now also known from New Mexico. Thus, the evolution of *Peraceras* is far more complex and interesting than was previously imagined.

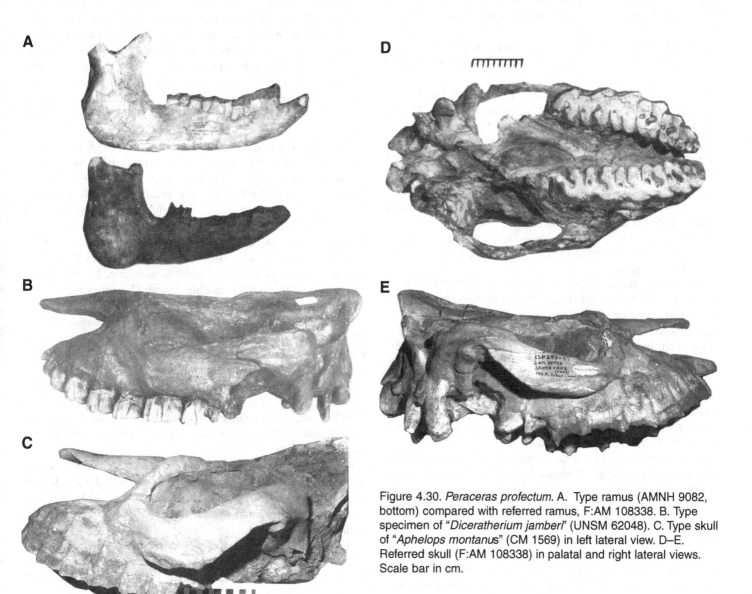

Figure 4.30. *Peraceras profectum*. A. Type ramus (AMNH 9082, bottom) compared with referred ramus, F:AM 108338. B. Type specimen of "*Diceratherium jamberi*" (UNSM 62048). C. Type skull of "*Aphelops montanus*" (CM 1569) in left lateral view. D–E. Referred skull (F:AM 108338) in palatal and right lateral views. Scale bar in cm.

Peraceras profectum (Matthew, 1899)

Aphelops meridianus Cope, 1877 (*non Rhinoceros meridianus*)
Aphelops jemezanus Cope, 1877 (in part)
Aceratherium profectum Matthew, 1899
Aphelops profectus Matthew, 1901
Aphelops profectus Osborn, 1904
Aphelops (*?Diceratherium*) *brachyodus* Osborn, 1904
Aphelops ceratorhinus Douglass, 1903
Aphelops montanus Douglass, 1908
Peraceras n. sp. Tanner, 1976
Diceratherium jamberi Tanner, 1977
Peraceras profectum Prothero and Manning, 1987

Figures 4.30, 4.31, Table 4.8

Holotype: AMNH 9082, a jaw, fragments of upper teeth, right squamosal region of a skull and atlas; possibly from the Barstovian Pawnee Creek Formation, Logan County, Colorado (*fide* Galbreath, 1953, p. 107).

Hypodigm: *Late Hemingfordian, Sheep Creek Formation, Sioux County, Nebraska*: Hilltop Quarry: F:AM 115001, back of skull.

Early Barstovian, Virgin Valley beds, Humboldt County, Nevada: F:AM 114921, left and right rami; F:AM 114922, mandible; UCMP 11396, humerus; UCMP 45715, radius.

Early Barstovian, Tesuque Formation, Skull Ridge Member, Santa Fe County, New Mexico: Northwest of White Operation: F:AM 114974, juvenile skull, partial ramus; Southeast of White Operation: F:AM 114975, femur; Between Old and New Santa Fe Road: F:AM 114976, ramus with m2–3; East Cuyamunque: F:AM 114978, right ramus.

Early Barstovian, Olcott Formation, Sioux County, Nebraska: New Surface Quarry: F:AM 114377, skull; West Sinclair Quarry: F:AM 114378, ramus.

Barstovian, Madison Valley Formation, Gallatin County, Montana: CM 857, partial skull with nasals, mandible and limb bones type specimen of *Aphelops ceratorhinus* Douglass, 1903.

Barstovian, Flint Creek l.f., Granite County, Montana: CM 1569, skull, mandible, two femora, humerus, and other skeletal fragments type specimen of *Aphelops montanus* Douglass, 1908.

Late Barstovian, Punchbowl Formation, Los Angeles County, California: LACM (CIT) 5310, right upper molar fragment.

Late Barstovian, Valentine Formation, Cornell Dam Member, Brown County, Nebraska: Norden Bridge Quarry: F:AM 114926, left ramus; F:AM 114927, right ramus.

Late Barstovian, Valentine Formation, Crookston Bridge Member, Brown County, Nebraska: Fairfield Creek #1: F:AM 114928, skull and jaws.

Late Barstovian, Valentine Formation, Devil's Gulch Member, Brown County, Nebraska: George Elliot Place, 4 miles north of Long Pine: F:AM 114929, mandible; F:AM 114930, humerus.

Latest Barstovian, unnamed formation, Cap Rock Member, Weld County, Colorado: H.C. Markman Collection, 0.5 miles southwest of Quarry Hill: F:AM 114923, skull and ramus; John Weiss Ranch, 8 miles northeast of Keota: F:AM 114924, two rami; F:AM 114925, ramus.

Barstovian/Clarendonian, Tesuque Formation, Pojoaque Member, Santa Fe County, New Mexico: North Fork Canyada Moquino: F:AM 114388, crushed skull, metatarsals; West Fork Canyada Piedra: F:AM 114386, partial maxilla, ramus; North Ceja del Rio Puerco: F:AM 114382, skull; Ceja del Rio Puerco: F:AM 114398, skull, mandible; Pojoaque Bluffs: F:AM 114379, skull; F:AM 114380, skull, mandible; F:AM 114383, partial skull; F:AM 114384, skull; F:AM 114385, rami; F:AM 114391, male skull, jaws; F:AM 114394, rami; F:AM 114395, palate; F:AM 114401, skull, mandible, humerus (axis, atlas); F:AM 114402, skull, mandible; F:AM 114403, skull; F:AM 114407, mandible; F:AM 114408, mandible. Camel Rock: F:AM 114404, partial skull, mandible; Tesuque Locality: F:AM 114948, forelimb; F:AM 114399, skull; Skull Ridge or Española Area: F:AM 114977, right ramus; Pojoaque Bluffs, Española Area: F:AM 114932, juvenile right ramus; F:AM 114955, tibia; F:AM 114931, ulna; F:AM 114934, juvenile radius, ulna, humerus; F:AM 114935, distal tibia, distal humerus; F:AM 114953, femur; F:AM 114958, radius; F:AM 114959, partial ulna; F:AM 114960, right ramus; F:AM 114964, left ramus; F:AM 114966, partial left ramus with m2–3; F:AM 114970, fore- and hindlimb elements; F:AM 114973, skull, ramus, vertebrae; North Pojoaque Bluffs, Española Area: F:AM 114954, humerus; F:AM 114956, tibia; F:AM 114972, juvenile maxillae with dP3–M1; Northwest Pojoaque Bluffs, Española Area: F:AM 114968, juvenile skull (see F:AM 114969); F:AM 114969, juvenile right ramus (see F:AM 114968); South Pojoaque Bluffs, Española Area: F:AM 114933, distal humerus; F:AM 114967, juvenile skull; Southwest Pojoaque Bluffs, Española Area: F:AM 114963, right ramus; West Pojoaque Bluffs, Española Area: F:AM 114957, juvenile ulna; F:AM 114961, left ramus; F:AM 114962, right ramus; F:AM 114971, maxilla; Lower Pojoaque Bluffs, Española Area: F:AM 114965, left ramus; District #1: F:AM 114406, mandible; San Ildefonso: F:AM 114390, skull, mandible; West Cuyamunque: F:AM 114389, skull, scapula.

Barstovian/Clarendonian, Zia Formation, Chamisa Mesa Member, Sandoval County, New Mexico: Canyada Moquino: F:AM 114943, crushed mandible; Arroyo Chamisa Prospect, left fork: F:AM 114942, crushed mandible.

Barstovian/Clarendonian, Tesuque Formation, Pojoaque Member, Sandoval County, New Mexico: Canyada del Zia: F:AM 114939, juvenile mandible; F:AM 114938, crushed skull and jaws; F:AM 114937, ulna, radius, carpals, tarsals; F:AM 114936, crushed mandible.

Barstovian/Clarendonian, Tesuque Formation, Chama–el-Rito Member, Rio Arriba County, New Mexico: Conical Hill Quarry: F:AM 114944, juvenile left ramus. District #2: F:AM 114947, left ramus, scapula, forelimb, femur; Chama–el-Rito: F:AM 114397, partial skull, mandible.

Barstovian/Clarendonian, Tesuque Formation, Pojoaque Member, Rio Arriba County, New Mexico: Santa Cruz 4th Wash: F:AM 114387, crushed skull; Santa Clara: F:AM 114392, skull

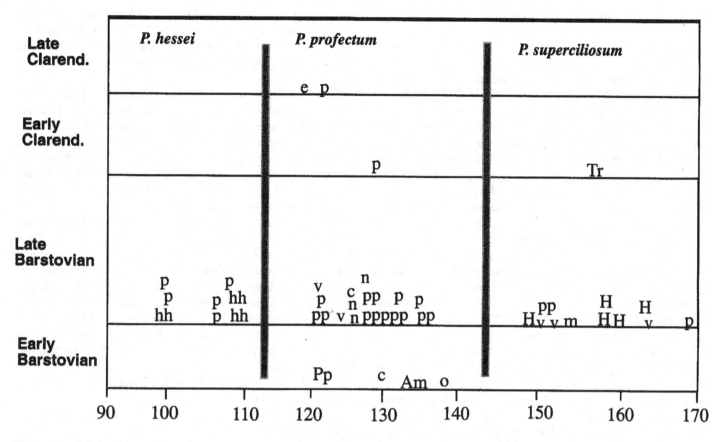

Figure 4.31. Histograms of *Peraceras* m1–3 lengths plotted over time. Symbols: Am = type jaw of "*Aphelops montanus*"; c = Pawnee Creek, Colorado; e = Exell l.f., Texas; H = Hotell Ranch, Nebraska; h = *P. hessei* from Texas; m = Madison Valley, Montana; n = Nevada; o = Olcott Formation, Nebraska; p = Pojoaque Member, Tesuque Formation, New Mexico; Pp = type jaw of *P. profectum*; Tr = type specimen of "*Peraceras troxelli*"; v = Valentine Formation, Nebraska.

(see ramus 89-52, F:AM 114393); F:AM 114393, ramus (see skull 88-52, F:AM 114392).

Early Clarendonian, Tesuque Formation, Pojoaque Member, Rio Arriba County, New Mexico: North Ojo Caliente, *Aelurodon* Wash: F:AM 114946, tibia; Mid Ojo Caliente, Prince Ranch Wash: F:AM 114945, right ramus with p4-m3.

Late Clarendonian, Tesuque Formation, Pojoaque Member, Rio Arriba County, New Mexico: Santa Cruz Area, 1st Wash: F:AM 114950, left maxilla; F:AM 114952, hindlimb elements; Santa Cruz Area, 2nd Wash: F:AM 114951, partial skull; Santa Cruz Area, 3rd Wash: F:AM 114949, right ramus.

Late Clarendonian, Mint Canyon Formation, Los Angeles County, California: LACM (CIT) 103522, left maxilla.

Known distribution: Late Hemingfordian to late Clarendonian, New Mexico, Colorado, Nebraska, Nevada, California, and Texas.

Diagnosis: Most primitive species of *Peraceras*, intermediate in size between *P. hessei* and *P. superciliosum* (M1–3 length = 117–142 mm; m1–3 length = 112–130 mm). Lateral narial notches have a sharp V-shaped border, rather than the gently curved U-shaped concavity of most rhinos. Distinguished from *P. hessei*

mainly by size and by greater robustness of skull and limbs. Distinguished from *P. superciliosum* by the lack of the derived, brachycephalic skull in *P. profectum*, and by its smaller size. Difficult to distinguish from primitive *Aphelops*, because they have overlapping ranges in size, time, and geography. The skull of *P. profectum* is distinct from *A. megalodus* in having a flat nasal-frontal profile, shorter more gracile nasals, and a slightly more flaring occiput. The jaws of *P. profectum* are distinct from *A. megalodus* in having slightly reduced lower premolars, an upturned symphysis (in females), and the diastema between the i2 tusk and the first premolar is relatively short. Most *Peraceras* jaws have better developed lingual cingula than those of *Aphelops*, and the labial cingula tend to be better developed as well. Most post-cranial elements of the two cannot be distinguished without associated cranial material.

Description: The most complete skulls referable to *P. profectum* are F:AM 108338 and UNSM 62048 (the type of "*Diceratherium jamberi*" Tanner, 1977). Both skulls are well preserved and nearly uncrushed, but the dentition is more worn in F:AM 108338. UNSM 62048 was illustrated by Tanner (1977), and F:AM 108338 was illustrated and described by Prothero and Manning

(1987, Fig. 9), but they are both shown here for reference (Fig. 4.30).

The skull of *P. profectum* is similar in size and general proportions to that of *A. megalodus*, except for several significant features: the nasal-frontal region is flat, rather than dorsally arched; the lambdoid crest flares more broadly; and the nasals are shorter and more pointed. These last two features are diagnostic of *Peraceras*, and separate it from any other aceratheriine from North America. Because *P. profectum* is very primitive in skull shape, and very different from the highly derived skull of *P. superciliosum*, many previous authors have misassigned material of *P. profectum* to other taxa (e.g., *Aphelops*, *Diceratherium*), basing their concept of *Peraceras* entirely on the most derived species. This newer conception of *Peraceras* demands that specimens be identified more carefully in the future.

The skull of *P. profectum*, as mentioned above, has a nearly flat nasofrontal skull profile, except for the normal dorsal flexure of the frontal-occipital region. The area between the orbits is actually somewhat concave. The nasals are short but pointed, without rugosities, and extend to the level of dP1. There is a roughened rugose area over the orbits that is more strongly developed than in *Aphelops*. This feature is highly variable, however, and not diagnostic of the genus. The flare of the orbits gives the skull a very diamond- or lozenge-like shape in dorsal view. The lateral broadening is further accentuated in the more brachycephalic *P. superciliosum*. The sagittal crest is a single ridge, and not the broad paired parasagittal crests of *Aceratherium* s.s. or *Chilotherium*. The occiput is nearly vertical, rather than anteriorly inclined, as in *P. superciliosum*. The lambdoid crest is broader than in *A. megalodus*, but not as broad as the condition in *P. superciliosum*. There are two prominent supraoccipital crests radiating dorsally from the foramen magnum, as in all rhinos.

The lateral view (Fig. 4.30) of *P. profectum* shows mostly primitive rhinocerotid features, except for the deep nasal incision, which reaches the level of posterior P4 (the primitive condition for aceratheriines). The nasal incision is a very sharp V-shaped notch, and not as deep or broad as in *Galushaceras* or *Aphelops*. This V-shaped notch is unique to primitive *Peraceras*, because the normal rhinocerotid condition is broader and more U-shaped. Because this condition does not occur in any other species of *Peraceras*, it appears to be autapomorphic for *P. profectum*. The infraorbital foramen is small and rounded in shape, located in a slight depression just below the nasal incision. The premaxillae of *P. profectum* are very reduced splints of bone that could not have supported an I1 chisel (true of all aceratheriines), and are broken in most specimens.

No complete skull of *P. profectum* has a well-preserved basicranium. Some features, however, are observable. The pterygoid flanges are very broad and posterolaterally flaring, as in *Aphelops*. There is also a distinct ridge of bone posterior to M3 on the maxilla, which occurs on many rhinos. The internal nares open at the level of the anterior M3. The anterior border of the internal nares is more acute and V-shaped than in any other rhino, although this may be an individual variation. At the basioccipital-basisphenoid juncture there is a strong raised boss, which thins posteriorly into a sharp ventromedial basioccipital ridge. The foramen ovale is well preserved but most of the ear region is badly damaged. However, the hypoglossal foramen is visible just anterior to the condyle.

The postglenoid processes are broader laterally and more robust than those of *Aphelops*, and this is greatly exaggerated in *P. superciliosum*. The paroccipital processes are long and slender, with a much more narrow triangular shape than the condition seen in *Aphelops*. In contrast, *P. superciliosum* has laterally broadened paroccipital processes, which are broader than in *Aphelops* and much broader than in *P. profectum*. All *Peraceras* characteristically have a deep pocket in the paroccipital process that separates it from the mastoid. The mastoid and postglenoid processes are fused along most of the length, as in the case of *Aphelops*. However, in most *P. superciliosum*, the mastoid and postglenoid processes are separate. Although the fusion of these processes was emphasized by Osborn (1898c), in fact they prove to be highly variable in fossil rhinos and of relatively little systematic value.

The upper dentition in F:AM 108338 is very worn and difficult to interpret. UNSM 62048 has a much less worn upper dentition, and was described in detail by Tanner (1977). The important features are as follows:

(1) P2–4 have strong lingual cingula closing the internal valleys, but M1–3 lack them.

(2) There is a faint labial cingulum on P2, but the other cheek teeth lack them.

(3) Crochets are strongly developed on all the cheek teeth, especially on M1–3.

(4) Antecrochets are present only as posterior swellings on the protoloph, and not as distinct as in the case of *Teleoceras*.

(5) Occasional weak cristae are present in some skulls (especially in P2–4), forming small medifossettes.

The mandible of F:AM 108338 (Fig. 4.30) indicates that it probably came from a female individual, because the i2 tusk has the short, blunt shape typical of female rhinos. The diastema is very short (diagnostic of *Peraceras*) and the symphysis is upwardly inflected (also diagnostic of *Peraceras*). There is a mental foramen below anterior p3. The posterior part of the jaw bears the typical rhinocerotid condition, with a broad medially sloping condyle subdivided into two distinct ridges. One of these ridges passes along the posterointernal margin of the ramus, a derived condition for rhinocerotids. The coronoid process is sharp and anteriorly pointed (not recurved), as is typical of many rhinos.

The lower dentition of F:AM 108338 is very worn, but less worn dentitions show the stereotypical rhinocerotoid pattern. Strong labial and lingual cingula are present in all the lower cheek teeth, a derived feature of *Peraceras*. The premolars are also somewhat reduced in length relative to the molars, a feature that frequently appears in *Peraceras*.

Although much postcranial material in the Frick Collection can be tentatively referred to *P. profectum*, almost none of it is definitely associated with diagnostic cranial remains. Elements that are associated are indistinguishable from those of *A. megalodus*.

Table 4.8. Cranial and dental measurements of *Peraceras* (in mm)

MEASUREMENT	*P. superciliosum*			*P. hessei*			*P. profectum*			*"D. jamberi"*
	Mean	SD	N	Mean	SD	N	Mean	SD	N	UNSM 62048
P2 to occiput	564	15	5	361	1	2	431	1	2	430
Lambdoid crest to nasals	385	—	1	364	22	2	433	28	2	453
Width @ zygoma	415	24	5	237	21	5	264	18	2	277
Width of occiput	168	30	4	131	20	5	186	—	1	165
Height of occiput	243	9	4	—	—	—	170	11	2	162
P2-M3 length	280	6	5	167	1	2	228	1	2	230
P2-4 length	128	3	5	76	3	2	111	14	3	110
M1-3 length	147	13	5	94	4	4	128	7	3	123
P2 length	38	1	5	24	1	4	29	4	4	29
P2 width	49	3	5	31	5	4	41	2	4	39
P3 length	45	1	5	27	1	5	34	3	4	32
P3 width	64	4	5	36	3	5	53	4	4	48
P4 length	47	3	5	30	2	5	35	4	4	33
P4 width	67	5	5	40	1	5	58	2	4	57
M1 length	51	2	5	31	3	6	43	5	4	40
M1 width	66	5	5	40	3	6	58	4	4	55
M2 length	55	4	5	35	3	6	45	4	4	45
M2 width	66	7	5	39	2	6	60	2	4	53
M3 length	48	6	5	34	2	6	45	2	4	35
M3 width	64	5	4	36	1	6	56	3	4	49
p2-m3 length	305	—	1	183	13	5	230	—	1	—
p2-4 length	117	—	1	—	—	—	113	—	1	—
m1-3 length	164	8	3	—	—	—	135	—	1	—
p2 length	31	—	1	23	2	3	31	2	2	—
p2 width	19	—	1	17	1	3	21	2	2	—
p3 length	41	1	1	29	3	6	38	1	2	—
p3 width	34	3	2	22	2	6	29	2	2	—
p4 length	45	1	3	30	2	6	38	1	2	—
p4 width	38	3	3	24	6	6	31	3	2	—
m1 length	50	3	3	29	3	6	43	1	2	—
m1 width	35	2	3	26	7	6	31	1	2	—
m2 length	52	9	3	31	5	7	45	1	2	—
m2 width	33	6	3	26	7	7	31	1	2	—
m3 length	58	4	3	32	6	7	44	1	3	—
m3 width	35	2	3	26	8	7	29	3	3	—

Indeed, much of the material listed under *A. megalodus* may in fact belong to *P. profectum*. Because there is no apparent postcranial difference between these two, the reader is referred to the better-known material of *Aphelops* described in Chapter 5.

Discussion: As outlined above, fragmentary material of *P. profectum* has been known for some time, but incorrectly assigned. The type specimen was originally placed in "*Aceratherium*" and has been referred to *Aphelops* since 1901. Because primitive *Peraceras* was unknown and misdiagnosed, Tanner (1977) incorrectly referred a skull to "*Diceratherium jamberi*" which he first (1976) thought was *Peraceras*. Prothero and Manning (1987) have since clarified the taxonomic status of *P. profectum* and shown that "*Diceratherium jamberi*" is a junior synonym.

Osborn (1904) described a poorly preserved cranium, palate, and upper cheek tooth row (AMNH 10873) as *Aphelops* (*?Diceratherium*) *brachyodus*. It was reportedly from the "Loup Fork Upper Miocene of the Little White River, South Dakota" although its exact location and horizon is unknown. However, the specimen compares best with the skulls of *Peraceras profectum* listed above, and so I consider it a junior synonym of *P. profectum*.

Aphelops ceratorhinus Douglass, 1903, and *Aphelops montanus* Douglass, 1908, were names given to rhino material from the Barstovian Madison Valley Formation, Gallatin County, Montana, and the Flint Creek l.f., Granite County, Montana (respectively). Matthew (1932, p. 420) synonymized *A. montanus* with *A. ceratorhinus*, because the only significant feature that distinguished them was the unique horn rugosities in *A. ceratorhinus*. These "horn rugosities" are in fact the only feature that makes these taxa distinct from *A. megalodus*, because they are otherwise within the *A. megalodus* or *Peraceras profectum* range of size and morphology. I have examined the specimens, and I do not attach much significance to the supposed "horn rugosities" in *A. ceratorhinus*. Small irregular roughenings that may or may not have indicated the presence of horns occur in many Tertiary rhinos. These include the "median horn" found only in the type specimen of *T. medicornutum* (AMNH 9832) but not on the many skulls of this taxon found since then, or the "paired horns" that led Tanner (1977) to misidentify a specimen of *Peraceras profectum* as "*Diceratherium jamberi*" (see below). When a large population sample of these species is examined, these supposed "horn rugosities" turns out to be mere individual variations. They may or may not have served as the base for a horn, but they are not taxonomically significant if they are found only in one specimen in a large, otherwise uniform population sample. Significantly, of these two known Montana Barstovian skulls, one has the rugosities and one does not—and the one without the rugosity, the type of *A. montanus*, also appears to be from a male, judging from the i2 tusk (thus ruling out sexual dimorphism). Unless a larger sample from this area consistently shows nasal rugosities (and none has turned up in a century of collecting), there seem to be no grounds for separating Barstovian *Aphelops* from Montana from *A. megalodus* found elsewhere in the western United States. Moreover, a close examination of the skulls show that they have nasal cross-sections more like those of *Peraceras profectum*, rather than *A. megalodus*.

Thus, I refer both taxas to *P. profectum* for the present until better skull material shows otherwise.

Peraceras profectum, although once poorly known, is now a very well-represented rhino. It is most abundant in the late Barstovian-Clarendonian Pojoaque Member of the Tesuque Formation in the Santa Fe Group, New Mexico. There it occurs in approximately equal numbers with *Teleoceras*, but only a few *Aphelops* are known. It is rare in the Hemingfordian, Barstovian, and Clarendonian of Colorado and Nebraska, but it is known from several localities in Nevada. Its earliest documented occurrence is from the late Hemingfordian Sheep Creek Formation of Nebraska, where a partial skull (F:AM 115001) clearly shows the broadened lambdoid of *P. profectum*. Much of the material referred to *A. megalodus* from the Sheep Creek Formation may in fact be from *P. profectum*. The rhino jaw from the late Clarendonian Exell l.f., Moore County, Texas (Dalquest and Hughes, 1966) is referable to *P. profectum*, as is the jaw from the early Clarendonian Ricardo Group of the Mojave Desert of California (Stock and Furlong, 1926). Thus, *P. profectum* has a predominately western and southwestern distribution, with only rare occurrences in the High Plains of Nebraska, Colorado, or Texas. It is unknown from most northern localities (Montana, South Dakota) and also absent from the Gulf Coastal Plain, where the dwarf *P. hessei*, *A. megalodus*, and two species of *Teleoceras* coexisted (Prothero and Manning, 1987). It is little wonder that *P. profectum* was so badly misunderstood, because it is rare in areas that are well studied and abundantly fossiliferous (especially the High Plains), and common in areas that are still poorly known (such as New Mexico or the Great Basin). Much of the New Mexico rhino material described by Cope (1874a) and referred to "*Aphelops jemezanus*" or "*Aphelops meridianus*" is probably *P. profectum*. Neither of these names can be used for New Mexican *Peraceras*, however, since the type specimen of *jemazanus* is indeterminate (see below), and the type specimen of *meridianus* is actually referable to *T. meridianum*, the dwarf *Teleoceras* from the Texas Gulf Coastal Plain (Prothero and Manning, 1987).

Peraceras hessei Prothero and Manning, 1987

Aphelops n. sp. (small form) Hesse, 1943
Diceratherium Quinn, 1955
Diceratherium sp. Patton, 1969
Aphelops sp. Prothero and Sereno, 1980
Peraceras sp. Prothero and Sereno, 1982
Peraceras hessei Prothero and Manning, 1987

Figures 4.31, 4.32, Table 4.8

Holotype: TMM 31219-228, a skull from the late Barstovian Cold Springs fauna, Coldspring, Texas.

Hypodigm: The Texas material was listed in Prothero and Manning (1987, p. 395). Additional material includes the following:
Barstovian/Clarendonian, Pojoaque Member, Tesuque Formation, Sandoval County, New Mexico: Canyada del Zia

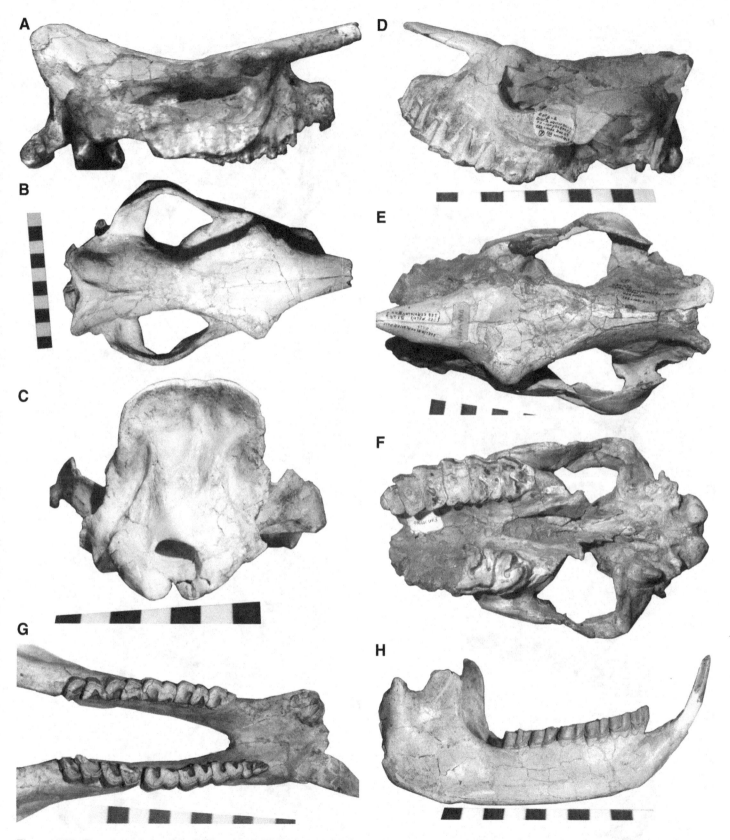

Figure 4.32. *Peraceras hessei*. A–C. Type skull (TMM 31219-228) in right lateral, dorsal and posterior views. D–F. Referred skull (F:AM 109360) in left lateral, dorsal, and palatal views. G–H. Referred mandible (TMM 31219-225) in occlusal and right lateral views. Scale bar = 2 cm increments.

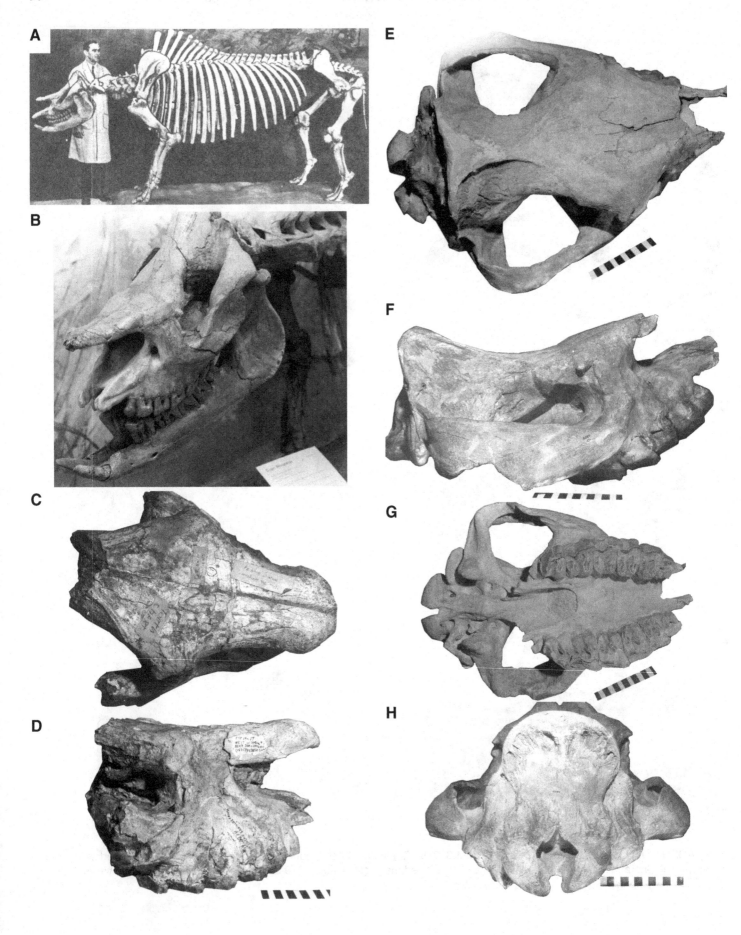

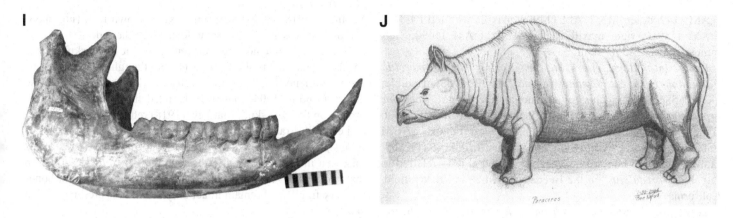

Figure 4.33. *Peraceras superciliosum.* A-B. Mounted composite skeleton (A) and close-up of skull (B) of UNSM 1235 from the Hotell Ranch (UNSM locality Bn-10) on display at the UNSM. C–D. F:AM 114410, a skull showing the broad, robust, possibly horn-bearing nasals. E–H. Type skull (AMNH 8380) in dorsal, right lateral, palatal, and posterior views. I. Lateral view of male mandible (UNSM 1237) from Hotell Ranch. Scale bars in cm. J. Restoration by Ben Nafus.

Locality: F:AM 114941, partial skull, forelimb; Rincon Quarry: F:AM 114940, crushed skull, partial skeleton; Near Chimayo, Española area: F:AM 114405, ramus.

Known distribution: Early and late Barstovian, Texas Gulf Coastal Plain; late Barstovian of New Mexico.

Diagnosis: Primitive dwarfed *Peraceras* (M1–3 length = 90–105 mm; m1–3 length = 98–110 mm) with relatively robust limb elements, reduced occiput, and lambdoid crest with a slight lateral flare.

Description: No further description needs to be added to that published by Prothero and Manning (1987). The New Mexico material adds one good uncrushed skull (F:AM 109360, Fig. 4.32) to the hypodigm, which was also described by Prothero and Manning (1987, figs. 5-4 to 5-7). The postcranial materials from New Mexico duplicate those already described from Texas by Prothero and Manning (1987).

Discussion: The dwarf *Peraceras* was first described and discussed based on material from the Texas Gulf Coastal Plain. Prothero and Manning (1987) briefly mentioned material referable to *P. hessei* from New Mexico, but this review has uncovered more such material from the late Barstovian-Clarendonian Pojoaque Member of the Tesuque Formation in the Santa Fe Group. However, *P. hessei* is still relatively rare compared to the abundant material of *P. profectum* from those same deposits. No *P. hessei* is known from the early Barstovian Skull Ridge Member of the Tesuque Formation, but *P. profectum* is abundant in these beds. *P. hessei* first appears in the early Barstovian of the Texas Gulf Coastal Plain, and may have secondarily migrated to New Mexico in the late Barstovian. If this is so, then the hypothesis of Prothero and Sereno (1982) that the dwarfing is related to the ecological limitations of the coastal plain forests is not falsified.

Peraceras superciliosum Cope, 1880

Peraceras superciliosum Cope, 1880
Teleoceras superciliosus [*sic*] Matthew, 1901
Aphelops superciliosus Hay, 1902
Peraceras superciliosum Osborn, 1904
Peraceras troxelli Matthew, 1918
Peraceras superciliosum Prothero and Manning, 1987

Figures 4.31, 4.33, Table 4.8

Holotype: AMNH 8380, a skull lacking nasals, possibly from the early Clarendonian Burge Formation equivalent, Hitchcock County, Nebraska (F.W. Johnson and R.H. Tedford, pers. comm.).

Hypodigm: *Late Barstovian, Fort Randall Formation, Charles Mix County, South Dakota:* South Bijou Hills: F:AM 114913, radius.

Late Barstovian, Valentine Formation, Burge Member, Brown County, Nebraska: June Quarry: F:AM 114911, tibia; F:AM 114912, astragalus.

Late Barstovian, Valentine Formation, Cornell Dam Member, Brown County, Nebraska: Norden Bridge Quarry: F:AM 114914, radius.

Late Barstovian, Valentine Formation, Devil's Gulch Member, Brown County, Nebraska: Elliot Place, 4 miles north of Long Pine: F:AM 114915, radius; F:AM 114916, nasal; F:AM 114917, humerus; F:AM 114918, Mc2, Mc3.

Late Barstovian, Valentine Formation, Crookston Bridge Member, Cherry County, Nebraska: Valentine Railroad Quarry A: F:AM 114813, magnum; F:AM 114902A–B, isolated right M1–2; F:AM 114903, isolated right M1, left P2; Ripple Quarry: F:AM 114904, left ramus with m1–3; Sawyer Quarry: F:AM 114905, maxilla with right P4–M2; F:AM 114907, right ramus with m3;

F:AM 114908, left M3; F:AM 114909, maxilla with left P4–M2; F:AM 114910, right maxilla with ?M1–2; F:AM 114906, left ramus; F:AM 114396, skull.

Early late Barstovian, Stewart Spring l.f., Esmerelda Formation, Esmerelda County, Nevada: UCMP 58943, mandible; UCMP 58628, astragalus; plus material referred to "*Aphelops cristalatus*" by Henshaw (1942), including LACM (CIT)2806 (type), left maxilla; LACM (CIT)2807, left ramus; LACM (CIT)762, ramus; LACM (CIT)2608, ramus; LACM (CIT)2809; ramus; plus limb bones; uncatalogued skull and jaws referred to "*Aphelops cristalatus*" in the Buena Vista Museum, Bakersfield, California.

Early late Barstovian, Madison Valley, Three Forks Basin, Jefferson County, Montana: CM 840, skull; CM 3304, 3307, 3202, partial skeletons; CM 818, 829, 831, 842, 3311, 841, 851, 854, 3305, skeletal fragments; CM 25378, skull and mandible; CM856, ramus.

Late Barstovian, Browns Park Formation, Douglass Mountain, Moffat County, Colorado: CM 11387, postcranials.

Barstovian/Clarendonian, Tesuque Formation, Pojoaque Member, Santa Fe County, New Mexico: Pojoaque Bluffs: F:AM 114400, tibia, fibula, footbones, femora; F:AM 114411, skull; West Cuyamunque: F:AM 114412, mandible. Near San Ildefonso: F:AM 114410, skull.

Late Barstovian, Ash Hollow Formation, Andrew Hotell Ranch, Banner County, Nebraska (Voorhies *et al.*, 1987) UNSM locality Bn-10: Large collection of specimens, including two mounted skeletons in the UNSM, and many disarticulated specimens.

Clarendonian, Ash Hollow Formation, Keya Paha County, Nebraska: Turkey Creek: F:AM 114409, skull.

Clarendonian, Exell l.f. (= Frick 4-way Pit), Ogalla Formation, Moore County, Texas (Dalquest and Hughes, 1966, p-. 85–86): MWU P174-38, jaw.

Early Clarendonian, Dove Spring Formation, Ricardo Group, Kern County, California: UCMP 27110, jaw; UCMP 27109, jaw (Stock and Furlong, 1926).

Early to middle Clarendonian, Valentine Formation, Burge Member, or Ash Hollow Formation, Cap Rock Member, Keya Paha County, Nebraska: Troxell's Rhino Quarry: F:AM 114919, scapula; F:AM 114920, astragalus.

Middle Clarendonian, Ash Hollow Formation, Cap Rock Member, Brown County, Nebraska: Quinn Ranch: F:AM 114901, calcaneum.

Known distribution: Late Barstovian to middle Clarendonian, South Dakota, Nebraska, New Mexico, and Montana.

Diagnosis: Largest species of *Peraceras* (M1–3 length = 143–165 mm; m1–3 length = 150–178 mm). Distinguished from all other aceratheriines from North America by its robust, broad brachycephalic skull with broad elevated lambdoid crest and anteriorly inclined occiput. Distinguished from similarly brachycephalic *Teleoceras* by its deep nasal incision, and brachydont teeth without strong antecrochets.

Description: *P. superciliosum* has already been described and figured by Osborn (1904), Cope and Matthew (1915), and Matthew (1918, 1932). Additional skulls shown here (Fig. 4.33) demonstrate essentially the same features as the type.

The only region not well preserved in the type skull is the nasals, which are broken. Cope (1880) thought that the nasals were "shortened" and considered that condition diagnostic of the genus. Osborn (1904), however, realized that they were simply broken on the type skull. Matthew (1918) returned to Cope's view and even illustrated (1918, fig. 9) the "difference" in the nasals of *Aphelops*, *Peraceras*, and *Teleoceras*. He may have gotten this mistaken impression because the type of *P. troxelli*, described in that same paper, also has broken nasals. Indeed, the nasal bones are very fragile and remain intact only in the best-preserved rhino skulls.

Such well-preserved material is now available in the Frick Collection. The nasals of *P. profectum* and *P. hessei* are shown elsewhere in this section. They are slender and pointed, like those of *Aphelops*, except that they are shorter and flat with no dorsal flexure. A skull of *P. superciliosum* (F:AM 114409) from the Burge Member of the Valentine Formation, Nebraska, shows a similar development of the nasals. They are long and slender, but the lateral margins are downturned, so they are U-shaped in cross-section. The tips are pointed but dorsoventrally thicker than in primitive *Peraceras profectum*. F:AM 114409 (Fig. 4.33C-D) has a slightly rugose dorsal tip on the nasals, which may have supported some sort of small horn. If this is so, then some *Peraceras* were not hornless (contrary to the name "aceratheriine"). Indeed, specimens of *P. superciliosum* from the Barstovian Andrew Hottell Ranch Quarry (Voohies *et al.*, 1987), Banner County, Nebraska (UNSM locality Bn-10) show very rugose areas at the tips of the nasals of male skulls, strongly suggesting that male *P. superciliosum* had a horn (Fig. 4.33).

Discussion: *Peraceras troxelli* was described by Matthew (1918, pp. 208–209) based on a skull (AMNH 11434) from the late Barstovian Burge Member of the Valentine Formation (Skinner and Johnson, 1984). Matthew distinguished it from *P. superciliosum* by the following definition: "Smaller than *P. superciliosum*, nasals more reduced, cut short to the narial notch above P4. Occiput narrower, occipital portion of cranium less sharply bent upwards. M3 smaller, less quadrate, crochet somewhat more developed on premolars and molars. A small crista on M3." (Matthew, 1918, p. 208)

Most of the dental features discussed above are not diagnostic, since they vary within populations of *Peraceras*. The M2 crista, for example, occurs in several skulls of *P. superciliosum*, including F:AM 114396. The nasals (as discussed above) are not reduced, but simply broken on the type skulls of *P. troxelli* and *P. superciliosum* and incorrectly restored. The skull proportions are difficult to determine. On close inspection, it appears that most of the dorsal and posterior portions of AMNH 11434 are plaster reconstruction. The small amount of original bone present could have been restored much more along the lines of typical *P. superciliosum*, with a broader, more dorsally inflected occiput. Thus, only size differentiates this skull of *Peraceras* from other Burge *P. superciliosum*, and even in this character (Fig. 4.31) it falls at the

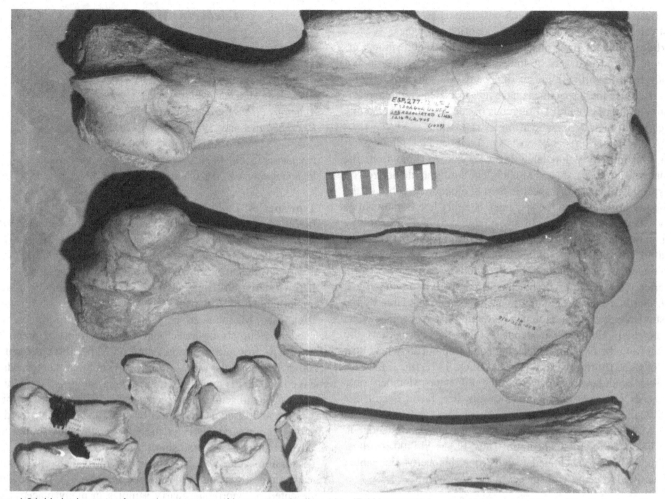

Figure 4.34. Limb elements of an unknown taxon of large aceratheriine rhino (F:AM 114400) from the Pojoaque Member of the Tesuque Formation. Scale bar in centimeters.

small end of the normal size range of *P. superciliosum.* F:AM 114409 which clearly has normal *P. superciliosum* skull features, is identical in size to *P. troxelli,* and even comes from the same deposits in Keya Paha County, Nebraska. Therefore, I consider *P. troxelli* to be a junior synonym of *P. superciliosum.*

P. superciliosum has a very restricted temporal and geographic distribution, unlike the other species of *Peraceras.* It is best known and most abundant from the late Barstovian and early Clarendonian Valentine Formation of Nebraska and South Dakota, but also occurs in the late Barstovian Madison River beds of Montana, and the late Barstovian to early Clarendonian Pojoaque Member of the Tesuque Formation of New Mexico. Early Clarendonian examples are rare, and the provenance is uncertain on all non-Barstovian *P. superciliosum.* There are no well documented late Clarendonian occurrences of *P. superciliosum.* It appears that *P. superciliosum* was a highly derived offshoot of *P. profectum* that briefly flourished in the late Barstovian, mostly in northern areas (Montana, South Dakota, Nebraska), although it was also sympatric in New Mexico with *P. profectum.* If it survived into the early Clarendonian, it was extremely rare, and was

definitely extinct by the late Clarendonian. This is the same time that *Teleoceras* becomes much more abundant and specialized in the same ecological niche occupied by *P. superciliosum* (judging from their convergence of skull features). As Prothero and Manning (1987) suggested, *P. superciliosum* appears to represent an unsuccessful attempt by an aceratheriine to compete with *Teleoceras* for the grazing niche. *P. superciliosum* is most common in the Burge and Crookston Bridge members of the Valentine Formation, and yet even there it is greatly outnumbered by *Teleoceras.* By the early Clarendonian, *Teleoceras* is almost completely dominant. As its great abundance in the Clarendonian and Hemphillian attest, it was very successful in its ecologic niche.

Large aceratheriine rhinoceros (genus uncertain)

Discussion: Because most rhinoceros taxa are diagnosed by the characters of the skull and teeth, postcranial material is frequently not identified. In this study (Chapter 5), I have attempted to document the detailed postcranial anatomy for the important North

American rhinocerotids for which postcranials are known. Thus, most rhino postcranials can now be identified to species.

While sorting the rhino postcranials in the Frick Collection, Earl Manning found two sets of limb elements that clearly do not belong to any North American rhinoceros known from postcranials. Because these specimens have no associated cranial material, they cannot be identified to genus or species. However, some indication of their systematic affinities is possible.

F:AM 114400 (Fig. 4.34) consists of a right and left femur, a left tibia, and most of a left tarsus of a very large rhino from the Pojoaque Bluffs area of New Mexico. The stratigraphic horizon on the specimen is no more specific than "Pojoaque Member of the Tesuque Formation," so the specimen is either late Barstovian or early Clarendonian in age. In size (Table 5.5) and morphology (Fig. 4.34), these elements closely resemble comparable elements of late Hemphillian *A. mutilus*, except that they are even larger than the largest aceratheriine rhino previously known from North America. They could easily be considered a very large individual of *A. mutilus*, except there are no aceratheriines this large by the Barstovian or early Clarendonian (assuming the provenience data are correct). Hemphillian deposits (Chamita Formation) are known from the Española Basin, but none crop out anywhere near the Pojoaque Bluffs area (Galusha and Blick, 1971, fig. 38). Of the possible candidates known from the Pojoaque Member, only *Peraceras superciliosum* approaches this size, but it is still disjunctly smaller. *A. megalodus* is much smaller (and very rare in New Mexico). These limbs are clearly too long and slender to have belonged to any teleoceratin. Thus, F:AM 114400 is probably from a large long-limbed aceratheriine rhinoceros, but cannot be identified to genus and species at the present time.

Tribe Teleoceratini Hay, 1902

Discussion: The teleoceratins are a distinctive, abundant, widespread, and well-known group of rhinos. They are recognized by their distinctive synapomorphies, particularly their short, stumpy limbs with robust, flattened carpals, tarsals, and metapodials. They are also characterized by having a very brachycephalic skull with a flaring lambdoid crest and broad zygomatic arches, and nasals that are U-shaped in cross-section, with or without a small terminal horn. Other synapomorphies include a nasal incision retracted to anterior P3 (not as far as in aceratheriines); a strong, lobal antecrochet on the upper molars; and an elongate calcaneal tuber. In general, most skeletal elements of a teleoceratin are easily distinguished from any other group.

The group is best known from the widespread Eurasian genus *Brachypotherium*, plus several other forms from the Miocene (*Diaceratherium*, *Brachydiceratherium*, *Prosantorhinus*, and *Aprotodon*). In North America, it is represented only by the widespread and abundant genus *Teleoceras*. Most of these genera are known from the Miocene, but they are first known from the late Oligocene of Europe (*Diaceratherium*) and flourished in the early Miocene of Eurasia before arriving in North America during the early Hemingfordian (Martin Canyon Formation of Colorado).

Teleoceras Hatcher, 1894

Rhinoceros Leidy, 1865 (non Linnaeus)
Aphelops Cope, 1878 (in part)
Aceratherium Marsh, 1887 (in part)
Teleoceras Hatcher, 1894
Aphelops (?*Peraceras*) Osborn, 1904
Paraphelops Lane, 1927
Teleoceras (*Mesoceras*) Cook, 1930
Peraceras Cook, 1930
Brachypotherium Yatkola and Tanner, 1979

Type species: *T. major* Hatcher, 1894
Included species: *T. americanum* (Yatkola and Tanner, 1979); *T. medicornutum* Osborn, 1904; *T. meridianum* (Leidy, 1865); *T. fossiger* (Cope, 1878); *T. proterum* (Leidy, 1885); *T. hicksi* Cook, 1927; *T. guymonense* n.sp. and *T. brachyrhinum* n.sp.

Known distribution: Early Hemingfordian to latest Hemphillian, North America.

Diagnosis: Medium- to large-sized teleoceratin rhinos with hypsodont teeth, strong antecrochets, greatly reduced premolars with dP1/p1 lost and occasional loss of P2/p2, thick cement on teeth, narrow nasals with strongly downturned lateral edges, enlarged premaxilla and I1, broad zygomatic arches, flaring lambdoid crests (skull semicircular in posterior view), a small terminal nasal horn and fused nasals, lower tusk (i2) shaped like a teardrop in cross-section, and characteristic teleoceratin body proportions of a barrel-shaped trunk and short, robust limbs.

Discussion: *Teleoceras* is one of the most distinctive, widespread, and easily recognized rhinoceroses in North America. It is represented by enormous quarry samples in many places. Even where it is known only from fragmentary materials, it is easily recognized. Because of the hypsodonty, skull shape, and distinctly short, robust limbs, it cannot be mistaken for material from any other North American genus.

Despite these facts, *Teleoceras* was recognized relatively late (1894). It was the last of the three major genera of Miocene North American rhinoceroses to be named. The first named species of the genus (*meridianum*) was described by Leidy in 1865, but referred to *Rhinoceros*. Cope (1878) referred his species *fossiger* to *Aphelops*, despite having a nearly complete skull that clearly did not resemble his other specimens referred to *Aphelops*. Leidy (1885, 1886) referred the large Mixson bone bed sample of *Teleoceras* to *Rhinoceros*, and Lucas (Leidy and Lucas, 1896) then referred it to *Aphelops*. Only when a skull with nasals was obtained did Hatcher (1894c) consider the taxon distinct from *Aphelops*. Indeed, the name *Teleoceras* ("end horn") refers to this terminal nasal rugosity. It is curious that the founders of American vertebrate paleontology took so long to recognize such a distinctive genus during a period of extreme typological oversplitting.

Once the name *Teleoceras* was proposed, authors began to recognize the taxon everywhere. In some cases, their enthusiasm for naming new species was excessive, so that four different species in three genera were proposed for a single quarry sample of

Teleoceras from the Hemphillian of Colorado (Lane, 1927; Cook, 1927, 1930).

Misconceptions about *Teleoceras* abound. Miscorrelation of the "Kimballian" (Schultz and Stout, 1961) with the post-Hemphillian (Breyer, 1981; Tedford *et al.*, 1987, demonstrated that it is early-middle Hemphillian) led Tanner (1975) to propose *T. schultzi* for "the largest, final stage of *Teleoceras* evolution." But *T. "schultzi"* (= *T. hicksi*) is actually *smaller* than the largest species, the early Hemphillian *T. fossiger*, and the true final species, *T. guymonense*, is actually a dwarf. As Prothero and Manning (1987) pointed out, *Teleoceras* is one of many generic lineages that reach their maximum size in the early Hemphillian and then became smaller in the late Hemphillian. Tanner (1959, 1967, 1975) reported again and again that *Teleoceras* is not known before the late Barstovian (Valentinian) when in fact *Teleoceras "thomsoni"* has been known from the early Barstovian since Cook described it in 1930 (and with the reassignment of Yatkola and Tanner's *"Brachypotherium americanus"* to *Teleoceras*, it is now known from the early Hemingfordian). In addition, the Frick Collection now has enormous early Barstovian and Hemingfordian samples of *Teleoceras*, so the gap between *"Brachypotherium americanus"* and late Barstovian *Teleoceras* suggested by Yatkola and Tanner (1979) is now completely filled. Instead, the Frick Collection gives a completely different picture of *Teleoceras* than exists in the literature before recent work (Prothero and Manning, 1987; Prothero *et al.*, 1989; Prothero, 1998).

There are many other misconceptions as well. Stanley (1977) suggested that *Teleoceras* arose suddenly by chondrodystrophic dwarfing. But this completely ignores the protracted development of the stumpy *Teleoceras* limbs (Prothero and Sereno, 1980) or the fact that almost the entire tribe (except for the most primitive genus, *Diaceratherium*) is characterized by these limb proportions (Prothero *et al.*, 1989). Even the man who erected the genus got it wrong. Hatcher (1894c) diagnosed *Teleoceras major* as "one third larger than *Aphelops fossiger*" when it is actually about one-third smaller.

Teleoceras americanum (Yatkola and Tanner, 1979)

Brachypotherium americanus [*sic*] Yatkola and Tanner, 1979
Brachypotherium americanum Prothero and Manning, 1987
Brachypotherium americanum Prothero, 1998

Figures 4.35, 4.36; Table 4.9

Holotype: KU 9857, skull with broken dentition, from the early Hemingfordian Martin Canyon Formation, Quarry A, Logan County, Colorado.

Hypodigm: *Middle Hemingfordian, Box Butte Formation, Box Butte County, Nebraska*: Above Dry Creek Prospect: F:AM 116093, back of juvenile skull; Dry Creek Prospect: F:AM 116096, ulna.

Late Hemingfordian? channel in Runningwater Formation, Dawes County, Nebraska: Perissodactyl Prospect: F:AM 115031, radius; F:AM 115032, distal femur; F:AM 115033, distal tibia;

F:AM 115034, footbones; F:AM 115035, left ramus.

Late Hemingfordian, Sheep Creek Formation, Dawes County, Nebraska: Ginn Quarry: F:AM 115009, left ramus; F:AM 115010, left ramus.

Late Hemingfordian, Sheep Creek Formation, Sioux County, Nebraska: Antelope Draw: F:AM 114420, skull, vertebrae, partial jaw; *Pliohippus* Draw: F:AM 115026, right ramus; F:AM 115027, right ramus with m2–3; F:AM 115037, tibia; F:AM 115039, distal tibia, unciform; F:AM 115040, tibia; East Ravine Quarry, Ranch House Draw: F:AM 115016, left ramus with m1–3; Hilltop Quarry, south of Antelope Draw: F:AM 115005, pelvis; F:AM 115011, left ramus; F:AM 115012, right ramus with m2–3; F:AM 115013, right ramus F:AM 115014, left ramus (edentulous); Greenside Quarry, Ranch House Draw: F:AM 114414, skull; F:AM 114415, skull; F:AM 114416, skull; F:AM 114417, palate; F:AM 115006, left ramus; F:AM 115007, left ramus; F:AM 115008, right ramus; F:AM 115038, tibia; Thompson Quarry, Stonehouse Draw, Kilpatrick Ranch: F:AM 114418, skull; F:AM 115019, mandible; F:AM 115020, left ramus; F:AM 115029, femur; F:AM 115036, tibia/fibula; Thistle Quarry, Antelope Draw: F:AM 114419, skull; F:AM 115015, right ramus; F:AM 115017, right ramus with m2–3; F:AM 115018, left ramus; F:AM 115028, femur; Long Quarry, Antelope Draw: F:AM 115002, right ramus; F:AM 115003, left ramus with m1–3; F:AM 115004, left ramus with m2–3; F:AM 115021, left ramus; F:AM 115022, right ramus with dp4–m2; F:AM 115023, left ramus; F:AM 115024, left ramus; F:AM 115025, left ramus; F:AM 115030, femur; Vista Quarry: F:AM 114421, partial skull.

Latest Hemingfordian, Massacre Lake l.f., Washoe County, Nevada: UCMP 61852, tibia.

Known distribution: Early to late Hemingfordian, Colorado, Nevada, and Nebraska.

Diagnosis: Small species of *Teleoceras* with slender post-tympanic process whose ventral edge is level with the postglenoid process; lingual cingula on upper premolars; and M2 is approximately equal in length to M3. Distal limb elements, especially the second metapodials, are longer and less robust than is typical of later *Teleoceras*. Like *Teleoceras*, however, the occiput is vertical and the dorsal skull profile is sharply inflected.

Description: The cranial material was described by Yatkola and Tanner (1979), and additional material is illustrated in Figure 4.35. The postcranials are described in Chapter 5.

Discussion: Yatkola and Tanner (1979) first described this skull as *Brachypotherium*, a common Miocene Eurasian genus of teleoceratin. Their identification was based on a skull (KU 9857) from the early Hemingfordian Quarry A in the Martin Canyon Formation, Logan County, Colorado (Galbreath, 1953; Tedford *et al.*, 1987; Tedford *et al.*, 2004). They compared this skull to the existing collections and literature, and found that it was much more primitive than any *Teleoceras* known until then (which apparently included only late Barstovian, Clarendonian, and Hemphillian specimens). They also made comparisons with the type skull of *Brachypotherium* (now *Diaceratherium*) *aurelianense* and other illustrated specimens, and concluded that the

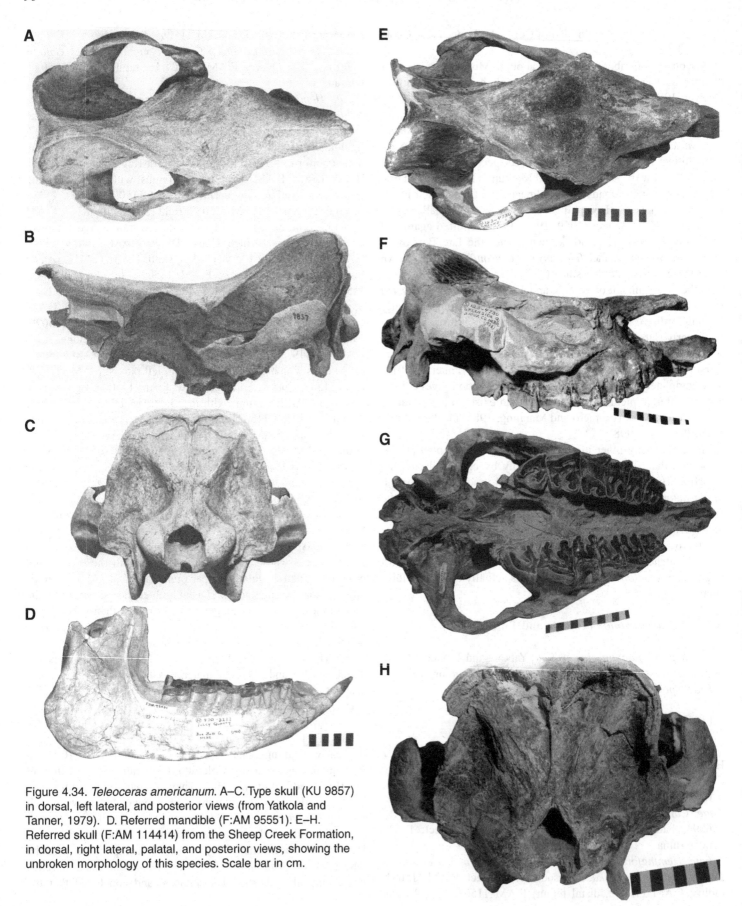

Figure 4.34. *Teleoceras americanum*. A–C. Type skull (KU 9857) in dorsal, left lateral, and posterior views (from Yatkola and Tanner, 1979). D. Referred mandible (F:AM 95551). E–H. Referred skull (F:AM 114414) from the Sheep Creek Formation, in dorsal, right lateral, palatal, and posterior views, showing the unbroken morphology of this species. Scale bar in cm.

Martin Canyon specimen was an American *Brachypotherium*.

The large collection of primitive teleoceratins in the Frick Collection has made most of their analysis obsolete. In particular, the primitive skulls from the middle and late Hemingfordian and early Barstovian bridge the gap between typical *Teleoceras* (as Yatkola and Tanner, 1979, understood it) and this species. Yatkola and Tanner (1979, p. 2) gave the following revised diagnosis based on their comparisons:

Foramen ovale and foramen lacerum medium separate, post-tympanic process light and just in contact with the post-glenoid process, M2 anteriopoterior length not much greater than M3, P1 larger, basilar mound on sphenoid small, greatest breadth across frontal lies above the anterior portion of the orbit, lacrimal expanded anteriorly, infraorbital foramen located outside of narial notch, teeth relatively low crowned, second metapodials more elongate than in *Teleoceras*.

This diagnosis is no longer valid when a larger series of skulls is examined, especially the skulls from the late Hemingfordian Sheep Creek Formation of Nebraska. For each of the "diagnostic" characters, the following was observed:

(1) <u>Foramina</u>: The foramen ovale and foramen lacerum medium are very close to each other in all teleoceratins and rhinocerotins (Groves, 1984; Prothero and Manning, 1987). The thin bony bridge that separates them is easily broken in fossils, and is highly variable when it is not. Most Hemingfordian teleoceratins still have this bony bridge, but so do some very derived late Tertiary *Teleoceras*, including the type specimen of *T. fossiger* (AMNH 8390). This character is too variable within populations, and too susceptible to breakage, to be valid.

(2) <u>Post-tympanic process</u>: In most primitive rhinos, this process is a very thin, wing-like flange, which nearly contacts and sometimes fuses with the postglenoid process. Its ventral edge extends no further ventrally than the postglenoid. In most of the Hemingfordian teleoceratins in the Frick Collection, this primitive condition is retained. Early Barstovian *Teleoceras* skulls from the Olcott Formation of Nebraska have more robust post-tympanic processes that extend ventrally below the level of the postglenoid process. This process also flares widely, resulting in a wide occiput. Thus, the character appears to be valid, but it is a symplesiomorphy of all primitive teleoceratins, so only the derived state can be used to distinguish post-Hemingfordian *Teleoceras*.

(3) <u>M2 length</u>: Primitively, the M2 length and M3 length are about equal, and this is true of nearly all the Hemingfordian teleoceratins (Table 4.9). Early Barstovian *Teleoceras* begins to show a lengthening of the M2, and the general lengthening of all the molars that typifies this genus. Again, however, this is a symplesiomorphy that does not justify placing the Hemingfordian specimens in *Brachypotherium*; it only differentiates them from post-Hemingfordian *Teleoceras*.

(4) <u>Basilar mound on sphenoid</u>: This character is highly variable within quarry samples and shows no consistent trends. Yatkola and Tanner (1979) could have considered it important only if they

compared only one or two type skulls without regard to populational variation.

(5) <u>Greatest breadth</u>: What Yatkola and Tanner (1979) meant by this character is unclear, for it is true of nearly all higher rhinocerotids, including *Aphelops* and *Peraceras*. It certainly does not distinguish *Brachypotherium* from *Teleoceras*.

(6) <u>Lacrimal exposure</u>: This character occurs in all higher rhinos as well. The shape of the lacrimal tends to be highly variable, and difficult to use as a systematic character.

(7) <u>Infraorbital foramen</u>: This feature is very difficult to use, because the infraorbital foramen is always very close to the narial notch. If there is crushing or distortion, it is almost impossible to tell whether it was "inside" or "outside." In undistorted skulls, the condition varies tremendously. There are several derived skulls of *Teleoceras* that have the infraorbital foramen outside the narial notch (e.g., F:AM 42978, a *T. major* from the Burge Member of the Valentine Formation), so this character does not distinguish *Brachypotherium* from *Teleoceras*.

(8) <u>Teeth relatively low crowned</u>: Most teleoceratins (except for the primitive *Diaceratherium* or *Prosantorhinus*) are higher crowned than primitive aceratheriines or rhinocerotins, but all three groups became more hyposodont during their evolution. The Hemingfordian teleoceratin skulls are lower crowned than the most derived Clarendonian and Hemphillian *Teleoceras*, but not noticeably lower crowned than those of the Barstovian.

(9) <u>Second metapodials</u>: This character is apparently valid, but symplesiomorphic. In fact, the entire late Hemingfordian sample of teleoceratin postcranials is less robust and shortened than the early Barstovian sample (see Chapter 5). From the early Barstovian on, the postcranials assume the stumpy, flattened, *Teleoceras* shape.

Since the time of Yatkola and Tanner's (1979) description of "*Brachypotherium americanus*" [*sic*], much has been learned. Originally, I (Prothero *et al.*, 1989; Prothero, 1998) followed their assignment of this species to *Brachypotherium*, but indicated (Prothero, 1998, p. 601) that this assignment was debatable. However, both their comparisons and my own were based on very limited access to the European material. Much additional new work has been done on the European Miocene rhinoceroses as well (Ginsburg *et al.*, 1981; Cerdeño, 1993; Heissig, 1999). "*Brachypotherium*" *aurelianense* has been transferred to the more primitive European Oligocene-Miocene genus *Diaceratherium*, so if the Hemingfordian teleoceratin were assigned to any European genus, it would be *Diaceratherium*, not *Brachypotherium*. The latter genus is a much more derived form (as currently defined) and in no way resembles the Hemingfordian teleoceratin. In addition, several European paleontologists, with their superior knowledge of the (largely unpublished) European material of *Diaceratherium*, have examined the Sheep Creek specimens in the Frick Collection, and agree with me that it should not be assigned to *Diaceratherium* or any European genus. Cerdeño (pers. comm.) found that *Diaceratherium aurelianense* from the locality of Neuville, France, has a much higher occiput which is less laterally excavated than the Hemingfordian teleoceratin. The occiput on

TABLE 4.9—Cranial and dental measurements of *Teleoceras* (in mm)

MEASUREMENT	*T. americanum*			*T.meridianum*			*T. medicornutum*			*"M. thomsoni"*
	Mean	SD	N	Mean	SD	N	Mean	SD	N	AMNH 82592
P2 to occiput	444	19	3	—	—	—	482	45	4	530
Lambdoid crest to nasals	435	38	3	—	—	—	488	47	4	525
Width @ zygoma	320	14	2	323	—	1	350	36	4	350
Width of occiput	163	9	2	170	—	1	200	—	1	240
Height of occiput	161	11	3	—	—	—	239	—	1	—
P2-M3 length	227	7	4	—	—	—	236	13	3	221
P2-4 length	97	3	4	—	—	—	94	8	3	89
M1-3 length	136	4	4	138	—	1	140	12	5	136
P2 length	30	2	4	—	—	—	28	3	3	26
P2 width	35	4	4	—	—	—	36	3	3	35
P3 length	34	2	4	—	—	—	32	3	5	29
P3 width	49	4	4	—	—	—	48	3	5	45
P4 length	37	2	4	—	—	—	36	3	5	33
P4 width	56	2	4	—	—	—	58	5	5	54
M1 length	46	6	4	46	—	1	41	3	6	46
M1 width	60	4	4	61	—	1	61	9	6	53
M2 length	51	7	4	49	—	1	50	5	6	46
M2 width	61	6	4	68	—	1	64	7	6	61
M3 length	53	11	4	47	—	1	50	4	6	52
M3 width	56	3	4	51	—	1	61	5	6	57
p3-m3 length	210	14	6	—	—	—	242	17	4	—
p3-4 length	72	6	6	—	—	—	72	7	4	—
m1-3 length	141	9	5	—	—	—	155	8	5	—
p2 length	24	1	3	—	—	—	26	2	2	—
p2 width	17	1	3	—	—	—	19	1	2	—
p3 length	33	2	6	24	2	2	35	3	5	—
p3 width	23	2	6	22	4	2	26	5	5	—
p4 length	39	4	5	31	1	2	41	3	6	—
p4 width	28	2	6	28	6	2	33	2	5	—
m1 length	43	4	6	38	—	1	48	5	6	—
m1 width	30	2	6	31	2	2	35	3	5	—
m2 length	49	4	6	44	3	2	55	3	6	—
m2 width	30	2	6	29	3	2	37	2	6	—
m3 length	53	3	6	50	1	2	57	2	6	—
m3 width	29	3	6	30	3	2	35	2	6	—

Diaceratherium aurelianense slopes anteriorly, while that on the Hemingfordian teleoceratin and most *Teleoceras* is vertical. The dorsal profile on *Diaceratherium aurelianense* is smoothly concave, while that on the Hemingfordian teleoceratin and *Teleoceras* is sharply inflected. In *Diaceratherium aurelianense,* the postglenoid and post-tympanic processes are in contact, while they are close together but separate in the Hemingfordian teleoceratin (and in juvenile specimens of *Diaceratherium aurelianense*), and fused in more advanced *Teleoceras.* As outlined in item 2 above, this character is not very diagnostic. The Hemingfordian teleoceratin and *Diaceratherium aurelianense* are similar in size, and the narial notch reaches the P3 in both specimens, but these are symplesiomorphic features.

Thus, nothing but symplesiomorphies unite the Hemingfordian teleoceratin with any European genus (*Diaceratherium* or *Brachypotherium*). Instead, it shares derived similarities in the occiput angle and dorsal skull profile with North American *Teleoceras.* For these reasons, I modify my taxonomic assignments of earlier papers (Prothero *et al.,* 1989; Prothero, 1998) and refer Yatkola and Tanner's (1979) species "*Brachypotherium americanus*" to *Teleoceras.*

Teleoceras medicornutum Osborn, 1904

Teleoceras medicornutus [*sic*] Osborn, 1904
Aphelops (?Peraceras) planiceps Osborn, 1904
Teleoceras bicornutus [*lapsus calami*] Osborn, 1905
Teleoceras (Mesoceras) thomsoni Cook, 1930
Teleoceras medicornutum Prothero and Manning, 1987

Figures 4.36, 4.37, Table 4.9

Holotype: AMNH 9832, a skull lacking occiput and jaws without an angle, from the "middle Miocene [Barstovian] of eastern Colorado, about twenty-four miles north of Pawnee Buttes" (Osborn, 1904, p. 319). Galbreath (1953, p. 108) considered it likely that the specimen came from the late Barstovian Vim-Peetz local fauna, although this cannot be determined with certainty.

Hypodigm: *Early Barstovian, Pawnee Creek Formation, Weld County, Colorado*: 1 mile west of Buttes: F:AM 115177, juvenile radius;Pawnee Quarry: F:AM 115174, radius; Ehmke (near Blick's Old Pit): F:AM 115175, tibia; F:AM 115176, radius; Big Spring Quarry: F:AM 115172A–B, two tibiae;150-200 yards South of Dark Cap: F:AM 115173, astragalus; Grover Side (Miocene): F:AM 115197, left radius; Rhino Quarry: F:AM 115198, radius; F:AM 115215A–B, two rami; F:AM 115216, ramus; F:AM 115217, mandible; F:AM 115218, partial left ramus; F:AM 115219, partial ramus; F:AM 115220, mandible; F:AM 115221, right ramus; F:AM 115222, left ramus; F:AM 115223, partial left ramus; Mastodon and Horse Quarry: F:AM 115058, left ramus; F:AM 115189, left ramus; F:AM 115190, right ramus; F:AM 115191, right ramus; F:AM 115192, four femora; F:AM 115193A–B, two femora; F:AM 115194A–E, five tibiae; F:AM 115195A–D, four fibulae; F:AM 115196A–F, six tibiae.

Early Barstovian, Olcott Formation, Dawes County, Nebraska: Survey Quarry: F:AM 114429, skull; F:AM 114430, skull; F:AM 115162, femur; F:AM 115163, scapula; F:AM 115164, humerus; F:AM 115165, humerus; F:AM 11516, ulna; F:AM 115167, left ramus with m2–3; F:AM 115168, left ramus; F:AM 115169, left ramus with m2–3; Observation Quarry: F:AM 115060, atlas; F:AM 115061, nasals; F:AM 115062A–C, three juvenile rami; F:AM 115141, femur; F:AM 115142, femur; F:AM 115143, humerus; F:AM 115144, radius; F:AM 115145, humerus; F:AM 115146, humerus; F:AM 115147, radius; F:AM 115148, distal humerus; F:AM 115157A–C, three patellae; Pioneer Quarry: F:AM 115171, ulna.

Early Barstovian, Olcott Formation, Sioux County, Nebraska: *Synthetoceras* Quarry, Kilpatrick Ranch, Ranch House Draw: F:AM 115140, femur; F:AM 114428, back of skull; Echo Quarry: F:AM 114427, skull, rami; West Sand Quarry: F:AM 114426, skull, limbs, mandible, vertebrae; Jenkins Quarry, Ashbrook Ranch, Sinclair Draw: F:AM 114422, partial skull; F:AM 114423, partial skull; F:AM 114424, skull; F:AM 114425, skull; F:AM 115067, right ramus; F:AM 115068, left ramus; F:AM 115089, thoracic vertebra; F:AM 115101, humerus; F:AM 115114A–E, five radii; F:AM 115116, radius; F:AM 115117A–C, three ulnae; F:AM 115125, Mc2; F:AM 115126A–C, three Mc4s; F:AM 115127A–B, two Mt4s; F:AM 115128A–C, three Mc3s; F:AM 115129A–D, four Mt3s; F:AM 115132, femur; F:AM 115133, femur; F:AM 115158A–B, two tibiae/fibulae; F:AM 115159A–B, two tibiae; F:AM 115160, tibia; F:AM 115161, tibia/fibula; Version Quarry, Sinclair Draw: F:AM 115120, ulna; Channel C, West Sinclair Draw: F:AM 115051, maxilla with M2–3; F:AM 115063, right ramus with m3; F:AM 115113C, radius; F:AM 115130, scapula; F:AM 115131, femur; East Sand Quarry, West Sinclair Draw: F:AM 115050, maxilla with M2–3; West Sand Quarry, West Sinclair Draw: F:AM 115115A–C, three radii; F:AM 115052, ramus with m2–3; F:AM 115100, distal humerus; West Surface Quarry, West Sinclair Draw: F:AM 115056, right ramus with m2–3; F:AM 115059, left ramus with m3; F:AM 115113A–C, three radii; F:AM 115118A–B, two ulnae; F:AM 115134, femur; F:AM 115135, femur; F:AM 115152, tibia; F:AM 115154, fibula; F:AM 115064, right ramus; New Surface Quarry, East Sinclair Draw: F:AM 115057, right ramus; F:AM 115102, humerus; F:AM 115112A–C, three radii; F:AM 115119A–B, two ulnae; F:AM 115136, femur; F:AM 115149, tibia; F:AM 115150, tibia; Quarry #2, East Sinclair Draw: F:AM 115053, back of skull; F:AM 115103, humerus; F:AM 115104, humerus; F:AM 115151, tibia; F:AM 115084, left ramus; F:AM 115085, right ramus; F:AM 115086, ramal fragment; Quarry #3, East Sinclair Draw: F:AM 115153, tibia; Quarry #4, East Sinclair Draw: F:AM 115105, humerus; F:AM 115106, humerus; Boulder Quarry, East Sinclair Draw: F:AM 115055A–B, juvenile left and right rami; F:AM 115111, radius; F:AM 115096, humerus; Humbug Quarry, Ranch House Draw: F:AM 115042, maxilla with P2–M2; F:AM 115043, maxilla with M1–3; F:AM 115044, maxilla with M1–3; F:AM 115045, maxilla with M2–3; F:AM 115046, maxilla with dP3–M2; F:AM 115047, maxilla with M2–3; F:AM 115048, max-

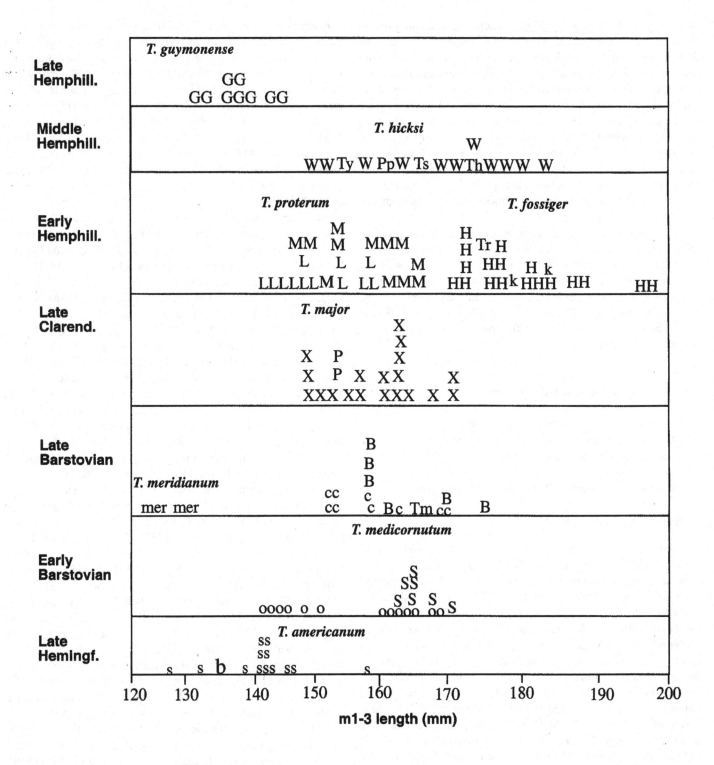

Figure 4.36. Histograms of *Teleoceras* m1–3 lengths through time. Symbols: B = Burge Formation, Nebraska; b = Box Butte Formation, Nebraska; c = Pawnee Creek Formation, Colorado; G = Guymon, Oklahoma; H = Hemphill beds, Texas (Box T Ranch, P.E. Pit); k = Long Island Rhino Quarry, Kansas; L = Love bone bed, Florida; M = Mixson's bone bed, Florida; mer = *T. meridianum*, Texas Gulf Coastal Plain; o = Olcott Formation, Nebraska; P = *T. brachyrhinum*, Pojoaque Member, Tesuque Formation, New Mexico; Pp = type of "*Peraceras ponderis*"; S = Skull Ridge Member, Tesuque Formation, New Mexico; s = Snake Creek Formation, Nebraska; Th = type of *T. hicksi*; Tm = type of *T. medicornutum*; Tr = type of "*Paraphelops rooksensis*"; Ts = type of *T. "schultzi"*; Ty = type of "*Paraphelops yumensis*"; W = Wray localities, Colorado; X = Xmas Quarry, Ash Hollow Formation, Nebraska.

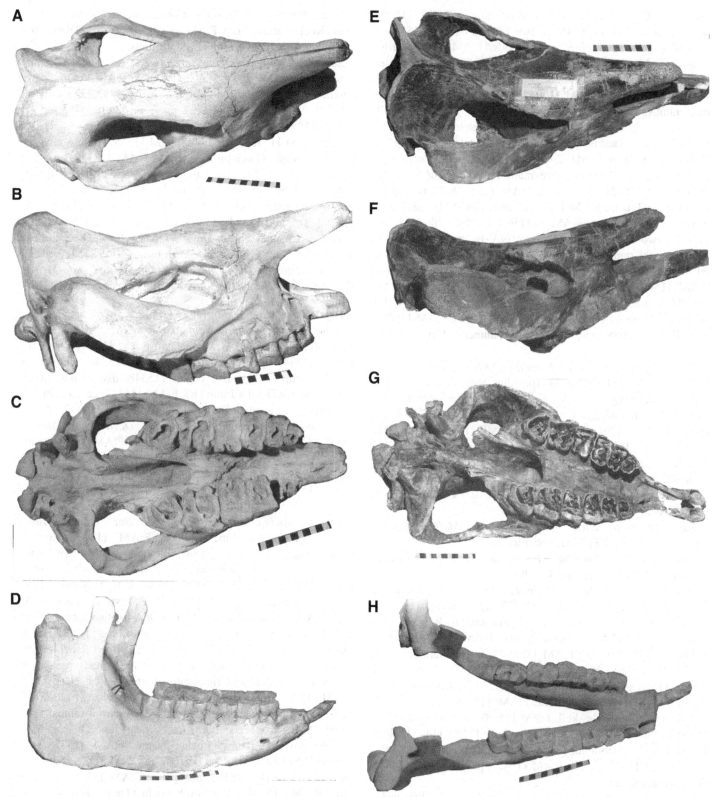

Figure 4.37. *Teleoceras medicornutum*. A–C. Type skull (AMNH 9832) in dorsal, right lateral, and palatal views. D. Type mandible in right lateral view. E–G. Type skull of "*T. (Mesoceras) thomsoni*" (AMNH 82592) in dorsal, right lateral, and palatal views, showing the aberrantly long premaxillae. H. Palatal view of type mandible (AMNH 9332). Scale bar in cm.

illa with dP2–4; F:AM 115076, right ramus; F:AM 115109, radius/ulna; F:AM 115110A–B, two radii; F:AM 115155A–D, four tibiae; F:AM 115069, right ramus; F:AM 115070, right ramus; F:AM 115071, left ramus with m3; F:AM 115072, left ramus; F:AM 115073, ramus; F:AM 115074, ramus; F:AM 115075, right ramus; F:AM 115077, right ramus; F:AM 115078, right ramus; F:AM 115093, humerus; F:AM 115094, juvenile humerus; F:AM 115095, humerus; Head of Antelope Draw: F:AM 115297, composite skeleton; Echo Quarry, Antelope Draw: F:AM 115049, maxilla with M1–2; F:AM 115054, right ramus with m1–3; F:AM 115107A–E, five radii; F:AM 115108A–E, five radii; F:AM 115121, radius/ulna; F:AM 115122A–B, two ulnae; F:AM 115123, ulna; F:AM 115137, femur; F:AM 115138, femur; F:AM 115139, femur; F:AM 115156A–C, three tibiae; F:AM 115041, two astragali; F:AM 115079A–B, two rami; F:AM 115080, right ramus; F:AM 115081, left ramus; F:AM 115082, left ramus; F:AM 115083, right ramus; F:AM 115091, humerus; F:AM 115092, distal humerus; F:AM 115097, humerus; F:AM 115098, humerus; F:AM 115099, humerus; F:AM 115181D, humerus; Campsite Echo Quarry, Antelope Draw: F:AM 115087, atlas; F:AM 115088, axis; F:AM 115090, humerus; F:AM 115124, ulna.

Early Barstovian, Carlin Formation, Elko County, Nevada: UCMP Loc. V-5264, UCMP 44716, radius, ulna, tibia, calcaneum.

Early Barstovian, Virgin Valley beds, Humboldt County, Nevada: UCMP 24293, left maxilla with P2–4; UCMP 24308, left ramus; UCMP 24294, right maxilla; UCMP 21304, tibia; UCMP 35585, teeth; UCMP 35584, calcaneum, foot bones; UCMP 11629, tusk and foot bones; UCMP 11619, teeth; UCMP 41054, right maxilla; UCMP 11615, tibia.

Early Barstovian, Mascall Formation, Grant County, Oregon: UCMP 1682, calcaneum, partial foot.

Early Barstovian, Sucker Creek Formation, Malheur County, Oregon: LACM (CIT) 5732, upper teeth.

Late Barstovian, Punchbowl Formation, Los Angeles County, California: LACM (CIT) 48543, Mt3.

Late Barstovian, Quartz Basin l.f., Deer Butte Formation, Malheur County, Oregon: UCMP 35282, femur, vertebrae.

Late Barstovian, Pawnee Creek Formation, Weld County, Colorado: Mastodon and Horse Quarry: F:AM 115178, scapula; F:AM 115179, humerus; F:AM 115180A–E, five humeri; F:AM 115181A–C, three humeri; F:AM 115182A–D, four humeri; F:AM 115183A–F, six radii; F:AM 115184A–C, three ulnae; F:AM 115185A–B, two femora; F:AM 115186, juvenile skull; F:AM 115187, juvenile skull; F:AM 115188, maxilla with M2–3; F:AM 114460, skull; F:AM 114462, skull; F:AM 114463, skull; F:AM 114472, mandible; F:AM 114473, skull; F:AM 114474, skull; F:AM 114475, skull; F:AM 114478, palate; F:AM 114482, skull and mandible.

Early late Barstovian, Stewart Spring l.f., Esmerelda Formation, Esmerelda County, Nevada: UCMP 58650, left ramus.

Early late Barstovian, Savage Canyon, l.f., Esmerelda Formation, Esmerelda County, Nevada: UCMP 58166, radius, vertebrae, humerus.

?*Late Barstovian, Pershing County, Nevada*: Home Station Pass, Rock Corral Wash: F:AM 115339, partial skeleton; Battle Mountain Area: F:AM 114461, poor ramus.

Late Barstovian, Valentine Formation, Devil's Gulch Member, Brown County, Nebraska: Williams Ranch: F:AM 114499, partial maxilla and ramus; Fairfield Creek: F:AM 115352, left ramus with m1–3; F:AM 115353A–C, three humeri; East Fork of Deep Creek: F:AM 115350, juvenile radius; Elliot Quarry: F:AM 115357, femur; F:AM 115358, mandible; F:AM 115359, radius, ulna, carpus; Rattlesnake Gulch on Pine Creek: F:AM 115349, skull fragment; F:AM 115351, partial tibia; Devil's Gulch Horse Quarry: F:AM 115355, juvenile left ramus; F:AM 115356A, tibia; F:AM 115360, juvenile skull; F:AM 115361, juvenile skull; F:AM 115363, scapula; F:AM 115364, astragalus; F:AM 115365, two calcanea; Channel above Devil's Gulch Horse Quarry: F:AM 115354, left ramus; F:AM 115356B, tibia; F:AM 115362, atlas.

Late Barstovian, Valentine Formation, Cornell Dam Member, Brown County, Nebraska: Norden Bridge Quarry: F:AM 115340, femur; F:AM 115341, ulna.

Late Barstovian, Valentine Formation, Crookston Bridge Member, Cherry County, Nebraska: Top of Devil's Jump Off exposures: F:AM 115343, calcaneum; F:AM 114344, astragalus; F:AM 115345, partial humerus; F:AM 115346, fibula; East Devil's Jump Off Quarry: F:AM 115348, distal tibia; Second Canyon above Devil's Jump Off: F:AM 115342, calcaneum; 0.75 miles west of Valentine Railroad Quarry: F:AM 115347, Mt2.

Late Barstovian, Valentine Formation, Devil's Gulch Member, Cherry County, Nebraska: Ripple Quarry: F:AM 115369, astragalus; Sawyer Quarry: F:AM 115368, femur; Niobrara River, south of Nenzel, upper zone: F:AM 115366, left and right rami; F:AM 115367, left and right rami; F:AM 115422, right ramus with m2–3.

Late Barstovian, Valentine Formation, Burge Member, Brown County, Nebraska: June Quarry: F:AM 115374, juvenile mandible; F:AM 115375, left juvenile ramus with m2 (m3 erupting); F:AM 115377, partial ramus; F:AM 115385A–B, two ulnae; F:AM 115386, humerus; F:AM 115387A–B, two radii; F:AM 115393, astragalus; F:AM 115394A–B, two calcanea; F:AM 115395A–C, three tibiae; F:AM 115396A–B, two femora; F:AM 115370, right ramus; F:AM 115371, right ramus; F:AM 115372, right maxilla; F:AM 115373, right maxilla. Beed Ranch on Pine Creek: F:AM 115421, ramus; F:AM 115423, ulna.Burge Quarry: F:AM 115376, left and right rami; Burge Sands, Deep Creek: F:AM 115389, Mc3. Middle fork, Deep Creek: F:AM 15388A–B, two atlases; Moore Creek: F:AM 115390, partial ramus; F:AM 115391A–B, two Mt3s; Plum Creek near June Quarry: F:AM 115392, juvenile ramus; Lucht Quarry: F:AM 115378, right ramus; F:AM 115380, juvenile right ramus; F:AM 115381A–C, three tibiae; F:AM 115382, astragalus; F:AM 115383, proximal ulna; F:AM 115384, Mt4; South Lucht Quarry: F:AM 115379, juvenile left and right rami.

Late Barstovian, Valentine Formation, Burge Member, Cherry County, Nebraska: 1/8 mile above June Quarry: F:AM 115417, ulna; South side of Niobrara River: F:AM 115411B, tibia; Whiteface Quarry: F:AM 115409, ulna. Midway Quarry: F:AM

115408, femur; F:AM 115410A, radius; Channel between Midway Quarry and Xmas Quarry: F:AM 115410B, radius; F:AM 115411B, tibia; 1 mile below mouth of Steer Creek: F:AM 115402, calcaneum; Deer Canyon: F:AM 115397, symphysis; F:AM 115399, left maxilla with M1–2; F:AM 114498, partial skull; Ewert Quarry: F:AM 115398, right ramus with m2–3; F:AM 115400, humerus; Boiling Spring Bridge: F:AM 115401A–B, two radii; F:AM 115403, right juvenile ramus; Burge Quarry: F:AM 115404, femur; F:AM 115405A–B, two astragali; F:AM 115406, Mt3; F:AM 115407, patella; F:AM 115412, right maxilla; F:AM 115413, juvenile left ramus; F:AM 115414, maxilla with zygoma; F:AM 115415, mandible; F:AM 115416, left ramus; F:AM 115418A–C, three radii; F:AM 115419A–B, two humeri; F:AM 115420, Mc4.

Late Barstovian, Valentine Formation equivalent, Dawes County, Nebraska: "N" Quarry: F:AM 115337, juvenile left ramus, humerus, femur; Surprise Quarry: F:AM 115338, calcaneum.

Early and late Barstovian, Texas Gulf Coastal Plain: dozens of catalogued specimens listed by Prothero and Manning (1987, pp. 409–410).

Early Barstovian, Red Basin fauna, Malheur County, Oregon (Shotwell, 1968): UO24135, skull and lower jaw; UO24160, partial upper dentition.

Known distribution: Early to late Barstovian, Colorado, Nebraska, Nevada, and Texas.

Diagnosis: Smaller species of *Teleoceras* (M1–3 length = 135–160 mm; m1–3 length = 140–170 mm). Differs from *T. major* in smaller size, weaker crochets and no cristae, especially on M2. The p2 is usually present. Distinguished from *T. americanum* by its greater size, M2 larger than M3, post-tympanic process robust and laterally extended, and shorter, more robust limbs.

Description: *T. medicornutum* was described and illustrated by Osborn (1904). Additional figures of well-preserved specimens are shown in Fig. 4.37. The postcranials are described in Chapter 5.

Discussion: Osborn (1904) based part of his diagnosis of *T. medicornutum* on what he believed to be a small frontal horn rugosity (hence the name *medicornutum*, "middle horned"). In fact, this is only a small broken area along the suture between the frontal bones, and has no resemblance to the true frontal rugosity of a tandem-horned rhino skull, such as that of *Diceros* or *Ceratotherium*. None of the many other skulls of *T. medicornutum* in the Frick Collection shows this feature, so it is clearly due to breakage on the type specimen. To my knowledge, no North American rhino shows evidence of having a frontal horn.

In a paper reviewing the recent progress in mammalian paleontology, Osborn (1905, p. 106, plate VII) briefly mentioned the name "*Teleoceras bicornutus*" in passing. He cited his 1904 decription of *T. medicornutum* immediately after giving the name *bicornutus*. He also illustrated an unnumbered specimen (plate VII, fig. 1) of "*Teleoceras bicornutus*" from the "middle Miocene of Colorado." The skull and jaws of "*Teleoceras bicornutus*" are clearly referable to *T. medicornutum*, but it is not the same specimen as the type, AMNH 9832. In his book *The Age of Mammals*, Osborn (1910, p. 252, fig. 127) demonstrated the skull he called "*Teleoceras bicornutus*," but here it was captioned as *T. medicornutus*. Hay (1930) suggested that these were in fact the same taxon. It seems clear from the evidence that "*Teleoceras bicornutus*" Osborn (1905) is a *lapsus calami*. Perhaps Osborn had used it as a manuscript name and changed it after his paper (not published until 1905) was delivered in Berne, Switzerland, in 1904. He never formally corrected this mistake, but is is clear (from Osborn, 1910, fig. 127) that "*Teleoceras bicornutus*" is a junior synonym of *T. medicornutum*.

Osborn (1904, pp. 321–322) also described the posterior portion of a skull (AMNH 9369) from the same Miocene Pawnee Buttes collection as the type specimen of *T. medicornutum*. He gave this specimen the name *Aphelops* (?*Peraceras*) *planiceps*. Despite Osborn's claims to the contrary, the occiput of *A. planiceps* is clearly shaped like many of the skulls of *Teleoceras medicornutum* now known. (In 1904, no occiput was known for *T. medicornutum*.) The skeleton (AMNH 9369) that was found near this skull is clearly referable to *T. medicornutum*. Osborn (1904, p. 322) remarked that this skeleton "indicates limbs somewhat longer than those of *T. fossiger*," which is indeed true of *T. medicornutum*. Matthew (1918, 1932) also considered *planiceps* to be a junior synonym of *T. medicornutum*.

Cook (1930) described a skull (AMNH 82592) he named *Teleoceras* (*Mesoceras*) *thomsoni* (Fig. 4.37E–G) from the "Pliocene Snake Creek beds of Sioux County, Nebraska." According to Skinner *et al.* (1977, p. 333), this skull is actually from the early Barstovian Olcott Formation. Cook compared this skull with *Aphelops, Teleoceras*, and *Peraceras*, and noted that it has a dentition more similar to *Aphelops* than to *Teleoceras*. Cook gave the following comparisons with *T. medicornutum* (Cook, 1930, p. 49): "As compared with *T.* (*Mesoceras*) *mediocornutus* [sic], the zygomatic arches are much deeper and heavier; the skull is relatively wide posteirorly, narrowing forward rapidly, giving the skull a very tapering, wedge-shaped effect. The molars and last two premolars are relatively smaller, and P3 relatively larger, than in *mediocornutus*[sic]." On this basis of this comparison, Cook placed *T. thomsoni* and *T. medicornutum* in his new subgenus *Mesoceras*, but gave no diagnosis or justification for this taxon.

Comparison of the type of *T.* (*Mesoceras*) *thomsoni* with the large Olcott sample in the Frick Collection shows that it falls within the normal range of variation of *T. medicornutum* (Fig. 4.36). The teeth are on the small end of the size spectrum, but the overall skull size is virtually identical with the type of *T. medicornutum*. The dental pattern is typical of primitive *T. medicornutum* as well. Cook was apparently thinking of very derived species of *Teleoceras* when he diagnosed this taxon. The other characters cited (heavy, deep zygomatic arches, and wedge-shaped skull) can be attributed to the fact that this skull is from a very large, robust, rugose old male individual, and these features are also within the range of variation of Olcott *T. medicornutum*. Only the unusually long premaxillae are unique to this specimen. Because no other rhino skull from the Olcott Formation or any other Barstovian locality shows such long premaxillae, this is probably an individual variation of no taxonomic significance. Thus, *T. (Mesoceras) thomsoni* has no validity and is a junior synonym of *T. medicornutum*.

Teleoceras brachyrhinum new species

Figures 4.36, 4.38, Table 4.10

Holotype: F:AM 114433, an adult skull from the early Barstovian Skull Ridge Member of the Tesuque Formation, Santa Fe County, New Mexico.

Hypodigm: *Early Barstovian, Tesuque Formation, Skull Ridge Member, Santa Fe County, New Mexico*: Tesuque Locality: F:AM 115224, articulated forelimb; South side of Big Wash: F:AM 114467, partial skull, mandible; F:AM 114468, skull, mandible; East Skull Ridge: F:AM 115235, left ramus; Skull Ridge: F:AM 115236, right ramus; F:AM 115225, mandible; F:AM 115226, atlas; F:AM 115229A–C, three humeri; F:AM 115230A–B, two radii; F:AM 115231, ulna; F:AM 115233, associated femora, tibia/fibula; F:AM 115237, right ramus; F:AM 115238, left juvenile ramus; F:AM 115239, femur, tibia, fibula, tarsals; F:AM 115240, humerus; F:AM 115301, right ramus with m1–3; F:AM 115308, humerus; F:AM 115309A,C, two radii; F:AM 114433, complete (? male) skull; F:AM 114435, partial female skull; F:AM 114436, young male skull and mandible; F:AM 114437, adult male mandible; F:AM 114438, mandible; F:AM 114439, complete male skull; F:AM 114440, skull; F:AM 114441, top of skull; F:AM 114442, partial maxilla; F:AM 114445, poor skull, mandible; F:AM 114446, partial skull, mandible; F:AM 114447, skull; F:AM 114453, skull, mandible, vertebrae; South Skull Ridge: F:AM 115227, femur; F:AM 115228, tibia; F:AM 115232, radius/ulna; F:AM 114443, battered skull; F:AM 114452, skull; West Skull Ridge: F:AM 114444, bottom of skull; White Operation: F:AM 115302, left ramus; F:AM 115303, left ramus; F:AM 115305, partial skull; F:AM 115306, associated rami, teeth, vertebrae, juvenile maxilla; F:AM 115309B,D, two radii; F:AM 115310, partial femur; F:AM 115311, associated hind limb bones; F:AM 114450, back of skull; F:AM 114451, female mandible; F:AM 114456, poor skull, mandible; F:AM 114458, skull; District 3 (3rd Division): F:AM 115312, mandible; F:AM 115313, maxilla, mandible, associated forelimb; F:AM 115314, mandible; F:AM 115315, ramus; F:AM 115316, femur, tibia; F:AM 114457, complete skull, F:AM 114459, partial ramus; Pojoaque, dark sand: F:AM 115304, left and right rami; Tesuque Pueblo: F:AM 115307, atlas, cervical vertebrae; East Fork, Joe Rak Wash: F:AM 114454, complete ?female skull; Between Joe Rak Wash and Nambe Creek: F:AM 115234, right ramus; Cuyamunque Terrace: F:AM

114455, juvenile skull; East Cuyamunque Locality: F:AM 115243, femur, radius, tarsals; F:AM 115244, associated forelimb; F:AM 115245, pelvis, limb bones, associated foot bones; Cuyamunque Wash: F:AM 115241, left ramus.

Early Barstovian, Tesuque Formation, Nambe Member, Santa Fe County, New Mexico: East Cuyamunque: F:AM 114431, poor female skull, mandible; F:AM 114432, skull; East Santa Fe: F:AM 114434, poor skull, mandible.

Late Barstovian/early Clarendonian, Tesuque Formation, Pojoaque Member, Rio Arriba County, New Mexico: Rhino District # 3: F:AM 115326, partial skull; Tesuque Locality: F:AM 114470, partial skull; F:AM 114476, partial juvenile skull; F:AM 115335, partial skeleton; F:AM 115336, mandible; Cuyamunque, Española Area: F:AM 115334, mandible; Pojoaque Hills (Bluffs), Española Area: F:AM 114479, skull, partial mandible, femur; F:AM 114490, partial pelvis; Pojoaque Bluffs: F:AM 115317, juvenile partial skull; F:AM 115318, vertebrae; F:AM 115319, hindlimb; F:AM 115320A–C, three tibiae; Tesuque Locality: F:AM 115332, atlas; F:AM 115329, articulated vertebrae; F:AM 115330, femur; F:AM 115331, radius; San Ildefonso Area: F:AM 115327, humerus; North Ceja del Rio, Puerco Locality: F:AM 115328, tibia, fibula; West Cuyamunque: F:AM 114481, damaged skull; Cuyamunque: F:AM 115321, mandible; F:AM 115322, mandible, teeth, associated limbs; West Cuyamunque: F:AM 115333, radius; North Tesuque: F:AM 115324, two femora, associated pelvis; Española Area: F:AM 115323, forelimb elements; District #1, Rhino Quarry #4: F:AM 15325, right ramus with p2–m1; Santa Cruz Area: F:AM 114485, palate, upper tusk.

Late Barstovian/early Clarendonian, Tesuque Formation, Chama–el-Rito Member, Rio Arriba County, New Mexico: Rio del Oso: F:AM 116090, astragalus; F:AM 116020, maxilla; Ojo Caliente: F:AM 114489, base of skull; F:AM 116015, calcaneum; Black Mesa: F:AM 114488, crushed partial skull.

Barstovian/Clarendonian, Tesuque Formation equivalent, Sandoval County, New Mexico: Arroyo Arenoso: F:AM 114471, partial skull.

Late Clarendonian, Tesuque Formation, Pojoaque Member, Santa Fe County, New Mexico: Santa Cruz Area: F:AM 115640, maxilla; Santa Cruz Area, 1st Wash: F:AM 115638, femur; F:AM 115639, humerus; F:AM 114469, complete mandible, skull; Santa Cruz Area, 4th Wash: F:AM 115636, tibia; F:AM 115637, ulna; Santa Cruz Area, south side of wash: F:AM 115643, palate; Santa Cruz Area, White Sand: F:AM 115642, partial skull; Santa Cruz Area, Red Layer: F:AM 115641, right ramus; San Ildefonso Locality: F:AM 114484, ramus; F:AM 114486, partial male skull, mandible; F:AM 114487, mandible.

Late Clarendonian, Tesuque Formation, Pojoaque Member, Rio Arriba County, New Mexico: Round Mountain Quarry: F:AM 115644, juvenile palate; F:AM 115645, mandible; F:AM 115646,

Figure 4.38. *Teleoceras brachyrhinum*, n.sp. (opposite page) A–D. Type specimen (F:AM 114433) in dorsal, right lateral, palatal, and posterior views. E. Referred specimen (F:AM 114435) in dorsal view, showing short nasals. F–G. Another referred skull (F:AM 114437) showing short nasals and robust premaxillae. H–I. Referred male mandible (F:AM 114439) in left lateral and occlusal views. Scale bar in cm.

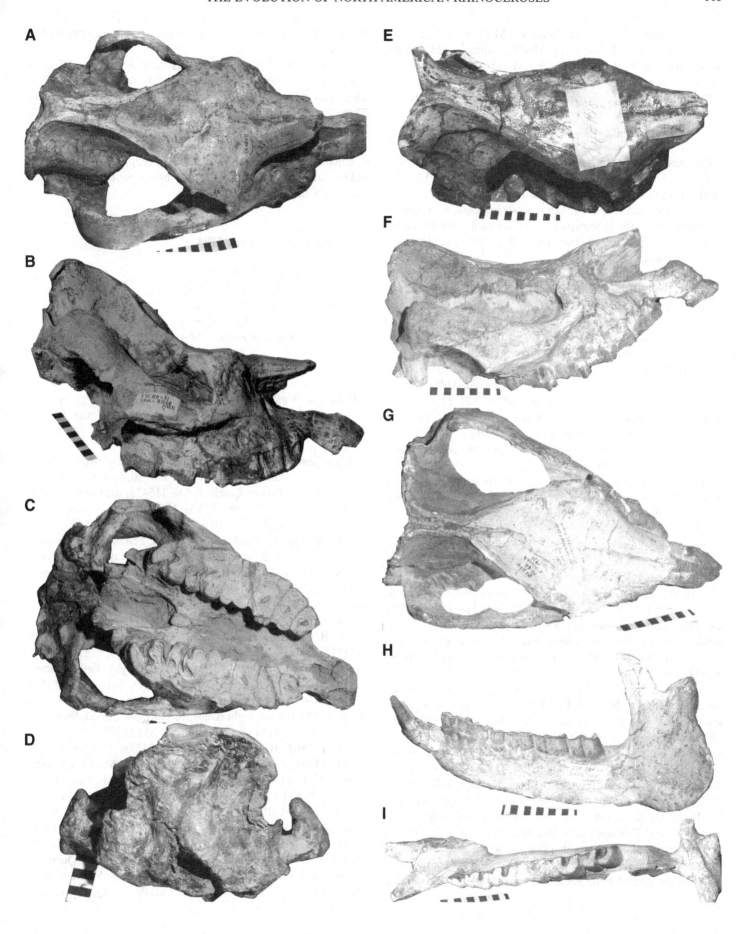

juvenile palate; F:AM 115647, femur; F:AM 115648, tibia; F:AM 115649, humerus and ulna; F:AM 115650, radius; F:AM 114483, skull, mandible; Santa Cruz Area: F:AM 115651A–B, two femora; North Santa Clara River: F:AM 115242, femur.

Etymology: Greek *brachyrhinum* "short nose" in reference to the unusually short and flat nasals.

Distribution: Early Barstovian to late Clarendonian, New Mexico.

Diagnosis: Similar to large *T. medicornutum* in size and morphology (M1–3 length = 150–160 mm; m1–3 length = 160–170 mm) except that the nasals are unusually short and flattened. The skull is more brachycephalic than other *Teleoceras*, with a broad flat frontal area. The premaxillae are robust with a distinct dorsal arch and a broad heavy anterior end for the I1 alveolus. The p2 is usually present.

Description: *Teleoceras brachyrhinum* is known from a large series of skulls from the Barstovian Skull Ridge Member, and from younger units in the Santa Fe Group as well. Although similar to *T. medicornutum* in size, these skulls uniformly differ from Colorado, Nebraska, or Texas specimens of *T. medicornutum* in having peculiarly brachycephalic skulls with unusually short, flattened nasals. This is true not only of the type specimen (Fig. 4.38A–D) but also of many other well-preserved skulls, such as F:AM 114437 (Fig. 4.38F–G). One skull, F:AM 114435 (Fig. 4.38E) carries this nasal shortening to extremes. This specimen has extremely short nasals, with a tiny nubbin at the tip in place of the normal *Teleoceras* horn boss. The *T. brachyrhinum* skull morphology is not only uniform within the New Mexico sample, but unknown outside New Mexico. In a group where few features are uniform in such highly variable population samples, such a difference suggests that the New Mexico *Teleoceras* is a distinct species.

Outside the skull features, there are few differences between *T. brachyrhinum* and *T. medicornutum*. The dentition is not as hypsodont or highly derived as in *T. major*, and few specimens have very strong crochets. In this respect, *T. brachyrhinum* is not as derived as *T. medicornutum*. The symphysis and post-i2 diastema are very short and upturned, which is probably related to the brachycephaly of the skull. The postcranial morphology (so far as it is known) closely matches *T. medicornutum*. Measurements are provided in Table. 4.10.

Discussion: Normally the variations in the length of the nasals would not seem to be a good character on which to base a new species. However, in the case of *T. brachyrhinum*, it is striking that all the *Teleoceras* skulls from the New Mexico Miocene have the shortened, flattened nasals with the tiny horn nubbin (or no sign of a horn base at all) and robust premaxillae, while all the *Teleoceras* skulls from other localities have the normal long nasals with the U-shaped cross-section. A few specimens of *T. hicksi* have similar shortening in the nasals, but they are not as flat as those of *T. brachyrhinum*. This consistent difference between large samples of undeformed skulls shows that the distinction is valid for systematics, and not just based on one aberrant skull (as is true for so many of the invalid rhino species synonymized in this volume).

Thus, I feel that the New Mexico sample deserve recognition as a new species.

This discovery shows that there is endemism in the Miocene faunas of New Mexico, something that has also been noted elsewhere (Galusha and Blick, 1971; Tedford, 1981; Tedford *et al.*, 2004; Tedford and Barghoorn, 1997). Although the Santa Fe Group yields other rhino species that are widespread in the western United States (all three species of *Peraceras*, plus *Aphelops megalodus*), only one species of *Teleoceras* appears in these beds, and it is an endemic. The implications of this pattern are discussed further in Chapter 6.

Teleoceras major Hatcher, 1894

Teleoceras major Hatcher, 1894
Teleoceras fossiger Matthew, 1918 (in part)
Teleoceras fossiger Hesse, 1935
Teleoceras proterum Quinn, 1955
Teleoceras proterum Patton, 1969

Figures 4.36, 4.39, Table 4.10

Holotype: YPM-PU 10645, a partial skull and right ramus; from the early Clarendonian Cap Rock Member, Ash Hollow Formation, Turtle Canyon, Sheridan County, Nebraska (Skinner *et al.*, 1968, p. 431).

Hypodigm: *Early Clarendonian, Clarendon beds, Donley County, Texas*: Blocker Ranch: F:AM 114530, poor skull; R. Farr Place, Bluff on Turkey Creek: F:AM 115425, mandible; R. Farr Land, Bluff Creek Quarry: F:AM 115426, right ramus; F:AM 114544, skull; North of Main Quarry: F:AM 115430, tibia; MacAdams Quarry: F:AM 115442, ulna; F:AM 114541, partial palate; Locality 5: F:AM 115437, symphysis; Locality 5, Quarry 2: F:AM 115438, left ramus; F:AM 114535, skull, mandible, vertebra; Locality 5, Quarry 3: F:AM 115429, right ramus with m2–3; F:AM 115439, radius; Quarry 4: F:AM 115444, left ramus; F:AM 115447, right ramus; F:AM 115448, occiput; F:AM 115449B, femur; F:AM 115450B–C, two tibiae; F:AM 115451A–B, two ulnae; F:AM 115453B, humerus; F:AM 114545, back of skull; F:AM 115452E, radius; Quarry 4, Golden Sand: F:AM 115449A, femur; F:AM 115450A,D, two tibiae; F:AM 115453A,C-D, three humeri; F:AM 115445, left ramus with m2–3; F:AM 115446, right ramus; F:AM 115452 A–D, four radii. Quarry 5: F:AM 115454, right ramus; F:AM 115455, right ramus; F:AM 115456A–B, two ulnae; F:AM 115457A–C, three humeri; F:AM 115458, radius; F:AM 115459, tibia; F:AM 115460, distal femur; F:AM 115461, fibula; F:AM 114532, skull; Locality 17, Quarry 1: F:AM 115427, juvenile left ramus; F:AM 115431, partial juvenile skull; F:AM 115433A–C, three radii; F:AM 115440A, femur; F:AM 115441, tibia; F:AM 115443, radius/ulna; F:AM 114531, ramus; F:AM 114534, mandible; F:AM 114536, skull; F:AM 114543, skull; F:AM 114562; Locality 17, Quarry 1, Yellow Sand: F:AM 115428, juvenile right ramus; F:AM 115434, left maxilla; F:AM 115440B, femur; Locality 17, Quarry 1, Grey

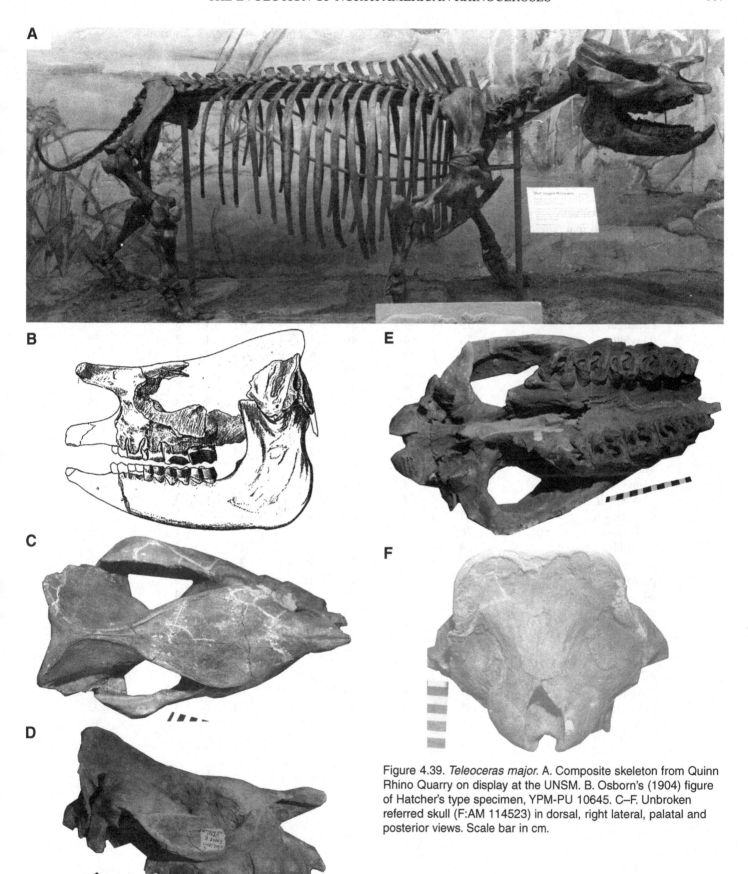

Figure 4.39. *Teleoceras major.* A. Composite skeleton from Quinn Rhino Quarry on display at the UNSM. B. Osborn's (1904) figure of Hatcher's type specimen, YPM-PU 10645. C–F. Unbroken referred skull (F:AM 114523) in dorsal, right lateral, palatal and posterior views. Scale bar in cm.

Table 4.10. Cranial and dental measurements of *Teleoceras* (in mm)

MEASUREMENT	*T. brachyrhinum*			*T.major*			*T. fossiger*			*T. brachyrhinum* F:AM 114433 Type
	Mean	SD	N	Mean	SD	N	Mean	SD	N	
P2 to occiput	428	39	2	471	38	4	—	—	—	—
Lambdoid crest to nasals	403	11	2	483	13	2	—	—	—	410
Width @ zygoma	353	4	2	333	18	4	355	7	3	355
Width of occiput	200	—	1	209	9	4	207	6	3	—
Height of occiput	180	—	1	203	28	4	217	20	3	—
P2-M3 length	256	6	4	243	17	4	305	—	1	251
P2-4 length	109	6	4	99	9	4	129	—	1	101
M1-3 length	149	3	4	143	8	5	178	7	3	151
P2 length	31	1	4	29	3	4	37	—	1	32
P2 width	40	7	4	32	4	4	32	—	1	46
P3 length	39	4	4	34	4	4	47	1	3	33
P3 width	58	2	4	46	3	4	59	1	3	59
P4 length	41	1	4	41	6	5	50	2	3	40
P4 width	66	4	4	59	3	5	63	7	3	69
M1 length	51	5	4	48	10	5	65	5	3	44
M1 width	68	4	4	63	4	5	71	9	3	65
M2 length	57	2	4	53	6	5	62	6	3	56
M2 width	68	3	4	63	2	5	76	6	3	73
M3 length	52	8	4	47	12	5	54	6	3	62
M3 width	63	5	3	57	4	5	71	4	2	68
p3-m3 length	243	4	4	220	5	7	250	9	3	245
p3-4 length	83	4	4	71	3	7	79	4	3	84
m1-3 length	165	3	4	148	4	7	173	5	4	167
p2 length	26	6	3	—	—	—	25	—	1	30
p2 width	20	1	3	—	—	—	15	—	1	20
p3 length	39	2	4	32	1	7	32	4	3	40
p3 width	31	3	3	22	5	7	24	3	3	30
p4 length	43	2	4	39	2	7	45	2	4	43
p4 width	33	2	4	28	1	7	34	7	3	32
m1 length	47	1	4	45	2	7	51	6	4	46
m1 width	37	3	4	29	1	7	32	5	4	38
m2 length	56	6	4	53	3	7	61	1	4	59
m2 width	34	2	4	30	2	7	33	4	4	35
m3 length	61	2	4	55	3	7	63	5	4	60
m3 width	35	3	4	38	2	7	31	6	4	33

Sandstone: F:AM 115432, right maxilla; Locality 22, Green Clay: F:AM 115435A, femur; Locality 22, Quarry 2, Stratum 2, Dilli Place: F:AM 115435B, femur; F:AM 115436, right ramus; Cliff Creek Pit: F:AM 114533, skull.

Early Clarendonian, Clarendon beds, Lipscomb County, Texas: Cole Highway Pit: F:AM 114561, palate; F:AM 115463, ramus; F:AM 115464, humerus.

Early Clarendonian, Lapara Creek fauna, Goliad Formation, Bee County, Texas: Berclair l.f. (= Farish Ranch l.f.), dozens of TMM specimens, including 31081-744, skull; Normanna l.f. (Bridge Ranch l.f.), several TMM specimens, including 31132-29, skull (see Prothero and Manning, 1987).

*Early Clarendonian, Ash Hollow Formation, Cap Rock Member, Brown County, Nebrask*a: Rhino Horizon #3 Quarry: F:AM 115470, right ramus with m1–3; F:AM 115471, right ramus; Martin Arent Ranch: F:AM 115511A–B, femur, tibia, footbones; 1 mile above Martin Arent Ranch: F:AM 115510, associated rami; 2 miles above mouth of Medicine Creek: F:AM 115539, right ramus; F:AM 115540, right ramus; Horsethief Canyon #1: F:AM 115504, juvenile right ramus; Horsethief Canyon #2: F:AM 114510, skull; Talus of Horsethief Canyon #3: F:AM 115506, distal humerus; East Clayton Quarry: F:AM 115507, humerus; F:AM 115508A–B, tibia, distal femur; F:AM 115509A–B, rami, skull roof; Head of C. Hurlburt: F:AM 115505, left ramus with m3; Cap Rock on Harvey Williams Ranch: F:AM 115502, right maxilla; F:AM 115503, right and left ramal fragments; Plum Creek: F:AM 115501, right ramus; F:AM 115538, distal femur; Quinn Rhino Quarry: F:AM 115474A–E, five humeri; F:AM 115472, scapula; F:AM 115473, scapula; F:AM 115474A–B, D–E; F:AM 115476A–F, six juvenile radii; F:AM 115477A–C, three juvenile ulnae; F:AM 115478, pelvis; F:AM 115479, pelvis; 115480A–D, four femora; F:AM 115481A–D, four femora; F:AM 115482A–C, three femora; F:AM 115483A–B, two femora; F:AM 115484A–C, three femora; F:AM 115485A–D, femora; F:AM 115486A–E, five juvenile femora; F:AM 115487A–C, three tibiae; F:AM 115488, tibia/fibula; F:AM 115489, composite skeleton; F:AM 114501, skull; F:AM 114502, skull, mandible; F:AM 114503, skull, mandible; F:AM 114504, skull, mandible; F:AM 114505, skull, mandible; F:AM 114508, skull, mandible; F:AM 114519, skull, mandible; F:AM 114520, skull, mandible; F:AM 115475A–C, three humeri; Quinn Rhino Quarry 2: F:AM 115466, left ramus; F:AM 115467, right ramus; F:AM 115468, left maxilla; F:AM 115469, left ramus; Quinn Rhino Quarry 3: F:AM 115465, palate, occiput; Deep Creek: F:AM 115512, femur; F:AM 115513, tibia; F:AM 115514, juvenile radius; F:AM 115515, proximal ulna; F:AM 115516, right ramus with m1–3; F:AM 115517, ramal fragment; F:AM 115518, distal humerus; F:AM 115519, juvenile left ramus; F:AM 115520, left ramus; F:AM 115521, symphysis; F:AM 115522, nasals; F:AM 115523, left ramal fragment; F:AM 115524, right ramal fragment; F:AM 115525, right ramal fragment; F:AM 115526, mandible; F:AM 114491, partial skull; Long Pine: F:AM 115529, humerus; F:AM 115495A, femur; F:AM 115497A–B, two ulnae; F:AM 115499, fibula; F:AM 115500A, tibia; Willow Creek: F:AM 115530, distal humerus; F:AM

115491, right ramus; F:AM 115493, right ramus; F:AM 115496B, femur; Phalen Ranch on Willow Creek: F:AM 114511, half of skull; F:AM 115490, right ramus; F:AM 115492, right ramus; F:AM 115494, right maxilla, zygoma; F:AM 115495, mandible; F:AM 115498, humerus; Quinn Quarry: F:AM 112216, skull, jaw.

Early Clarendonian, Ash Hollow Formation, Cap Rock Member, Keya Paha County, Nebraska: 6 miles northeast of Springview: F:AM 114507, partial skull.

Early Clarendonian, Ash Hollow Formation, Cap Rock Member, Sherman County, Nebraska: Davis Creek: F:AM 115424, juvenile right ramus.

Early Clarendonian, Ash Hollow Formation, Cap Rock Member, Sheridan County, Nebraska: Deer Creek: F:AM 115634, rami; F:AM 115635, juvenile partial skull.

Early Clarendonian, Snake Creek Formation, Johnson Member, Sioux County, Nebraska: *Hipparion affine* channel in *Merychippus* Draw: F:AM 115989, humerus; F:AM 115990, right M3; F:AM 115991, right ramus; F:AM 115992, Mc3; F:AM 115993, astragalus.

Early Clarendonian, Esmerelda Formation, Mineral County, Nevada: UCMP 11640, skull and radius; UCMP 222284, Mt3; Fish Lake Valley: UCMP 33929, humerus; Cedar Mountain l.f.: UCMP 58172, disarticulated skull and jaws.

Middle Clarendonian, Clarendon beds, Potter County, Texas: North end, above Cap: F:AM 115462, right M1–2.

Middle to late Clarendonian, Ash Hollow Formation, Cap Rock Member, Cherry County, Nebraska: Johnson Rhino Quarry: F:AM 115544, left ramal fragment; F:AM 115545, right ramal fragment; F:AM 115546, vertebra; Horn Quarry: F:AM 115531, left ramus; Mouth of Hay Creek: F:AM 114493, partial skull, ramus; John's Ranch near mouth of Hay Creek: F:AM 115543, tibia; Snake River: F:AM 115527A,C, two tibiae; Niobrara River: F:AM 115527B, D, two tibiae; F:AM 115528, left ramus; F:AM 115532, left ramus; Schlagle Creek: F:AM 115537, juvenile maxilla; F:AM 115533, right ramus; F:AM 115534, right ramus; F:AM 115535, humerus; F:AM 115536, fibula; Morton Ranch: F:AM 114492, posterior skull; South Kat Quarry: F:AM 114496, back of skull; Deer Canyon: F:AM 114495, partial skull; Snake River: F:AM 114494, skull; F:AM 114514, juvenile skull.

Middle to late Clarendonian, Ash Hollow Formation, Cap Rock Member, Keya Paha County, Nebraska: Pat Herron Hill: F:AM 115547, ramus; F:AM 115548, ulna; F:AM 115549, distal humerus; Cook Ranch: F:AM 114512, skull; F:AM 115558, left ramus; F:AM 115559, right premaxilla; F:AM 115564, juvenile tibia; North of Springview: F:AM 115541, symphysis; F:AM 115550, right ramus; F:AM 115551, pelvic fragments; F:AM 115552, juvenile humerus; F:AM 115553, juvenile radius; South of Burton: F:AM 115542, left ramus; F:AM 115560, juvenile ulna; F:AM 115562, distal humerus; West of Norton: F:AM 115561, distal femur; F:AM 115563, juvenile femur; Turkey Creek: F:AM 115554, scapula; F:AM 115555, tibia; F:AM 115556, partial right ramus; F:AM 115557, symphysis; North Side Highway 183: F:AM 114506, skull.

Middle to late Clarendonian, Ash Hollow Formation, Cap Rock

Member, Brown County, Nebraska: Frank Lessig Canyon: F:AM 114513, right side of skull, mandible.

Middle to late Clarendonian, Ash Hollow Formation, Cap Rock Member, Todd County, South Dakota: Near top of butte, northeast of Haystack Butte: F:AM 115587, tibia; F:AM 115588, humerus.

Middle to late Clarendonian, Ash Hollow Formation, Merritt Reservoir Member, Cherry County, Nebraska: Rosebud Agency Quarry: F:AM 114516, skull, mandible; F:AM 115586, radius; Chief Hollow Horn Bear Quarry: F:AM 114517, skull; F:AM 114518, skull; F:AM 115583, juvenile skull; F:AM 115584, left maxilla; F:AM 115585, humerus; Mensinger Quarry: F:AM 115576, distal femur; F:AM 115577, humerus; Bolling Quarry: F:AM 115578, juvenile humerus; Bear Creek Quarry: F:AM 115573, left ramus, right ramus; Talus of Bear Creek: F:AM 115575, Mc3; Spring Canyon: F:AM 115574, partial hindlimb; Gallup Gulch Quarry: F:AM 115579, scapula; F:AM 115580, distal humerus; F:AM 115581, mandible; Gallup Springs Prospect: F:AM 115582, right ramus, left ramus.

Middle to late Clarendonian, Ash Hollow Formation, Merritt Reservoir Member, Brown County, Nebraska: L.B. Ross Place on Plum Creek: F:AM 115565, scapula; F:AM 115567, femur; F:AM 115568A–B, two radii; F:AM 115569, ulna; F:AM 115570, partial juvenile skull; F:AM 115571, right maxilla; F:AM 115572, pelvis; North of Johnstown: F:AM 115566, distal humerus; Pratt Quarry: F:AM 114529, skull; Pratt Slide: F:AM 115591, radius; F:AM 115592, tibia; Jonas Wilson Ranch: F:AM 115589, toothless jaw; F:AM 115590, right ramus; E.E. Beed Place: F:AM 115293C, atlas.

Middle to late Clarendonian, Ash Hollow Formation, Merritt Reservoir Member, Sheridan County, Nebraska: Hay Springs: F:AM 115630, femur; Dierex Brothers Ranch: F:AM 115631, tibia; Turtle Canyon: F:AM 115628, nasals; F:AM 115629, juvenile left ramus.

Late Clarendonian, Snake Creek Formation, Murphy and Laucomer Members, Sioux County, Nebraska: Olcott Quarry: F:AM 115632, right ramus; Kilpatrick Ranch: F:AM 115633, distal humerus.

Late Clarendonian, Clarendon beds, Moore County, Texas: Exell l.f. (= Frick 4 Way Pit): F:AM 114537, skull.

Late Clarendonian, Ash Hollow Formation, Merritt Reservoir Member, Cherry County, Nebraska: Morton Ranch: F:AM 115594, left ramus; Platybelodon Marl: F:AM 115593, right ramus; F:AM 115595, Mt3, magnum; Balanced Rock Quarry: F:AM 115596, ulna; F:AM 115597A–B, two radii; F:AM 115598, two tibiae/fibulae; Machairodus Quarry: F:AM 115253, humerus; F:AM 115254, femur; F:AM 115255, partial juvenile skull; F:AM 115256, mandible; F:AM 115260C, radius; F:AM 114522, skull, mandible; Leptarctus Quarry: F:AM 115257A–B, two humeri; F:AM 115258A–B, two femora; F:AM 115259, ulna; F:AM 115260A–B, two radii; F:AM 115261, left ramus; F:AM 115262, right ramus; F:AM 115263, right ramal fragment; F:AM 115264, right juvenile ramus; F:AM 114521, skull; Steer Creek High Channel: F:AM 115249, Mc3; F:AM 115250, posterior ramal fragments; F:AM 115251, atlas; Hans Johnson Place: F:AM

115252, mandible; F:AM 115293E, atlas; Wade Quarry: F:AM 115246A–B, two femora; F:AM 115247A–B, two ulnae; F:AM 115248, tibia; F:AM 115599, juvenile partial skull; F:AM 115600, right maxillary fragment; Xmas Quarry: F:AM 115601A–D, four humeri; F:AM 115602A–E, five radii; F:AM 115603A–E, five radii; F:AM 115604A–F, six ulnae; F:AM 115605A–G, seven ulnae; F:AM 115606, pelvis; F:AM 115607, pelvis; F:AM 115608A–D, four femora; F:AM 115609A–D, four femora; F:AM 115610, fibula; F:AM 115612A–E, five tibiae; F:AM 115611A–D, four tibiae; F:AM 115265, partial skull; F:AM 115266, right ramus, left ramus; F:AM 115267A–B, right ramus, left ramus; F:AM 115268A–B, right ramus, symphysis, left ramus; F:AM 115269, left ramus; F:AM 115270, right ramus; F:AM 115271, left ramus; F:AM 115272, right ramus; F:AM 115273, right ramus; F:AM 115274, right ramus; F:AM 115275, mandible; F:AM 115276, mandible; F:AM 115277, left ramus; F:AM 115278, left ramus; F:AM 115279, left ramus; F:AM 115280, left ramus; F:AM 115281, right ramus; F:AM 115282, right ramus; F:AM 115283, left ramus; F:AM 115284, right ramus; F:AM 115285, right ramal fragment; F:AM 115286, left ramus; F:AM 115287, right ramal fragment; F:AM 115288, left ramus; F:AM 115289, mandible; F:AM 115291, right ramus; F:AM 115292A–E, five atlases; F:AM 115293A–B, D, F-G, five atlases; F:AM 115294A–D, four atlases; F:AM 115296, two scapulae; F:AM 115298A–C, three scapulae; F:AM 115299A–B, two scapulae; F:AM 115300A–D, four humeri; F:AM 114523, skull, mandible; F:AM 114524, skull; F:AM 114525, palate; F:AM 114526, skull, mandible; F:AM 114527, skull; F:AM 114528, complete skull; Xmas Quarry Extension: F:AM 115295, scapula; Xmas Quarry Tunnel Rock Locality: F:AM 115290, right ramus; Kat Quarry: F:AM 115617, juvenile ramus; F:AM 115623E, radius; Kat Quarry Channel Deposits: F:AM 115622, left ramus; F:AM 115625B, tibia; F:AM 114540, skull; Quarterline Kat Quarry Channel Deposits: F:AM 115615, juvenile right maxilla; F:AM 115623A, radius; West Line Kat Quarry Channel Deposits: F:AM 115620, juvenile right maxilla; F:AM 115621, juvenile left ramus; F:AM 115624, ulna; F:AM 115623B–D,F, three radii; F:AM 114539, back of skull; F:AM 114542, skull; East Kat Quarry Channel Deposits: F:AM 115613, juvenile skull; F:AM 115614, right ramus, left ramus; F:AM 115618, right ramus; F:AM 115619, right ramus; F:AM 115627, humerus; F:AM 115625A, tibia; F:AM 114538, skull, mandible; Connection Kat Quarry Channel Deposits: F:AM 115626, tibia; Trail Side Kat Quarry Channel Deposits: F:AM 115616, right ramus.

Late Clarendonian, Humboldt Formation, Elko County, Nevada: CCC Camp: UCMP 35381, calcaneum, metatarsals; UCMP 38437, astragalus; UCMP 38279, calcaneum; UCMP 38281, Mc3; UCMP 38281, humerus.

Late Clarendonian, Truckee Formation, Churchill County, Nevada: UCMP 88565, mandible, carpals.

Clarendonian, Ellensburg Formation, northwest of Selah, Yakima County, Washington (LACM locality 6438): LACM 10661, humerus; LACM 10625, tibia.

Known distribution: Early to late Clarendonian, South Dakota, Nebraska, Kansas, Texas, and Nevada.

Diagnosis: Slightly larger in size than *T. medicornutum* (M1–3 length 145–165 mm; m1–3 length = 150–175 mm) with a more brachycephalic skull and slightly shorter nasals (but longer than those in *T. brachyrhinum*). Teeth more derived than those in *T. medicornutum* in having well-developed crochets, especially on M2, that occasionally show branches or spurs. Occasional cristae may be present. The p2 is absent. Postcranial skeleton more stumpy and robust than *T. medicornutum*. Distinguished from *T. fossiger* by its much *smaller* size (*contra* Hatcher, 1894a, 1894b; Osborn, 1904). Distinguished from *T. proterum* by its less derived dentition and less robust limbs. Distinguished from *T. meridianum* by its much larger size.

Description: Hatcher (1894a, 1894b) and Osborn (1904) briefly described the type specimen of *Teleoceras major*. Many of the features listed in Hatcher's diagnosis are erroneous. *T. major* is not "one-third larger than *T. fossiger*," but up to one-third smaller. The absence of the crochet is simply due to extreme wear on the teeth of the type skull.

The cranial material of *T. major* is now known from a much more complete and well preserved series of skulls in the Frick Collection. Some of these specimens are shown in Fig. 4.40. In most respects, *T. major* differs little from *T. medicornutum*. However, the unworn upper dentitions from the Clarendonian of Nebraska and Texas consistently show more derived features than is typical of *T. medicornutum*. The crochet is commonly very long, particularly on M2–3. In some specimens, the tip of the crochet is curved labially, and may even bifurcate. F:AM 114525 (Fig. 2.9) even shows a bifurcate crochet on the right M2 and a long, highly recurved crochet on the left M2 (thus showing the individual variability of this character). The antecrochet is typically much larger and more developed than in *T. medicornutum*, frequently touching the crochet. On worn upper premolars, the antecrochet closes the valley between the protoloph and metaloph. Cristae are occasionally developed in some individuals. F:AM 114525 (Fig. 2.9) shows an aberrantly complex pattern on P4, with several cristae and other enamel folds joining to form a series of small fossettes. In most other features, except size, *T. major* is difficult to distinguish from *T. medicornutum*. The limbs (see Chapter 5) are more robust on average than *T. medicornutum*, but not significantly so. The distinction between the two species is one of the most difficult within the entire genus.

Discussion: Hatcher's type specimen of *T. major* is too poorly preserved with badly worn teeth to observe many important diagnostic features and his diagnosis was erroneous in many ways and useless in others. The type locality has produced very little additional material, and the sample from the Cap Rock Member is too small to form a useful topotype. Nevertheless, Clarendonian *Teleoceras* seems distinct from *T. medicornutum* and easily distinguished from *T. fossiger* (*contra* Matthew, 1918). Large quarry samples from the late Clarendonian Merritt Dam Member of the Ash Hollow Formation (Skinner and Johnson, 1984) occur in many places in Nebraska. The enormous samples from Quinn

Rhino Quarry (one skull illustrated by Matthew, 1932), Xmas Channel Quarry, and Kat Channel Quarry, and many smaller quarries provide a full range of variation of the species. From these samples, it is clear that *Teleoceras major* is a distinct species, and not synonymous with *T. fossiger* or *T. medicornutum*, as suggested by Skinner *et al.* (1968, p. 431).

Teleoceras major is now also known from complete articulated specimens in death position at Poison Ivy Quarry, Ashfall Fossil Bed State Park, Antelope County, Nebraska. Preliminary descriptions have been published by Voorhies (1981, 1985, 1992), and more complete descriptions of the hundreds of skeletons of *T. major* are anticipated in the future.

Teleoceras meridianum (Leidy, 1865)

Rhinoceros meridianus Leidy, 1865
Aphelops meridianus Cope, 1875
Teleoceras minor Olcott, 1909
Teleoceras meridianum Prothero and Manning, 1987

Figures 4.36, 4.40, Table 4.9

Holotype: USNM 3177, a broken ?M2
Hypodigm: Listed in Prothero and Manning (1987, p. 411), plus additional specimens from the late Barstovian of Nebraska ("*T. minor*") mentioned below.

Early Clarendonian, Dove Spring Formation, Ricardo Group, Kern County, California: UCMP 19442, partial calcaneum; LACM 59613, right maxilla with M2–3, mandible.

Known distribution: Early to late Barstovian, Texas and Nebraska.

Diagnosis: Dwarf *Teleoceras* (M1–3 length = 120–-135 mm; m1–3 length = 120–135 mm), which is similar to *T. medicornutum* in most features except for its smaller size and its proportionately more robust limbs.

Description: Most of the known material of *T. meridianum* was described by Prothero and Manning (1987).

Discussion: The history of the misinterpretations of Leidy's type specimen of "*meridianus*" was reviewed by Prothero and Manning (1987). Larger samples of this rhino from the Texas Gulf Coastal Plain made it clear that the type specimen of *meridianus* came from a dwarf species of *Teleoceras*.

Specimens of a dwarf *Teleoceras* from the Barstovian of Nebraska also appear to be referable to *T. meridianum*. Olcott (1909) described but did not illustrate a skull (YPM-PU11553) of a small *Teleoceras* from the "Loup Fork beds of the Niobrara River Valley, Cherry County, Nebraska," and named it *Teleoceras minor*. Skinner and Johnson (1984, p. 268) mentioned an immature skull (F:AM 109518) and left ramus (F:AM 109520) from Norden Bridge Quarry (late Barstovian, Cornell Dam Member of the Valentine Formation) which they referred to *T. minor*. In size and morphology, these specimens match the Texas specimens of *T. meridianum* quite well, especially F:AM 108306, a partial skull of *T. meridianum* from the late Barstovian McMurray Pit, San

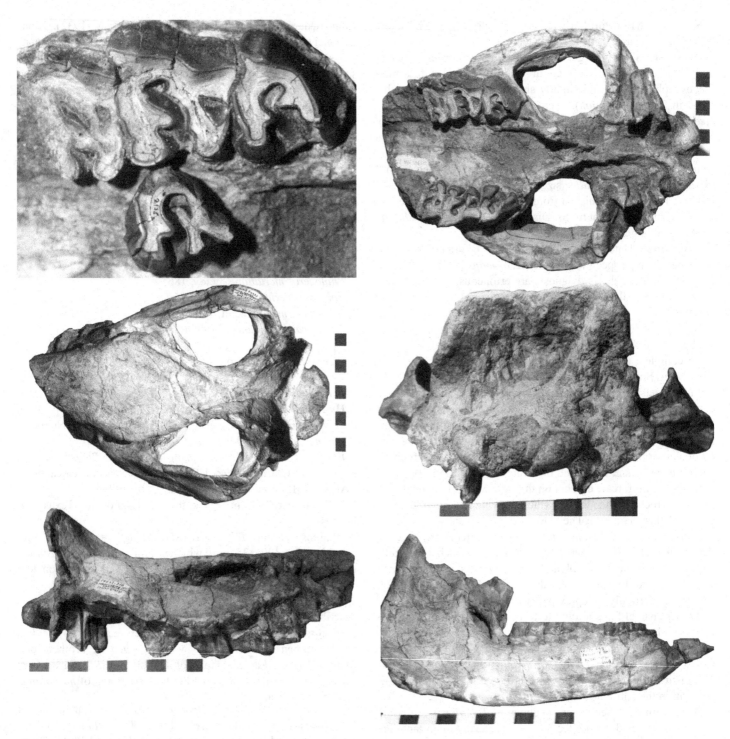

Figure 4.40. *Teleoceras meridianum*. A. Comparison of type specimen, an isolated upper molar (USNM 3177, bottom) with referred upper dentition (F:AM 108306). B–E. Referred skull (F:AM 108306) in dorsal, right lateral, palatal, and posterior views. F. Referred lower jaw (F:AM 108839). Scale bar in cm.

Jacinto County, Texas. This strongly suggests that *T. minor* is a junior synonym of *T. meridianum*.

Teleoceras fossiger (Cope, 1878)

Aphelops fossiger Cope, 1878
Aceratherium acutum Marsh, 1887
Teleoceras fossiger Osborn, 1898
Aceratherium fossiger Toula, 1902
Teleoceras fossiger Osborn, 1904

Figures 4.36, 4.41, Table 4.10

Holotype: AMNH 8390, a skull missing the left maxilla and much of the anterodorsal skull roof, from the early Hemphillian Beaver Creek area, Decatur County, Kansas (according to F.W. Johnson, pers. comm., who researched Hill's 1877 field notes in the AMNH archives).

Hypodigm: *Early Hemphillian, Phillips County, Kansas*: Large sample in the AMNH, USNM, and KU collections from Long Island Rhino Quarry. Selby Ranch: F:AM 115653, right ramus.

Early Hemphillian, Deer Lodge Valley, Powell County, Montana: 2 miles northeast of Warm Springs: CM 9178, left ramus with m1–3; CM 1390, postcranials; CM 3309 postcranials.

Early Hemphillian, Smith Valley Fauna, Coal Valley Formation, Lyon County, Nevada: Smith Valley: F:AM 114556, skull, jaws; F:AM 115853, palate, skull fragments, nasals, rami, femur; LACM (CIT) 17282, Mt3.

Early Hemphillian, Thousand Creek Formation, Humboldt County, Nevada: 2 miles east of Railroad Ridge: F:AM 114580, skull, mandible; Due south of Railroad Ridge, South Point: F:AM 115841, astragalus, tarsals; 6 miles southwest of camp: F:AM 115842, right maxilla; F:AM 115843, lower tusk (I2); 2.5 miles south of camp: F:AM 115844, ramus; northwest of Blick Quarry: F:AM 115845, patella, astragalus; F:AM 115846, lower molars; 2.5 miles north of "E" Quarry: F:AM 115847, right ramus; Thousand Creek Quarry: F:AM 115848, humerus; F:AM 115849, radius; F:AM 115850, ulna; F:AM 115852, patella; Thousand Creek Fields: F:AM 115851, fibula; UCMP 75206, metapodial; UCMP 70205, astragalus; UCMP 22901, nasals; UCMP 31393, 31407, 31411, 31416, fragments; LACM (CIT) 90774, tibia; LACM (CIT) 91927, astragalus, many other catalogued specimens in the LACM (CIT) collections.

Early Hemphillian, Elko County, Nevada: Hi Level Quarry, Carlin Area: F:AM 115838, humerus; Lake Quarry: F:AM 115839, juvenile right ramus; F:AM 115840, juvenile left ramus; Goose Creek: F:AM 115854, footbones.

Early Hemphillian, Rattlesnake Formation, Grant County, Oregon: UCMP 23177, P2; UCMP 23177, dP2; UCMP 23175, M3; UCMP 23181, Mc3; UCMP 32860, scaphoid; UCMP 23178, lunar; UCMP 23179, calcaneum; UCMP 32182, Mt3; UCMP 29970, 29972.

Early Hemphillian, Drewsey Formation, Stinking Water Creek l.f., Harney County, Oregon: LACM (CIT) 93030, Mc3; LACM (CIT) 93026, humerus; LACM (CIT) 94622, calcaneum; LACM (CIT) 94739, astragalus; many other catalogued specimens in the LACM (CIT) collections.

Early Hemphillian, Alturas Formation, Modoc County, California: UCMP 31017, distal metapodial; UCMP 31005, Mt2.

Early Hemphillian, Mehrten Formation, Stanislaus County, California: UCMP 44378, mandible, radius, humerus; LACM (CIT) 61668, jaw, tibia; LACM (CIT) 52749, foot bones; LACM (CIT) 61875, humerus

Early Hemphillian, Kern River Formation, Kern County, California: LACM (CIT) 1206, right and left male rami.

Early Hemphillian, Gracias Formation, Honduras: Walker Museum 1790 (in the Field Museum collections), ramus (Olson and McGrew, 1941).

Early Hemphillian, Ogallala Formation, Ellis County, Oklahoma: Port of Entry Pit: F:AM 114558, partial palate; F:AM 114563, partial skull; F:AM 115654, partial ramus; F:AM 115655, right ramus; F:AM 115656, partial ramus; F:AM 115657, ramus; F:AM 115658, right ramus; F:AM 115659, left ramus; F:AM 115660, right ramus; F:AM 115660, right ramus; F:AM 115661, right ramus; F:AM 115662, left ramus; F:AM 115663, symphysis; F:AM 115664, right ramus; F:AM 115665, right ramus; F:AM 115666, right ramus; F:AM 115667, right ramus; F:AM 115668, partial left ramus; F:AM 115669, left ramus; F:AM 115670, right ramus; F:AM 115671, left ramus; F:AM 115672, left ramus; F:AM 115673, right ramus; F:AM 115686, atlas; F:AM 115687A–D, four radii; F:AM 115688, fibula; F:AM 115689A–B, two tibiae; F:AM 115690A–B, two tibiae/fibulae; F:AM 115701, right maxilla with P4–M2; F:AM 114584, ramus; Northeast of Pit 1: F:AM 114570, mandible; Old Holcum Place: F:AM 115679, Mc3; Adair, north of Port of Entry Pit: F:AM 115680, Mt3; Fritzler Place: F:AM 115692, femur; F:AM 115693, humerus; Fritzler Lower Pit: F:AM 115691, associated hind limb; Fritzler South Pit: F:AM 115694, associated hindlimb.

Early Hemphillian, Ogallala Formation, Clark County, Kansas: Jack Swayze Quarry: F:AM 114552, skull; F:AM 115652, axis; Arens Ranch: F:AM 114554, skull; F:AM 115984D, tibia; Rhino Quarry, Young Brothers Ranch: F:AM 114546, skull; F:AM 114547, skull; F:AM 114548, skull; F:AM 114549, skull; F:AM 114551, skull; F:AM 115953, ramus; F:AM 115955, symphysis; F:AM 115967, right ramus; F:AM 115972, right ramus; F:AM 115973, left ramus; F:AM 115974A–C, three humeri; F:AM 115975A–D, four humeri; F:AM 115976A–D, four radii; F:AM 115977A–F, six radii; F:AM 115978A–E, five ulnae; F:AM 115979A–C, three ulnae; F:AM 115980, pelvis; F:AM 115981A–C, three femora; F:AM 115982A–B, two femora; F:AM 115983A–F, six tibiae; F:AM 115931, maxilla; F:AM 114550, skull; F:AM 114553, skull; F:AM 114555, skull; F:AM 114557, skull; F:AM 115926, maxilla; F:AM 115927, lambdoid crest; F:AM 115928, maxilla; F:AM 115929, M1–2; F:AM 115930, M1–2; F:AM 115932, maxilla; 115933, partial maxilla; F:AM 115934, left ramus; F:AM 115935, right ramus; F:AM 115936, left ramus; F:AM 115937, right ramus; F:AM 115938, right ramus; F:AM 115939, juvenile right ramus; F:AM 115940, right

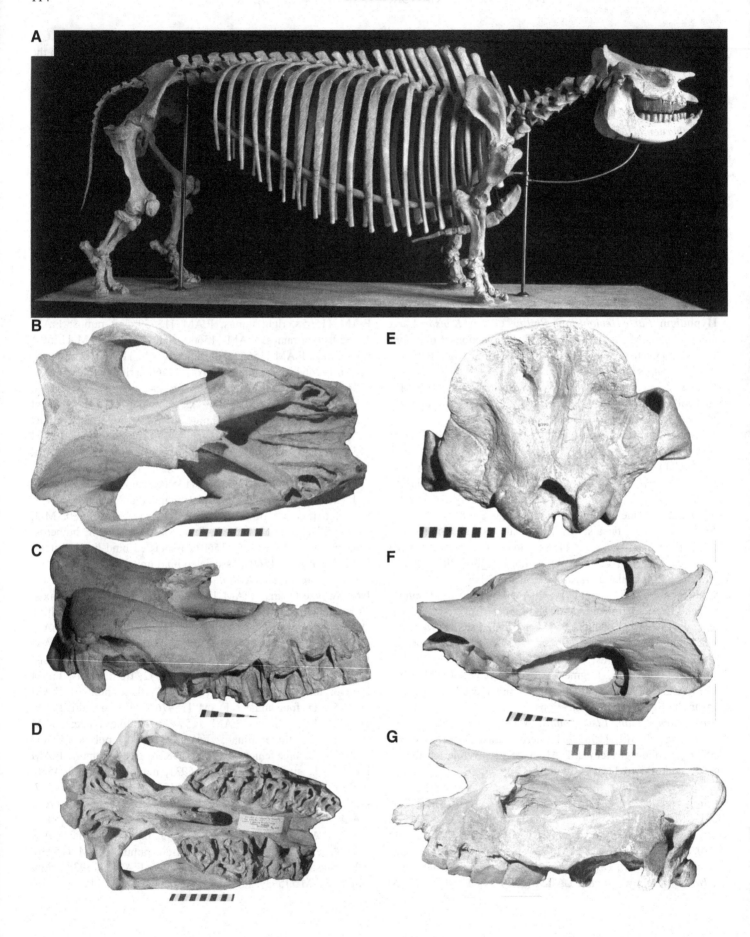

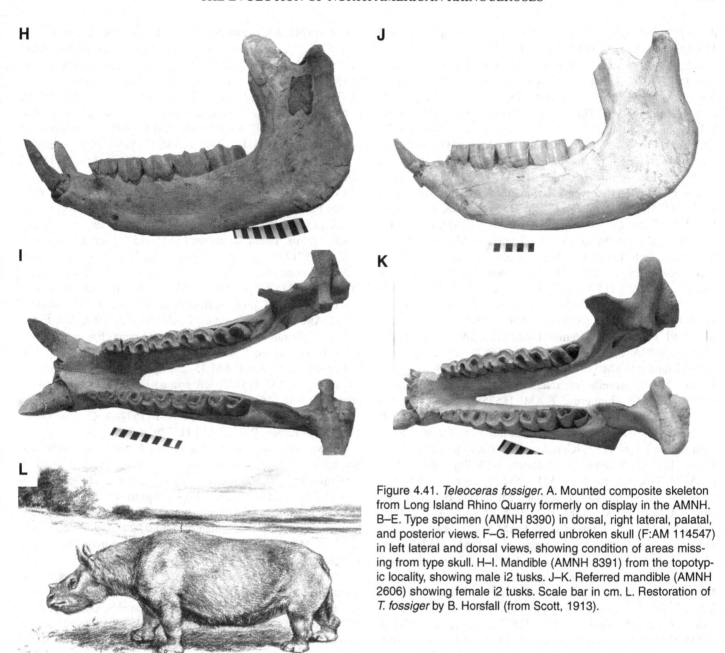

Figure 4.41. *Teleoceras fossiger*. A. Mounted composite skeleton from Long Island Rhino Quarry formerly on display in the AMNH. B–E. Type specimen (AMNH 8390) in dorsal, right lateral, palatal, and posterior views. F–G. Referred unbroken skull (F:AM 114547) in left lateral and dorsal views, showing condition of areas missing from type skull. H–I. Mandible (AMNH 8391) from the topotypic locality, showing male i2 tusks. J–K. Referred mandible (AMNH 2606) showing female i2 tusks. Scale bar in cm. L. Restoration of *T. fossiger* by B. Horsfall (from Scott, 1913).

ramus; F:AM 115941, juvenile right ramus; F:AM 115942, right ramal fragment; F:AM 115943, juvenile right ramus; F:AM 115944, left ramus; F:AM 115945, juvenile right ramus; 115946, juvenile right ramus; F:AM 115947, right ramus; F:AM 115948, juvenile left ramus; F:AM 115949, juvenile left ramus; F:AM 115950, right ramus; F:AM 115951, left ramus; F:AM 115952, ramus; F:AM 115954, left ramus; F:AM 115956, right ramus; F:AM 115957, left ramus; F:AM 115958, left ramus; F:AM 115959, left ramus; F:AM 115960, mandible; F:AM 115961, right ramus; F:AM 115962, right ramus; F:AM 115963, edentulous ramus; F:AM 115964, right ramus; F:AM 115965, right ramus;

F:AM 115966, left ramus; F:AM 115968, left ramus; F:AM 115969, right ramus; F:AM 115970, left ramus; F:AM 115971, right ramus; F:AM 115984A–C,E–G, six tibiae; F:AM 115985A–F, six fibulae; F:AM 116094, atlas, axis.

Early Hemphillian, Ogallala Formation, Lipscomb County, Texas: Martin Sebit Ranch: F:AM 115683, juvenile right ramus; F:AM 115684, ramus; F:AM 115685, tibia; Nation Gravel Pit: F:AM 115682, symphysis; Geiger Ranch: F:AM 115681, left M2; F:AM 115695, left maxilla with M1–3; Bradney Pit 1: F:AM 115678, right ramus; F:AM 115837, left ramus; Stewart Pit 1: F:AM 115675, humerus; F:AM 115676, radius; F:AM 115677, distal humerus; Box T Ranch: F:AM 114566, mandible; F:AM 114568, skull; F:AM 114573, skull; F:AM 114577, skull; Box T Ranch, Pit 1: F:AM 114560, partial palate; F:AM 114576, skull; F:AM 115730, shattered juvenile skull; F:AM 115731, right maxilla; F:AM 115733, right juvenile maxilla; F:AM 115734, juvenile right maxilla; F:AM 115735, juvenile left maxilla; F:AM 115736, left juvenile maxilla; F:AM 115737, left ramus; F:AM 115738, left ramus; F:AM 115740, right ramus; F:AM 115741, left ramus; F:AM 115742, right ramus; F:AM 115743, juvenile left ramus; F:AM 115744, juvenile right ramus; F:AM 115745, juvenile right ramus; F:AM 115746, juvenile left ramus; F:AM 115749, scapula; F:AM 115750, scapula; F:AM 115751, thoracic vertebra; F:AM 115752A–C, three humeri; F:AM 115753A,B,D, three humeri; F:AM 115754A,C,D, three humeri; F:AM 115755A–C, three humeri; F:AM 115756A–F, six ulna; F:AM 115757A–D, four ulnae; F:AM 115758A–D,F,G, six radii; F:AM 115759A–G, seven radii; F:AM 115760A–B, two femora; F:AM 115761A–B, two femora; F:AM 115763A–B, two juvenile femora; F:AM 115764A–E, five tibiae (E also has fibula); Box T Ranch across from Pit 1: F:AM 115831, femur; F:AM 115832A–B, two humeri; F:AM 115833A–C, three radii; F:AM 115834, humerus; F:AM 115835, femur; F:AM 115836, tibia; Box T Ranch, north of Pit 1: F:AM 115732, juvenile left ramus; F:AM 115753C, humerus; F:AM 115754B, humerus; F:AM 115758E, radius; F:AM 115762A–B, two femora; F:AM 115764F, tibia; Box T Ranch, northeast of Pit 1: F:AM 115747, ramal fragment; F:AM 115748, right ramus; F:AM 115759C, radius; Box T Ranch, west of Pit 1: F:AM 115739, left ramus; Box T Ranch, 1st Valley, west of Pit 1: F:AM 115716, right ramus; F:AM 115827A, tibia; Box T Ranch, West Valley: F:AM 115674, partial right ramus; Box T Ranch, Bridge Creek: F:AM 115827B-E, four tibiae; F:AM 114564, ramus; F:AM 114569, top of skull; F:AM 114571, ramus; F:AM 115765, ramus; F:AM 115766, right ramus; F:AM 115767, right ramus; F:AM 115768, right ramus; F:AM 115769, left ramus; F:AM 115770, right ramus; F:AM 115771, left maxilla with M2–3; F:AM 115772, right maxilla with P3-M2; F:AM 115773, left maxilla; F:AM 115774, right ramus; F:AM 115775, left ramus; F:AM 115776, right ramus; 115777, right ramus with m3; F:AM 115778, right ramus with m3; F:AM 115779, right ramal fragment; F:AM 115780,juvenile left ramus; F:AM 115781, juvenile right ramus; F:AM 115817, femur; F:AM 115818A–C, three humeri; F:AM 115819A–C, three humeri; F:AM 115823A–E, five ulnae; F:AM 115824A–E, five radii; F:AM 115826A–E, five tibiae; F:AM 115828, fibula; F:AM 115830A–D, four femora; Box T Ranch, West Draw: F:AM 114559, badly crushed skull; F:AM 114572, skull; F:AM 114575, broken mandible; F:AM 114578, 114579, ramus; F:AM 114581, ramus; F:AM 114583, skull; F:AM 115785, left ramus; F:AM 115790, right ramal fragment; F:AM 115801, juvenile right maxilla; F:AM 115802, juvenile left maxilla; F:AM 115803, juvenile right maxilla; F:AM 115808, right ramus; F:AM 115810, right maxilla with M1-2; F:AM 115816A–C, three tibiae/fibulae; F:AM 115820A–F, six radii; Box T Ranch, West Draw, Topsoil: F:AM 115815, humerus; Box T Ranch, West Draw, Sand: F:AM 115789, left rami, right rami;

F:AM 115794, left ramus; Box T Ranch, West Draw, Sandy Clay: F:AM 115782, right ramus, left ramus; 115783, left ramus; F:AM 115784, right ramus; F:AM 115786, left ramus, right ramus; F:AM 115787, right ramus; F:AM 115788, left ramus; F:AM 115791, right ramus; F:AM 115792, right ramus; F:AM 115793, right ramus; F:AM 115800, juvenile left ramus, juvenile right ramus; F:AM 115807, left ramus; F:AM 115809, left maxilla with P4-M2; F:AM 115811, palate; F:AM 115825A–B, two femora; V.V. Parker Ranch: F:AM 114567, partial skull; F:AM 115696, right maxilla with M2–3; F:AM 115697, left maxilla with M2–3; F:AM 115698, right maxilla with M2–3; F:AM 115699, right ramus; F:AM 115700, ramal fragment; F:AM 115705, left ramus; F:AM 115706, left ramus; F:AM 115707, right ramus with m3; F:AM 115708, ramal fragment; F:AM 115709, ramal fragment; F:AM 115719, left ramus; F:AM 115723, right ramus; F:AM 115724A, humerus; F:AM 115726B, radius; V.V. Parker Ranch, Pit 1: F:AM 115702, maxilla; F:AM 115703, juvenile right maxilla; F:AM 115704, juvenile right maxilla; F:AM 115710, symphysis; F:AM 115711, left ramus; F:AM 115712, left ramus; F:AM 115713, right ramus; F:AM 115714, right ramal fragment; F:AM 115715, left ramus; F:AM 115717, symphysis; F:AM 115718, right ramal fragment; F:AM 15720, right ramus; F:AM 115721, left ramus; F:AM 115722, left ramus; F:AM 115724B,C, three humeri; F:AM 115725A–B, two ulnae; F:AM 115726A, radius; F:AM 115727A–B, two femora; F:AM 115728A–C, three tibiae; V.V. Parker Ranch, Pit 4: F:AM 115729, right ramus.

Known distribution: Early to middle Hemphillian, Nebraska, South Dakota, Kansas, Oklahoma, Texas, and Nevada.

Diagnosis: Largest species of *Teleoceras* (M1–3 length = 160–190 mm; m1–3 length = 170–200 mm). Cristae well developed in P3–4, fusing with crochet to form medifossettes. Very long crochets on M1–3, which occasionally fuse with long antecrochets. M2 much larger than M1. Teeth very hypsodont. Skull very brachycephalic.

Description: *T. fossiger* is perhaps the best-known species of *Teleoceras*, since mounted skeletons are on display in several museums. It was fully described and illustrated by Osborn (1898a, 1898b) and by Cope and Matthew (1915). Further illustrations are provided in Figure 4.43, and the postcranial skeleton is described in Chapter 5.

Discussion: Although there is some uncertainty about the exact age and provenance of the type specimen of *T. fossiger*, it is well represented by a series of skulls and skeletons from the nearby early Hemphillian Long Island Rhino Quarry, Phillips County, Kansas. From this material, it is clear that early Hemphillian *T. fossiger* is the largest and one of the most derived species of *Teleoceras* (*contra* Tanner, 1975). It was very widespread in the early Hemphillian river channels of Kansas, Oklahoma, and Texas, so that enormous samples are now available from many localities. Especially large samples are known from Long Island Rhino Quarry (AMNH, USNM, KU, and several other collections), Jack Swayze Quarry, Clark County, Kansas (F:AM collection; Sternberg Memorial Museum, Fort Hays, Kansas, collections), the Frick Port of Entry (P.E.) Pit, Ellis County, Oklahoma

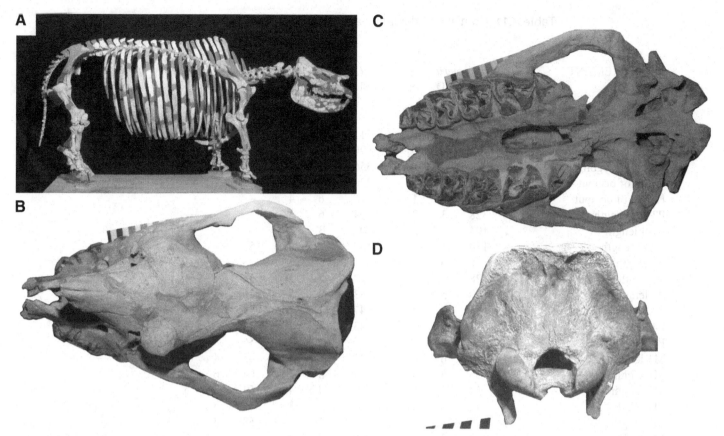

Figure 4.42. *Teleoceras proterum*. A. Composite skeleton from Love bone bed on display at the Florida Museum of Natural History, University of Florida, Gainesville. B-D. Referred skull (UF 9078) from Mixson's bone bed in dorsal, palatal, and posterior views. Scale bar in cm.

(F:AM collections), and several localities in adjacent Lipscomb County, Texas, especially the F:AM collections from Box T Ranch and V.V. Parker Ranch (discussed in Schultz, 1977, pp. 56–70). These large samples make *T. fossiger* one of the best-known North American fossil mammals, but also offer unparalleled opportunities for the study of population variation.

Because of the distinctiveness of this species, very few invalid synonyms have been proposed. Marsh (1887) named a skull from the Long Island Rhino Quarry *Aceratherium acutum* (possibly to spite Cope), but Osborn (1904) clearly demonstrated that it is a junior synonym of *T. fossiger*.

Teleoceras proterum (Leidy, 1885)

Rhinoceros proterus Leidy, 1885
Aphelops fossiger Leidy and Lucas, 1896
Teleoceras fossiger var. *T. proterum* Osborn, 1904
Teleoceras fossiger Harrison and Manning, 1983

Figures 4.36, 4.42, Table 4.11

Holotype: USNM 3190, an isolated M3, from the early Hemphillian Mixson's (also spelled Mixon's) Bone Bed, Alachua Formation, Levy County, Florida.

Hypodigm: Thousands of mostly uncatalogued specimens from Mixson's Bone Bed are found in the Frick Collection. Among the catalogued specimens are: F:AM 112239, left maxilla; F:AM 112240, skull; F:AM 112242, skull; F:AM 116095, composite skeleton (axis, atlas); F:AM 116098, juvenile skull; F:AM 116099, composite upper dentition. In addition, over 2000 specimens of *T. proterum* are catalogued from Love Bone Bed, Alachua County, Florida (Webb *et al.*, 1981) and from McGehee Farm in the Florida Museum of Natural History collections. A composite mounted specimen of this species is on display at the Florida Museum of Natural History (Fig. 4.42A).

Known distribution: Latest Clarendonian to early Hemphillian of Florida.

Diagnosis: Smaller species of *Teleoceras* (M1–3 length 140–150 mm; m1–3 length = 140–165 mm) with much thinner crests and stronger cingula on P3–4. Crochets and antecrochets thinner and shorter than the typical bulbous condition seen in other species of *Teleoceras*. Limb elements heavier, shorter, and more robust than those of any other species of *Teleoceras*.

Description: *T. proterum* was originally described by Leidy and Lucas (1896). Although much additional material from Mixson's Bone Bed is now available in the Frick Collection and the University of Florida Museum of Natural History, it is all highly fragmentary (as is all the material from this locality). The skulls

Table 4.11. Cranial and dental measurements of *Teleoceras* (in mm)

MEASUREMENT	T. proterum			T.hicksi			T. guymonense			T. "schultzi"
	Mean	SD	N	Mean	SD	N	Mean	SD	N	UNSM 5800
P2 to occiput	475	—	1	511	8	4	—	—	—	—
Lambdoid crest to nasals	—	—	—	490	7	3	—	—	—	—
Width @ zygoma	—	—	—	356	27	4	—	—	—	—
Width of occiput	184	—	1	226	13	4	—	—	—	—
Height of occiput	210	—	1	198	6	4	—	—	—	—
P2-M3 length	328	3	2	254	6	3	—	—	—	—
P2-4 length	104	16	3	100	3	2	—	—	—	—
M1-3 length	150	6	3	152	5	4	159	—	1	—
P2 length	36	6	4	28	1	3	28	2	4	—
P2 width	28	3	3	37	2	2	32	1	4	—
P3 length	39	5	4	33	2	3	36	3	4	—
P3 width	43	8	4	47	11	2	53	6	4	—
P4 length	41	7	4	43	8	3	44	6	4	—
P4 width	49	8	4	67	1	2	56	8	4	—
M1 length	47	6	4	45	3	4	45	4	5	—
M1 width	55	5	4	73	3	3	64	6	5	—
M2 length	54	3	4	53	4	4	53	3	5	—
M2 width	56	5	4	71	5	3	65	3	5	—
M3 length	49	7	4	52	5	4	50	3	5	—
M3 width	42	13	3	60	7	3	62	3	5	—
p3-m3 length	242	16	3	241	16	5	—	—	—	238
p3-4 length	82	7	3	72	7	5	—	—	—	73
m1-3 length	159	8	4	172	12	6	140	4	4	165
p2 length	—	—	—	—	—	—	—	—	—	—
p2 width	—	—	—	—	—	—	—	—	—	—
p3 length	39	4	3	33	4	5	—	—	—	36
p3 width	23	3	3	29	5	3	—	—	—	23
p4 length	43	3	3	42	5	7	35	2	2	44
p4 width	27	5	3	32	4	5	25	3	2	37
m1 length	51	5	4	41	2	7	40	4	4	49
m1 width	29	4	4	35	2	7	27	4	4	38
m2 length	54	2	4	59	5	7	47	5	4	60
m2 width	26	2	4	36	3	7	28	3	4	37
m3 length	55	2	4	63	6	7	53	2	4	59
m3 width	26	4	4	33	3	7	26	1	4	40

and most of the jaws are broken and crushed, so few complete skulls are known. The most complete cranial materials are shown in Fig. 4.44 and measurements are given in Table 4.11. Postcranial remains were described by Harrison and Manning (1983) and also are discussed in Chapter 5.

Discussion: Most authors (Leidy and Lucas, 1896; Osborn, 1904; Harrison and Manning, 1983) have considered the Mixson *Teleoceras* as a subspecies or variant of *Teleoceras fossiger*. However, it is completely disjunct in size (Fig. 4.36), and slightly smaller than *T. major*, and also remarkably distinct in morphology. Although it is obviously closely related to its High Plains contemporary, *T. fossiger*, I feel that the distinctive morphology, non-overlapping size distributions, and geographical separation justify retaining a separate species for early Hemphillian *Teleoceras* from Florida.

The Frick sample of *Teleoceras proterum* is enormous. At present, there are 156 trays of uncatalogued specimens, and a large number of skulls in crates that have never been unpacked. Harrison and Manning (1983) calculated that there were a minimum number of 111 individuals in this collection, and Mihlbachler (pers. comm.) calculated that there were at least 117 individuals represented. Every element is represented by dozens of specimens, allowing an unparalleled opportunity for the study of population variation. Harrison and Manning (1983) have discussed just one particularly interesting aspect of this sample.

Teleoceras hicksi Cook, 1927

Teleoceras hicksi Cook, 1927
Paraphelops rooksensis Lane, 1927
Paraphelops yumensis Cook, 1930
Peraceras ponderis Cook, 1930
Teleoceras yumensis Matthew, 1932
Teleoceras yumensis Frye *et al.*, 1956
Teleoceras schultzi Tanner, 1975
Teleoceras ocotensis Dalquest and Mooser, 1980
Teleoceras fossiger Carranza-Castañeda, 1989

Figures 4.36, 4.43, Table 4.11

Holotype: DMNH 304, a complete skull from the middle Hemphillian Wray local fauna, Yuma County, Colorado.

Hypodigm: *Middle to late Hemphillian, Ogallala Formation, Yuma County, Colorado*: Hundreds of specimens from the Wray area in the DMNH collections, plus the following F:AM specimens: Spencer Canyon: F:AM 115919, femur; Eight Mile Canyon: F:AM 115920, maxilla; F:AM 115921, partial ramus; East Fork, Haskel Canyon: F:AM 115906, maxilla; F:AM 115907, maxilla with M3; F:AM 115918, scapula; Locality C, Wray Area: F:AM 115170, mandible; Wray Area: F:AM 115922, maxilla; F:AM 115923, maxilla; F:AM 115924, partial ramus; Wray Area (surface): F:AM 115925, maxillary fragment; Wray Locality A: F:AM 114597, right ramus; F:AM 115893A, tibia; 115908, edentulous rami; F:AM 115909, right ramus; F:AM 115910, right ramus; F:AM 115911, ramus; F:AM 115912, femur; F:AM 115913A–B, two ulnae; F:AM 115914, humerus; F:AM 115915, tibia; F:AM 115916, articulated vertebrae; F:AM 115917, atlas; Wray Locality B: F:AM 114598, partial skull; F:AM 115894, maxilla; F:AM 115895, right ramus; F:AM 115896, right ramus; F:AM 115897, left ramus; F:AM 115898, left ramus; F:AM115899, ramus; F:AM 115900A–B, two radii; F:AM 115901, proximal ulna; F:AM 115902, fibula; F:AM 115903, left ramus; F:AM 115904, right ramus; F:AM 115905, ramus; Wray Locality C: F:AM 114585, mandible; F:AM 114586, edentulous skull; F:AM 114587, skull; F:AM 114588, skull; F:AM 114589, skull; F:AM 114595, humerus; F:AM 114596, femur; F:AM 114599, skull; F:AM 114600, partial skull; F:AM 114601, mandible; F:AM 115855, left ramus; F:AM 115856, left ramus; F:AM 115857, right juvenile ramus; F:AM 115858, right ramus; F:AM 115859, right ramus; F:AM 115860, right ramus; F:AM 115861, left juvenile ramus; F:AM 115862, left ramus; F:AM 115863, juvenile left ramus; F:AM 115864, right ramus; F:AM 115865 juvenile right ramus; F:AM 115866, right ramus; F:AM 115867, right ramus; F:AM 115868, right ramus; F:AM 115869, left ramus; F:AM 115870, right ramus; F:AM 115871, left ramus; F:AM 115872, right ramus; F:AM 115873, left ramus; F:AM 115874, left ramus; F:AM 115875, left ramal fragment; F:AM 115876, left ramus; F:AM 115877, left ramus; F:AM 115878, edentulous left ramus; F:AM 115879, mandible; F:AM 115882, mandible; F:AM 115883, left ramus; F:AM 115884, ramus; F:AM 115885, mandible; F:AM 115886A–B, two humeri; F:AM 115887, scapula; F:AM 115888A–D, four ulnae; F:AM 115889A–D, three radii; F:AM 115890, humerus; F:AM 115891A–B, two femora; F:AM 115892A–B, two fibulae; F:AM 115893B, C, two tibiae; F:AM 115880, mandible; F:AM 115881, right ramus.

Middle Hemphillian, "Kimball" Formation, Frontier County, Nebraska: UNSM locality Ft-40, *Amebelodon fricki* Quarry: UNSM 5800, left ramus, type of *T. schultzi*; UNSM 62099, mandible, referred to *T. schultzi*; UNSM 62103 and 62098, paratypes of *T. schultzi*; plus many other additional specimens listed in Tanner (1975).

Middle to late Hemphillian, Chamita Formation, Rio Arriba County, New Mexico: 4 miles up Rio Grande from San Juan: F:AM 116089, astragalus; East Alcalde, Santa Fe: F:AM 116018, calcaneum; Opposite Alcalde, Santa Cruz: F:AM 116016, maxilla; Four miles up Rio Grande: F:AM 116014, radius; Two miles north of San Juan: F:AM 116013, edentulous ramus; East Black Mesa, Santa Fe: F:AM 116017, calcaneum/astragalus; Seventeen miles up Rio Grande: F:AM 116021, ulna.

Middle to late Hemphillian, Hemphill beds, Lipscomb County, Texas: Box T Ranch, 1st Valley, West Pit 1: F:AM 115829A–B, two tibiae; F:AM 115821, humerus; F:AM 115822A–B, two ulnae; F:AM 115812, left ramus; F:AM 115813, left ramus; F:AM 115814, right ramus; F:AM 115804, right juvenile ramus; F:AM 115805, right juvenile ramus; F:AM 115806, juvenile mandible; F:AM 115796, left ramus; F:AM 115797, right ramus; F:AM 115798, left ramus; F:AM 115799, right ramus; Box T Ranch, 1st

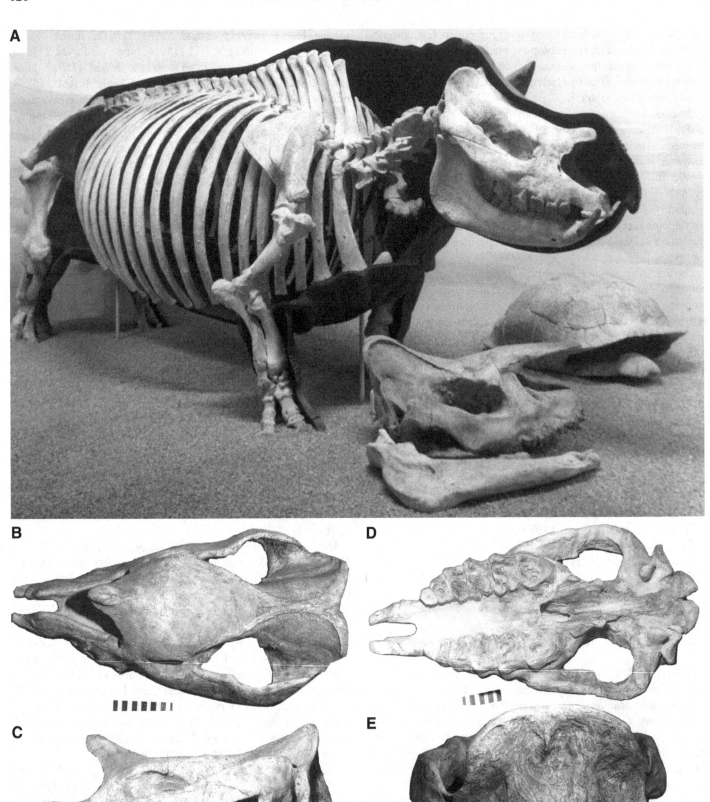

F

G

H

Figure 4.43. T*eleoceras hicksi.* A. Mounted composite skeleton from the Wray locality, including the type skull (DMNH 304). B–E. Referred skull from the Wray locality (DMNH 715) in dorsal, left lateral, palatal, and posterior views. F–H. Referred skull from the Snake Creek Formation, Nebraska (F:AM 114590), showing the variation in nasals and other skull proportions within this species. Scale bars in cm.

Valley, West Pit 1, Bridge Creek: F:AM 115795, left ramus; Box T Ranch, East Draw Sand: F:AM 115988, left M2; Box T Ranch, 1st Valley: F:AM 114565, ramus; F:AM 114574, partial skull.

Middle Hemphillian, Hemphill beds, Armstrong County, Texas: McGehee Place Quarry: F:AM 115987, radius; Armstrong Place: F:AM 115986, radius.

Late Hemphillian, Hemphill beds, Hartley County, Texas: Rentfro Pit 1: F:AM 116027, juvenile rami. Sanford Pit: F:AM 116029, left M2; Burson Ranch Pit: F:AM 116028, maxilla.

Late Hemphillian, Hemphill beds, Hemphill County, Texas: Coffee Ranch Quarry: MWU 6942, i2 tusk.

Late Hemphillian, Ash Hollow Formation, Brown County, Nebraska: Lake bed on Willow Creek: F:AM 116012, palate.

Late Hemphillian, Snake Creek Formation, Johnson Member, Sioux County, Nebraska: *Aphelops* Draw: F:AM 114590, skull.

Late Hemphillian, Panaca Formation, Lincoln County, Nevada: Panaca General Field: F:AM 116022, patella (Stock, 1921; Lindsay *et al.*, 2002).

Late Hemphillian, Mount Eden Formation, Mount Eden l.f., San Timoteo Badlands, Riverside County, California (Albright, 1999a): LACM (CIT) 72070, left ramus; several uncatalogued teeth in the F:AM collection.

Late Hemphillian, Shutler Formation, McKay Reservoir l.f., Umatilla County, Oregon (Shotwell, 1956, 1968): dozens of catalogued specimens in the UCMP, LACM (CIT), and UO collections.

Late Hemphillian, Rome Formation, Dry Creek l.f., Malheur County, Oregon: LACM (CIT) 6731, astragalus.

Late Hemphillian, Chalk Buttes Formation, 2 miles southeast of

Harper, Malheur County, Oregon: LACM (CIT) 90739, calcaneum; LACM (CIT) 92039, Mc3; LACM (CIT) 91643, Mt3; LACM (CIT) 88832, astragalus; LACM (CIT) 88841, scapula; many other catalogued postcranial fragments.

Late Hemphillian, Ringold Formation, Franklin County, Washington: UW 52685, mandible (Gustafson, 1977).

Late Hemphillian, Pinole Tuff, Contra Costa County, California: UCMP 29710, skull, isolated teeth.

Late Hemphillian, Coal Valley Formation, Yerington, Lyon County, Nevada: numerous fragments in the UCMP collections.

Late Hemphillian, Bidahochi Formation, Keams Canyon, Navajo County, Arizona: Unknown Location: F:AM 116030, humerus, ulna; Gentry Ranch: F:AM 116031, calcaneum, two astragali. Jeddito: F:AM 116023, edentulous left ramus; F:AM 116032, distal ulna.

Late Hemphillian, Upper Bone Valley Formation, Polk County, Florida: Payne Creek Mine: UF 26691, right rami; UF 24690, mandible; UF 24692, left maxilla with dP3–4, M1; UF 24721, symphysis; UF 24608, humerus; UF 24680, astragalus; UF 24673, partial pelvis; UF 2467, partial humerus; UF 53980, right calcaneum; UF 24696, P2; UF 24683, left ramus with p2-m2; Palmetto Mine: UF 10390, tooth fragment; UF 14792, lower molar; UF 14795, lower molar; UF 14791, lower molar; UF 14787, left lower molar; UF 14804, lower molar; UF 14789, right lower molar; UF 14796, right lower molar; UF 14790, lower molar; UF 14794, lower molar; UF 14788, lower molar; UF 58400, distal fibula; UF 14801, right M3; UF 14800, left M3; UF 14798, M2; UF 14799, Mc3; UF 14802, upper molar; UF 14793, lower molar; UF 14484, upper molars; UF 13227, upper molar; Tiger Bay Mine: UF

14485, upper molar; Noralyn Mine: UF 12053, left M3.

Late Hemphillian, Rancho el Ocote l.f., Guanajuato, Mexico: Material listed by Dalquest and Mooser (1980, p. 5) as hypodigm of *Teleoceras ocotensis*.

Late Hemphillian, Yepomera l.f., Chihauhau, Mexico: LACM (CIT) 30236, upper molar.

Known distribution: Middle to late Hemphillian, Texas, Colorado, Nebraska, Kansas, Nevada, Florida, New Mexico, Arizona, and Guanajuato, Mexico.

Diagnosis: Medium-sized *Teleoceras* (M1–3 length = 150–165 mm; m1–3 length = 155–180 mm) with a more brachycephalic skull than *T. fossiger*. Nasals shortened and upturned in many specimens, but not flattened as in *T. brachyrhinum*. The p2 is lost.

Description: *T. hicksi* was described by Cook (1927). Additional well preserved material in the Frick Collection is illustrated in Figures 4.45 and measurements are given in Table 4.11.

Discussion: The *Teleoceras* sample from Wray, Colorado, has been the most misinterpreted of all rhino quarry samples. It is fairly homogeneous in size (Fig. 4.36) and morphology, and yet Lane (1927) and Cook (1927, 1930) managed to name no fewer than five species in three genera for this single population. Only the incredible oversplitting of Gregory and Cook (1928) exceeds the oversplitting of this example for typological, rather than biological, thinking.

Cook (February 1927) named the first (and only valid) new taxon from the Wray locality, *Teleoceras hicksi*, based on a complete skull. Lane (September 1927) created a new genus and species, *Paraphelops rooksensis*, for a very robust lower jaw of Wray *Teleoceras*. Nearly all the features he cited are clearly due to individual variation common in rhinos, as a much more competent paleontologist, W.D. Matthew (1932, p. 417) realized. Cook (1930) compounded Lane's errors by erecting two new species for perfectly normal *Teleoceras* jaws from the same quarry. ?*Peraceras ponderis* is based on a lower jaw of *T. hicksi* (DMNH 732) which is a robust female. It is clearly not a *Peraceras* jaw, however, since it fails to show the upward inflection of the symphysis or short diastema characteristic of that genus. It also lacks p2 (a derived *Teleoceras* character), so it should be referred to *Teleoceras*. All of the features cited by Cook (1930) are simply due to normal individual variation. Similarly, another jaw from the same quarry (DMNH 731) was the basis for Cook's (1930) species *Paraphelops yumensis*. Other than its slightly smaller size (Fig. 4.38), it is within the normal range of variation of Wray *T. hicksi*. Cook's reference of this jaw to Lane's "*Paraphelops*" is even more curious, because it is the least similar to "*Paraphelops rooksensis*" of all the invalid species named for the Wray *T. hicksi* sample.

Tanner (1975) named another new species, *Teleoceras schultzi*, based on a left ramus (UNSM 5800) from UNSM locality Ft-40, the *Amebelodon fricki* Quarry, which produces the Cambridge l.f. Tanner thought that this "Kimballian" specimen was post-Hemphillian, but as has been shown elsewhere (Breyer, 1981; Tedford *et al.*, 1987), this locality is actually early middle Hemphillian in age. The erroneous age assignment apparently influenced Tanner in his decision to name a new species, because the morphology is remarkably close to *T. hicksi*. Tanner distinguished the two species by the slightly larger size of *T. schultzi*, but the type specimen falls well within the range of variation of Wray *T. hicksi* (Fig. 4.36). The other features (more brachycephalic skull; wider basioccipital region; larger and more massive postcranials) are all due to normal individual variation, and are well matched by several specimens from Wray in the Frick Collection. Tanner presented no statistics to support his claim that the two can be distinguished. Indeed, he apparently compared only referred material of *T. hicksi* from Nebraska, and never mentions examining material from the type locality in Colorado. The sample of the Wray *T. hicksi* in the Frick Collection clearly shows that *T. schultzi* cannot be distinguished from *T. hicksi*, and thus is a junior synonym. Perhaps Tanner's conception that the "Kimballian" rhino must be the "latest and last" species of *Teleoceras* was influenced by his belief that the "Kimballian" was post-Hemphillian, so he grasped at straws in attempting to justify his new taxon. Ironically, *T. schultzi* is considerably smaller than *T. fossiger* (NOT the largest species, as Tanner implied), so it is neither the largest nor the last of the species of *Teleoceras*. Instead, many genera (*Aepycamelus, Yumaceras, Tapirus, Calippus, Nimravides, Pliohippus, Neohipparion, Epicyon, Leptarctus, Macrogenis, Ilingoceras, Barbourofelis, Indarctos*, and *Prosthennops*) exhibit their largest species during the early Hemphillian, and each of these lineages then had smaller species in the late Hemphillian (Prothero and Manning, 1987).

In spite of the confusion about the "Kimballian," for decades many authors (e.g., Matthew, 1932) thought that *Teleoceras* was extinct before the late Hemphillian. However, it is only rare in comparison to the great abundance of *Aphelops mutilus* in the classic late Hemphillian quarries, such as Coffee Ranch, Texas, and Edson, Kansas; each of these localities produce only scraps of *T. hicksi* (see hypodigm above). *T. hicksi*, although relatively rare, was widespread in many late Hemphillian localities (listed in the hypodigm above).

Dalquest and Mooser (1980) described isolated teeth from the late Hemphillian Rancho el Ocote l.f. as *Teleoceras ocotensis*. They based their identification on Tanner's (1975) misconceptions about *T. fossiger, T. hicksi*, and *T. "schultzi"*, and decided their species was more advanced than "Kimballian" *T. schultzi*. Carranza-Castañeda (1989) described and figured additional *Teleoceras* material from Rancho el Ocote, including skulls, mandibles, and partial skeletons. He compared this material only to *T. fossiger* material in the UNSM collections (much of which is mislabeled, and is actually *T. hicksi*). On this basis, he synonymized *T. ocotensis* with *T. fossiger*. However, the measurements given by Dalquest and Mooser (1980) and Carranza-Castañeda (1989) are too small for *T. fossiger* as currently conceived, but instead match the known size range of *T. hicksi* (Table. 4.11). Thus, *T. ocotensis* should be considered a junior synonym of *T. hicksi*, not *T. fossiger*. This is consistent with the late Hemphillian age of the fauna as well.

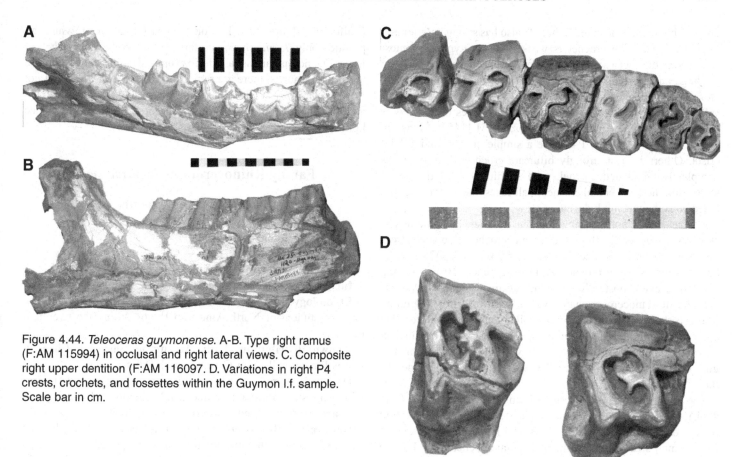

Figure 4.44. *Teleoceras guymonense*. A-B. Type right ramus (F:AM 115994) in occlusal and right lateral views. C. Composite right upper dentition (F:AM 116097. D. Variations in right P4 crests, crochets, and fossettes within the Guymon l.f. sample. Scale bar in cm.

Teleoceras guymonense new species

Teleoceras cf. *fossiger* Schultz, 1977

Figure 4.44, Table 4.11

Holotype: F:AM 115994, a right ramus with p4–m3, and part of the coronoid and angle.

Hypodigm: *Late Hemphillian, Ogallala Formation, Oklahoma County, Texas*: Guymon Area: F:AM 114592, partial skull, mandible; F:AM 114593, partial skull; Clay Flats, Guymon Area: F:AM 114591, partial skull; F:AM 114594, ramus; F:AM 116008, mandible; Sand Flats, Guymon Area: F:AM 116088, tibia; F:AM 116091, tibia; F:AM 116097, composite upper dentition; F:AM 116000, atlas; Railroad Sand, Guymon Area: F:AM 116087, right ramus; F:AM 115997, left ramus; F:AM 115998, ramus; Highway Sand, Guymon Area: F:AM 115994, right ramus; F:AM 115995, left ramus; F:AM 115996, right ramus; F:AM 116003, ulna; F:AM 116004, radius; F:AM 116005, left ramus; F:AM 116010, right ramus; F:AM 116011, eleven footbones; Highway Bottom Sand: F:AM 115999, right ramus; F:AM 116001, right ramus; Highway Clay, Guymon Area: F:AM 116092, radius; B.F. Guymon Area: F:AM 116002, left ramus; D.F. Guymon Area: F:AM 116006, left ramus; F:AM 116007, partial skull; Topsoil Sand, Guymon Area: F:AM 116009, rami.

Late Hemphillian, Ogallala Formation, Sherman County, Kansas: Edson Quarry: F:AM 116025A–B, two unciforms.

Late Hemphillian, Hemphill beds, Hemphill County, Texas: Coffee Ranch: F:AM 116026, right Mt4; F:AM 116019, cuboid; F:AM 116024, right ramus.

Late Hemphillian, Chamita Formation, Rio Arriba County, New Mexico: Santa Cruz: F:AM 112215, palate.

Etymology: In reference to its occurrence in the Guymon area, Oklahoma.

Known distribution: Latest Hemphillian of Oklahoma, Kansas, Texas and New Mexico.

Diagnosis: Very small (m1–3 length = 135–150 mm) species of *Teleoceras* with very short, robust limbs. M3 with ribs on lingual side of crochet. Otherwise similar to *T. hicksi*.

Description: *T. guymonense* is known from a large sample from the late Hemphillian Optima (= Guymon) local fauna, Texas County, Oklahoma (Schultz, 1977, pp. 81–82). Unfortunately, most of the sample is badly crushed and very poorly preserved. Several skulls are known from this locality, but they are so crushed that little can be said about them, other than that they are clearly *Teleoceras*. Isolated upper teeth of *T. guymonense* are abundant, and in most respects they resemble those of *T. hicksi* except that they are much smaller. P2 is strongly bilophodont, with strong anterior, lingual, and posterior cingula. A faint crochet is present, but there is no antecrochet. P3 also has strong anterior and lingual cingula. The crochet is weaker than is typical of most *Teleoceras*, and occasional cristae are present. In most examples, the large antecrochet has merged with the metaloph during wear, closing the valley between the lophs.

P4 is highly variable, as it is in many examples of *Teleoceras* (see, for example, the palate of *T. major*, F:AM 114525, shown in Fig. 2.9). Some isolated P4s have a simple crochet and a faint crista. Others have a strongly bifurcate crochet. One tooth has completely crenulated enamel, with additional folds developed along with the crochet and cristae. In all cases, the antecrochet has fused with the metaloph during wear.

M1 has a weak anterior cingulum, but no lingual cingulum. The crochets are generally short and the antecrochet is separate from the metaloph until extreme wear stages are reached. M2 typically has a short simple crochet and antecrochet. Broad flaring anterolingual and posterolingual cingulae are also present. M3 has the typical rhinocerotid shape with a weak anterior cingulum. Many isolated M3s in the Frick Collection show features that appear to be unique to *T. guymonense*. There is a short but distinct rib on the lingual face of the crochet, which produces a short lingual spur on the wear surface. I have seen no other rhinoceros molar with this feature.

The lower jaw is best described from the type specimen (Fig. 4.44A–B). It shows the characteristic stereotypical rhinocerotid pattern on the premolars and molars, and is unremarkable except for its small size. There is a mental foramen beneath p4, but the jaw anterior to this tooth is unknown. The angular process is badly broken, but appears to have a typical thickened ventral margin. The anterior portion of the coronoid process is preserved, also. It sweeps up broadly and gradually behind the m3, with a wide anterior face. Isolated lower premolars and i2 tusks look like those from any other *Teleoceras* except for their smaller size.

The known postcranial elements of *T. guymonense* are described in Chapter 5.

Discussion: Despite the widespread notion that *Teleoceras* was extinct before the late Hemphillian, there are actually two species present at that time. Most specimens fit within the normal size range of *T. hicksi*, and can be referred to that species. But the Frick Collection sample from the Optima (= Guymon) l.f. of Texas County, Oklahoma, contains no *Aphelops*, but two species of *Teleoceras*, *T. hicksi* and the dwarf species, *T. guymonense*. Although the fossils are poorly preserved, they are clearly much smaller than any other post-Barstovian species of *Teleoceras*. Indeed, it appears to be the culmination of the size reduction trend of *Teleoceras* during the Hemphillian, from the large *T. fossiger* to *T. hicksi* to *T. guymonense*. It is also the second example of dwarfing demonstrated in *Teleoceras*, after the late Barstovian dwarf *T. meridianum*.

T. guymonense has a distinct distribution. It occurs in the Panhandle of Texas (Coffee Ranch) and Oklahoma (Optima), in Kansas (Edson), and in the surrounding montane regions of Mexico (Ocote l.f.) and New Mexico (Chamita Formation). However, it does not occur as far west as Arizona (Jeddito l.f.) or

California (Mount Eden l.f.) or Nevada (Panaca l.f.), which all produce only *T. hicksi*, nor as far north as Nebraska (*Aphelops* Draw l.f.) nor as far east as Florida (upper Bone Valley fauna). In all of those faunas, there are remains referable to *T. hicksi* but not to *T. guymonense*. Only the Chamita Formation of New Mexico and the type locality of Optima (= Guymon) l.f. seem to produce both species.

Family Rhinocerotidae *incertae sedis*

Woodoceras new genus

Figure 4.45, Table 4.12

Type species: *Woodoceras brachyops* new species.
Included species: Type species only.
Etymology: In honor of Horace Elmer Wood II (1901-1975), premier student of North American rhinos during the twentieth century.
Known distribution: Early Arikareean Blue Ash Channel, Harris Ranch, Fall River County, South Dakota.
Diagnosis: Large short-faced rhino (m1–3 length = 121 mm) with very short diastema, reduced p2, compressed cheek tooth row, and m3 metalophid reduced to a single loph.
Description: *Woodoceras brachyops* is based on a badly broken mandible lacking the anterior teeth and the portion posterior to m3. However, in its preserved parts, it is so peculiar that it clearly cannot be referred to any known genus or species of rhinocerotid. The alveoli for the broken i2 tusks are very large (27 mm in diameter) and face anteriorly, suggesting that this specimen was a male. The symphysis is broad and robust, but very short. There is almost no diastema between the alveolus for i2 and p2, and certainly no room for a dp1. There are two mental foramina, one anterior to the p2, the other immediately below p3.

The alveolus for p2 is only a single small hole, so evidently it was the locus of an unusually short, possibly knob-like or cylindrical tooth. The preserved teeth are so worn that almost no crown pattern remains. Clearly, this specimen was an old individual, so its peculiarities are not due to it being a juvenile. The p3 and p4 have a more normal rhinocerotid shape, but they too are extremely reduced relative to the molars. The m1–2 are so worn that little can be said about them. However, the hypolophid of m3 is very peculiar in that its anterolabial portion is very reduced. As a result, the hypolophid is a single transverse lophid, without the typical anterolabial "hook" seen in all other rhinocerotids. In this respect, it seems to be approaching the bilophodont condition seen in tapirs.

The mandible appears deep-jawed because of its relative shortness. The depth of the mandible is proportional to the m1–3 length, so it is not deep-jawed in the absolute sense.

Discussion: The peculiar features of *W. brachyops* are clearly all autapomorphies, because they are seen in no other rhinoceros. The short-faced bilophodont condition suggests that this rhino was

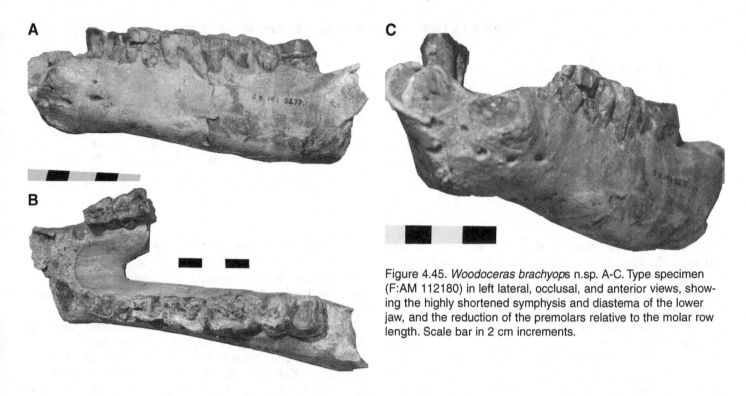

Figure 4.45. *Woodoceras brachyops* n.sp. A-C. Type specimen (F:AM 112180) in left lateral, occlusal, and anterior views, showing the highly shortened symphysis and diastema of the lower jaw, and the reduction of the premolars relative to the molar row length. Scale bar in 2 cm increments.

Table 4.12. Dental measurements of *Woodoceras* and *Diceratherium* (in mm)

MEASUREMENT	*D. tridactylum*			*D. armatum*			*D. annectens*			*D. niobrarense*			*Woodoceras*
	Mean	SD	N	Mean	SD	N	Mean	SD	N	Mean	SD	N	F:AM 112180
i2 width	29	1	2	22	4	5	20	4	3	26	7	3	27
Diastema length	60	9	2	54	4	4	45	11	3	44	7	3	11
p2-m3 length	189	16	13	218	15	5	170	14	3	233	10	3	172
m1-3 length	113	6	13	134	6	5	104	6	5	140	2	4	121
m1-3 /p2-m3ratio	0.59			0.59			0.61			0.60			**0.70**
p2 length	27	2	2	26	1	3	18	2	2	26	4	3	—
p2 width	17	1	2	17	1	3	12	1	2	18	1	3	—
p3 length	32	1	2	30	2	4	23	1	4	34	3	4	24
p3 width	19	1	2	23	2	4	18	1	4	22	2	4	21
p4 length	32	1	2	33	4	5	28	2	5	38	2	4	29
p4 width	24	2	2	27	2	5	22	1	5	27	2	4	24
m1 length	31	3	13	41	2	5	30	5	5	43	4	4	34
m1 width	24	2	13	29	2	5	23	2	5	29	2	4	27
m2 length	38	3	13	45	2	5	35	2	5	45	3	4	37
m2 width	25	2	13	30	2	5	24	2	5	30	2	4	28
m3 length	45	1	2	58	4	5	38	4	5	51	2	4	47
m3 width	29	1	2	30	2	5	23	2	5	29	3	4	30
Length symphysis	103	—	1	95	25	5	60	10	5	91	5	2	63
Depth jaw below p2	70	—	1	62	7	5	50	6	5	63	3	2	62
Depth jaw below m2	74	—	1	72	5	5	62	7	5	77	2	2	64

Table 4.13. Dental measurements of *Diceratherium* (in mm)

MEASUREMENT	*D. tridactylum*			*D. armatum*			*D. annectens*			*D. niobrarense*			*D.matutinum*
	Mean	SD	N	Mean	SD	N	Mean	SD	N	Mean	SD	N	
m2 length	38	3	13	45	2	5	35	2	5	45	3	4	48
m2 width	25	2	13	30	2	5	24	2	5	30	2	4	30
astragalus length	59	1	2	72	2	8	58	1	9	68	4	6	64
astragalus width	56	2	2	62	2	8	49	1	8	61	3	6	58

taking on a browsing mode of life more like that of a tapir. Unfortunately, there are no other materials to help us better understand this rhino, but the known mandible is so different from any other taxon known that it clearly requires a new genus and species.

Mihlbachler (pers. comm.), however, is of the opinion that this is a pathological specimen, which lost some anterior teeth and then resorbed some bone in the jaw. This may be known in brontotheres, but I have never seen a similar occurrence in thousands of fossil rhino jaws, many of which lack anterior teeth, so I am reluctant to dismiss it as an aberration. If later workers find that this is indeed the case, they can synonymize the taxon with *Diceratherium armatum*.

The type specimen is from a locality known as the Blue Ash Channel in the Harris Ranch sequence, Fall River County, South Dakota (Simpson, 1985). This locality also yields specimens of *D. annectens* and *D. armatum*, as well as specimens of the rodents *Leidymys* and *Geringia*, advanced leptauchenine oreodonts, and the horse *Miohippus* that suggest an early Arikareean age.

Woodoceras brachyops new species

Holotype: F:AM 112180, a partial mandible, from the early Arikareean Blue Ash Channel, Harris Ranch, Fall River County, South Dakota.

Hypodigm: Type specimen only.

Etymology: The name *brachyops* is Greek for "short face" in reference to the very shortened anterior portion of the skull.

Known distribution: Type locality only.

Diagnosis: Same as for genus.

Description: Same as for genus.

Diceratherium matutinum (Marsh, 1870)

Rhinoceros matutinus Marsh, 1870
Diceratherium matutinum Wood, 1939

Holotype: YPM 11873, astragalus and right m2; from the Miocene (Burdigalian, early Hemingfordian) Kirkwood Formation, near Farmingdale, Monmouth County, New Jersey.

Hypodigm: Types and AMNH 48980, protocone of left M1 or M2; AMNH 104645, anterior lateral fragment of left upper molar; YPM-PU21680, protocone fragment of right M1 or M2. All from the Shark River, west of Garden State Parkway exit 100A, at junction with Route 33, New Shrewsbury Township, Monmouth County, New Jersey.

Known distribution: Early Miocene (Burdigalian, based on elasmobranch teeth, *fide* G. Case, 1982) of New Jersey.

Discussion: The affinities of the *Diceratherium* astragalus and m2 from the Kirkwood Formation of New Jersey are difficult to determine. The astragalus compares favorably (Table 4.13) with either *D. armatum* or *D. niobrarense*, although it is shorter than any known astragalus of *D. armatum*. The m2 compares more favorably with *D. niobrarense* than it does with *D. armatum*. Based on such limited material, it is impossible to show conclusively that *D. matutinum* is a senior synonym of any other species of *Diceratherium*. Because the New Jersey dicerathere is also very geographically isolated from High Plains specimens, I retain the name *matutinum* until more complete material is found which will definitely show whether it is distinct or not. The material is not even complete enough that reference to *Diceratherium* is certain; it could also pertain to *Menoceras*. Hence, it is placed in "Rhinocerotidae *incertae sedis*" until better material is found.

Gulfoceras westfalli Albright, 1999

Figure 4.46

Holotype: LSUMG V-2622, right M3.

Hypodigm: *Late Arikareean, Fleming Formation, Toledo Bend l.f., Newton County, Texas* (Albright, 1999b): LSUMG V-2249, left M3; LSUMG V-2621, right M3; LSUMG V-2574, left astragalus.

Discussion: Described by Albright (1999b) from the late Arikareean Toledo Bend l.f., this taxon is based on several isolated M3s and a single astragalus which are too tiny to refer to any other late Arikareean rhinoceros from North America. Albright (1999b) compared the specimens to *Hyracodon*, but noted that it has the typical rhinocerotid M3 shape, with no metastyle. He also compared it to *Subhyracodon*, but noted that is was much smaller than this taxon, and also has a ridge that ascends the anterior surface of the hypocone, and a prominent lingual cingulum that is

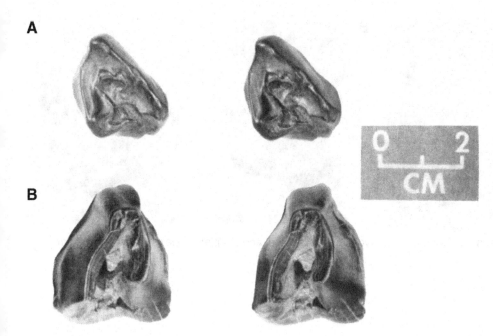

Figure 4.46. *Gulfoceras westfalli*. A. Left M3, LSUMG V-2249. B. Right M3, V-2622 (type specimen), showing the crochet on such small teeth. (After Albright, 1999b, fig. 5).

continuous with the anterior cingulum, both features missing in *Subhyracodon*. More importantly, it has a large and well-developed crochet, which rules out affinities with any other small, primitive rhino, such as *Penetrigonias* (which is a good match in size, nonetheless). Even though the material is barely adequate to diagnose a new taxon, I will retain it for the present. As Albright (1999b) notes, there is no other taxon to which it can be referred, and it probably represents a dwarf of some Miocene form not yet described from elsewhere.

The Haughton Astrobleme rhino

This taxon comes from the filling of an ancient meteorite crater on Devon Island in the Canadian Arctic, and is currently being described by Mary Dawson (pers. comm.). Originally mentioned in Hickey *et al.* (1988) and Whitlock and Dawson (1990), it is clearly early Miocene in age, based on fission-track dates on impact debris from the base of the deposit, and the associated fauna. Nonetheless, in many features this specimen resembles much more primitive rhinos from the middle Eocene through Oligocene. The upper premolars are only partially molarized, and closely resemble those of *Trigonias* or primitive specimens of *Subhyracodon*. The M3 does not have the typical triangular rhinocerotid pattern, but retains a metastyle, a characteristic of amynodonts and more primitive ceratomorphs. However, it does have the chisel-like I1 and the tusk-like i2, so it is a rhinocerotid, not an amynodont. Its limbs, however, are relatively short and stumpy like those of a teleoceratin, and the femora lack a third trochanter, a feature found in almost no other perissodactyl. Like higher rhinocerotines, however, it has a contact between the scaphoid and lunar.

This peculiar combination of features is seen in no other rhi-

noceros that has ever been described, and certainly merits a new genus and species. Its exact systematic position cannot be assessed until the full decription is published.

Giant Arikareean Postcranials

Figures 4.47, 4.48

Discussion: Like the giant aceratheriine bones described earlier (page 93, Fig. 4.34), the Frick Collection also yields another set of postcranials which clearly belong to an unknown and undescribed taxon, but we have no cranial material from this taxon yet. In this case, the specimens (F:AM 111846A-C) consist of three huge astragali (Fig. 4.47) and a humerus (Fig.5.15A) from two localities in the late Arikareean Harrison Formation in eastern Wyoming: 77 Hill Quarry, and Horse Creek. At 83 mm wide and 92 mm in height, the astragali are significantly larger than the astragali of any rhinoceros from the Arikareean (Fig. 5.33A–B, Table 5.3). As such, they certainly belong to a previously undescribed giant rhinoceros taxon (the largest known from the late Oligocene), but it is hard to determine what taxon to name it, based only on astragali. Their size is comparable to that of the smaller indricotheres such as *Juxia* from the late Eocene and early Oligocene. However, it is not possible to determine whether these astragali are dicerathere or indricothere, since they are similar in most features. Very large indricotheres, such as *Paraceratherium*, have rounded trochleae on their astragali, and in this respect, the specimens look more like a diceratherine, although the specimens of *Juxia borrisiaki* do not show this rounding of the trochleae. For the present, we must leave these specimens as "large rhinocerotoid, indeterminate" until better material turn up.

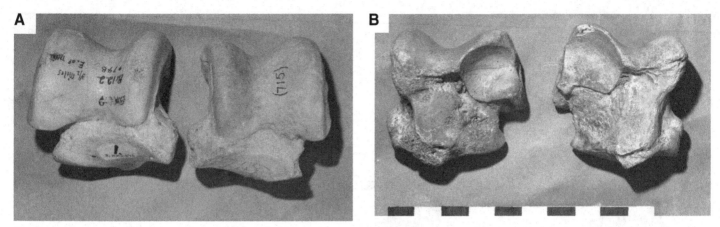

Figure 4.47. Giant astragali (F:AM 114846) from the Arikareean of Wyoming in (A) dorsal and (B) plantar views. Scale bar = 2 cm increments.

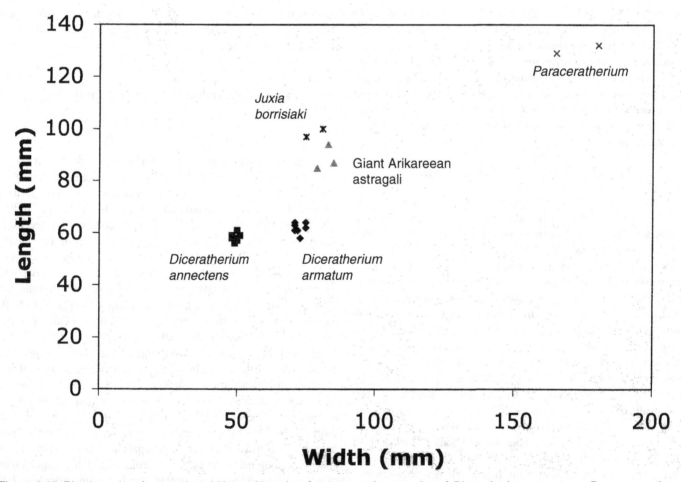

Figure 4.48. Bivariate plot of astragalar widths and lengths of representative samples of *Diceratherium annectens, D. armatum*, the three giant Arikareean astragali (triangles), and the indricotheres *Juxia borrisiaki* and *Paraceratherium*.

Invalid Taxa Based on Indeterminate Specimens

Rhinoceros crassus Leidy, 1858

Rhinoceros crassus Leidy, 1858
Aceratherium crassum Cope, 1874
Aphelops crassus Osborn, 1890
Rhinoceros crassus Osborn, 1904
?*Aphelops crassus* Matthew, 1918
Peraceras crassus Voorhies, 1990

Discussion: The first name applied to a North American Miocene rhinoceros was *Rhinoceros crassus*, described and named by Leidy (1858) from some fragmentary rhino remains collected by Ferdinand Hayden from the "sands of the Niobrara River, Nebraska," In 1869, Leidy (pp. 228–229) further described and illustrated the specimens (USNM 775), a right maxillary fragment with a broken M3, a right dP3, and a fragmentary left ramus with ?dp1, and a left I1. Leidy (1869, p. 228) stated that "it is uncertain whether all the specimens belong to the same species." If the chisel-like I1 is indeed associated with the rest of the material, then the name *crassus* would be referred to *Teleoceras*, since all other post-Hemingfordian North American rhinos are aceratheriines, which have lost their I1. The isolated M3 best matches specimens of *Teleoceras medicornutum* from the Valentine Formation (particularly a maxilla, F:AM 115372, from the June Quarry in the Burge Member), but it is also a reasonable match for specimens of *Peraceras superciliosum* (e.g., F:AM 114396) from Sawyer Quarry in the Devil's Gulch Member of the Valentine Formation. Most *Aphelops* from the late Barstovian and Clarendonian of northern Nebraska are too small to match the M3 of *crassus*. Thus, the lectotype M3 could pertain to either *Teleoceras* or *Peraceras*, and the stratigraphic and locality data given by Leidy is too vague to be of much help. Osborn (1904, p. 308) considered the name *crassus* indeterminate "owing to the uncertainty as to the type," and Matthew (1931, p. 420) considered it indeterminate "until adequate topotypes are known." Even with adequate topotypes from the "sands of the Niobrara River" (most likely the Valentine or Ash Hollow formations, according to Skinner and Johnson, 1984), the type material is still too incomplete to determine whether it pertains to *Teleoceras* or *Peraceras*. If the I1 can be proven to be associated with the lectotype M3, then it would be a senior synonym of *T. medicornutum*. However, since no such association is proven, I consider the name *crassus* a *nomen dubium* since the type is inadequate to decide to which genus it it belongs.

Rhinoceros oregonensis Marsh, 1873

Rhinoceros oregonensis Marsh, 1873
Aceratherium oregonensis Matthew, 1899
Diceratherium oregonensis Loomis, 1908
Diceratherium armatum Fremd *et al.*, 1994

Discussion: *Rhinoceros oregonensis* was described by Marsh (1873) based on material from the early Barstovian Mascall Formation of central Oregon. The type material is a fragmentary upper molar (YPM 10002, a right ?M2 without an ectoloph). Matthew (1899) referred it to *Aceratherium*, but Loomis (1908) referred it to *Diceratherium*. Peterson (1920, p. 412) considered it indeterminate, because it is so fragmentary. As Peterson pointed out, it could be referred to *Teleoceras* as easily as *Diceratherium*, and the same could be said of any other genus of North American Miocene rhino. Matthew (1932, pp. 418–419) suggested that the presence of a medifossette agrees with *Teleoceras*, but there are now specimens of *Aphelops* and *Peraceras* with medifossettes. Matthew (1932) considered the type specimen to be indeterminate until more complete topotypes could be found, but even with much larger collections (including a few additional specimens from the Mascall Formation), the type specimen is so poor that it will always be indeterminate and the name invalid. Fremd *et al.* (1994) considered the specimen to be referable to *Diceratherium armatum*, but it is not clear that it came from the John Day Formation, let alone that it is well enough preserved to be referred to that species.

Aphelops jemezanus Cope, 1875

Aphelops jemezanus Cope, 1875
Aphelops jemezanus Matthew, 1932

Discussion: Cope (1875a, 1875b, 1875c) described *Aphelops jemezanus* from "near the town of Santa Clara, on the west side of the Rio Grande." The type material consists of a fragmentary right ramus (USNM 115) with very worn m1–3, the angular process, and the base of the coronoid (figured in Cope, 1875c, plate 73, fig. 3; plate 74, fig. 4). Most of the features Cope used to distinguish this specimen from "*Aphelops*" *meridianus* (actually *Teleoceras meridianum*, according to Prothero and Manning, 1987) were slight variations in the posterior and ventral margins of the ramus that are not diagnostic when a large sample of jaws is considered (as noted by Matthew, 1932).

The teeth of the type are so worn and poorly preserved as to be virtually useless for identification. They lack lingual cingula, so they probably are not referable to *Peraceras*, the most common rhino from the Miocene of New Mexico. They compare well in size and morphology to rami of both *Teleoceras* and *Aphelops* from the late Clarendonian of the Santa Fe Group. Galusha and Blick (1975, table 3) list *A. jemezanus* as probably derived from the upper Pojoaque Member of the Tesuque Formation, or possibly from the Chamita Formation. The material mentioned above is from the upper Pojoaque Member, but there are also several specimens from the Chamita Formation in the Frick Collection that match the type of *A. jemezanus*. Even with much larger topotypic material, it is clear that the type of *A. jemezanus* is inadequate to determine to which genus it should be referred, and thus the name is indeterminate.

Anchisodon tubifer Cope, 1879

Discussion: In 1879, Cope placed several fragmentary specimens from the John Day region in his new genus *Anchisodon*, which included a new species *Anchisodon tubifer*. This taxon was based on an indeterminate M2 that was never figured, was referred to *Aceratherium* by Merriam and Sinclair (1907), and then to *Caenopus* by Matthew (1909), but has never been adequately discussed. According to Merriam and Sinclair (1907), the type was originally in the Cope Collection at the American Museum, and now appears to be lost. Based on its size (as given by Cope) and locality, it may be a junior synonym of Marsh's taxon *Diceratherium annectens*. Because the type is lost, and was never adequately described or figured, I consider *Anchisodon tubifer* to be a *nomen dubium*.

Eusyodon maximus Leidy, 1886

Eusyodon maximus Leidy, 1886
Teleoceras fossiger Osborn, 1904
Teleoceras fossiger Matthew (*in* Cope and Matthew, 1915)

Discussion: Leidy (1886) proposed the name *Eusyodon maximus* for a tusk from the early Hemphillian Mixson's Bone Bed of Florida, which he originally thought came from an extinct boar. The next year Leidy realized his mistake. "Several more characteristic specimens prove that he had formerly committed a blunder in referring the fragments of a tooth to an extinct boar with the name *Eusyodon maximus* which is only part of the lower tusk of a rhinoceros" (Leidy, 1887, p. 309). Osborn (1904, p. 313) and Matthew (*in* Cope and Matthew, 1915) treated *Eusyodon maximus* as a synonym of *Teleoceras fossiger*. However, if this is so, then *Eusyodon* Leidy, 1886, would have priority over *Teleoceras* Hatcher, 1894.

Because the name *Eusyodon* had long been forgotten and *Teleoceras* is well established, it seems appropriate that the name *Eusyodon* be suppressed. However, such a procedure is unnecessary. The type, USNM 3288, is a badly broken left i2. It matches several i2s of *Teleoceras proterum* from the Mixson's Bone Bed sample quite well. It also matches some tusks that are referable to *Aphelops malacorhinus* from Mixson's Bone Bed. The most diagnostic feature would be the presence of a thegosis facet that would indicate whether an I1 was present (*Teleoceras*) or absent (*Aphelops*). USNM 3288 has its tip broken off, so it is impossible to tell whether such a facet was present. Thus, *Eusyodon maximus*, although most likely representing a tusk of *Teleoceras proterum*, cannot be assigned to any taxon with certainty, and so must remain indeterminate.

Caenopus simplicidens Cope, 1891

Aceratherium simplicidens Osborn and Wortman, 1894
Caenopus (= Subhyracodon) simplicidens Osborn and Matthew, 1899
Caenopus simplicidens (of doubtful validity) Troxell, 1921
Amphicaenopus? simplicidens Wood, 1927

Discussion: The type specimen, AMNH 10708, consists only of a left M3 and part of the M2. The M2 has no lingual cingulum, and M3 has a weak lingual cingulum. As Osborn (1898c), Troxell (1921a), and Wood (1927) pointed out, in size and morphology it could match *Subhyracodon*, *Trigonias*, or *Amphicaenopus*, so it is indeterminate.

Aceratherium persistens Osborn, 1904

Discussion: Osborn (1904) described "*Aceratherium persistens*" based on a partial skull from the "Loup Fork beds, Pawnee Creek, Colorado." According to Galbreath (1953, p. 107), this locality could be either in the Martin Canyon area or in the Sand Canyon area, which have faunas ranging in age from early Hemingfordian (Martin Canyon l.f.) to early Clarendonian (Sand Canyon l.f.), and several Barstovian faunas in between. Wood (1927, p. 72) referred this specimen to *Diceratherium*, but Matthew (1932, p. 418) compared it to the Chinese Pliocene form "*Diceratherium*" *palaeosinense*. In morphology, it compares favorably with a female *D. niobrarense* but it could also be a female skull of *Menoceras barbouri*. The type of *persistens* is too incomplete to be confident of taxonomic assignment, so I regard "*Aceratherium persistens*" as a *nomen dubium*.

Teleoceras felicis Freudenberg, 1922

Aphelops sp. Felix and Lenk, 1891
Teleoceras felicis Freudenberg, 1922

Discussion: *Teleoceras felicis* was proposed by Freudenberg (1922) based on the distal end of a humerus originally described by Felix and Lenk (1891) as *Aphelops* from the ?Pliocene Tequixquiae locality of Mexico. Felix and Lenk (1891, table 30, no. 9) illustrate a bone that is neither *Teleoceras* nor *Aphelops*. It is approximately the right size for a rhino, but the entepicondylar process is far too slender for any late Tertiary North American rhino (see Chapter 5). It is too small for any mid-Tertiary rhino. If it is indeed a rhinoceros, it does not yet match any yet described from North America. For the present, I consider the name *Teleoceras felicis* to be indeterminate.

Caenopus yoderensis Schlaikjer, 1935

Discussion: Schlaikjer (1935) described a small lower jaw from the early Chadronian Yoder l.f. of Wyoming that he named *Caenopus yoderensis*. The type material (MCZ 2097) is a broken ramus with m1–3. It is too fragmentary to be very diagnostic, but in size it most closely matches *Teletaceras radinskyi*. However, the small posterior cingulum on m3, and the relatively small m1, are not completely within the published range of variation of *T. radinskyi*. In addition, it is such an incomplete specimen and its size is so small that it could also be from a hyracodont. I regard the *Caenopus yoderensis* as indeterminate based on its type materials.

5. Postcranial Osteology

INTRODUCTION

The study of many fossil mammals frequently concentrates on teeth to the exclusion of other features. In many cases, this is necessary, because the majority of fossil mammal genera are known primarily or exclusively from their teeth. In the case of rhinoceroses, this is not so. Most of the North American genera are known from articulated skeletons or quarry samples where most elements are known. Based on large samples from the White River Group, Scott (1941) described the postcranial skeletons of *Trigonias* and *Subhyracodon*. Similarly, the availability of the huge sample from Agate Springs Quarry allowed Peterson (1920) to describe the postcranials of *Menoceras*. Osborn (1898c) and Harrison and Manning (1983) described some elements of the skeleton of *Teleoceras*, and a number of the papers cited in this book (e.g., *Teletaceras* by Hanson, 1989; *Floridaceras* by Wood, 1964; Prothero and Manning, 1987, for some species of *Peraceras* and *Teleoceras*) described isolated elements of some rhino skeletons. Of the rhinos for which postcranials are known, however, there is still no published description of the complete skeletons of *Diceratherium, Aphelops,* or *Peraceras*, nor are there comprehensive comparative measurements published for any of these rhinos. Even where a comprehensive description of an individual skeleton (e.g., Scott, 1941), there are no element-by-element comparative descriptions and illustrations that would allow identification of isolated skeletal elements.

In this chapter, I discuss, compare, and illustrate a series of each element from each genus (where known). Tables of measurements (Fig. 5.1, Tables 5.1-5.12) make it possible to compare the size of the different species within each genus. Due to the current lack of adequate description and measurement in the literature, most rhino postcranials in museums around the world remain unidentified. These comparative descriptions, illustrations, and measurements should make it possible to identify almost any North American rhino fossil.

The terminology used herein follows Scott (1941) and Sisson and Grossman (1975).

AXIAL SKELETON
Vertebral column

The vertebral column of rhinoceroses is one of the most stereotyped and conservative parts of the anatomy (as it is in most mammals). Even in rhinos with unusual limb proportions, like the teleoceratins, there are few distinctive features of the vertebral column. Consequently, most vertebrate collections have not bothered to identify or sort out individual vertebrae. In addition, the fragile spines and processes of vertebrae are easily broken off, so that most fossilized vertebrae are simply the centrum and part of the neural arch. In such cases, it would be pointless to try to sort out the vertebrae.

However, there are a number of mounted skeletons of rhinos on display in museums across the country. The vertebral columns are mostly composed of loose but associated vertebrae found in the same quarry. Because these mounts are usually permanent, it is difficult to study individual vertebrae. Occasionally, articulated specimens are found, but these are seldom prepared so that each vertebra can be seen from all sides or studied individually. Consequently, it is difficult in most collections to even find vertebrae that can be identified and sorted out. One such collection, the UNSM sample of *Menoceras barbouri* from Bridgeport Quarries, contains at least a drawer full of each individual vertebra. The following descriptions and figures are based on this material and amplify the descriptions of *M. arikarense* by Peterson (1920). Where differences between *Menoceras* and other rhinos occur (other than those due to size), they have been noted.

Atlas (Fig. 5.2): In anterior view, all ceratomorph atlases have deeply concave anterior articular surfaces (which contact the occipital condyles) separated medially by deep dorsal and ventral notches. However, the ventral notch is weakest in *Menoceras* and higher rhinos. The neural arch of rhinos is generally a low, semiconical ridge, lower than that found in tapirs. The transverse processes are broad and flaring and in a horizontal plane with the centrum. A short hypophysial tubercle occurs in the posteroventral end of the centrum. All rhino atlases have a distinct alar notch on the anterior edge of the transverse processes, a feature not found in tapirs. The posterior articular surfaces (which contact the axis) are broad and gently saddle-shaped, with nearly flat surfaces. Primitively, they are nearly circular in posterior view, but in *Subhyracodon* and higher rhinos they become dorsoventrally compressed.

Most of the foramina found in the atlases of other mammals are found in rhinos as well. All have a deep suboccipital foramen lateral to the neural arch, and facing laterally. The intervertebral canal, whose posterior end lies lateral to the posterior articular sur-

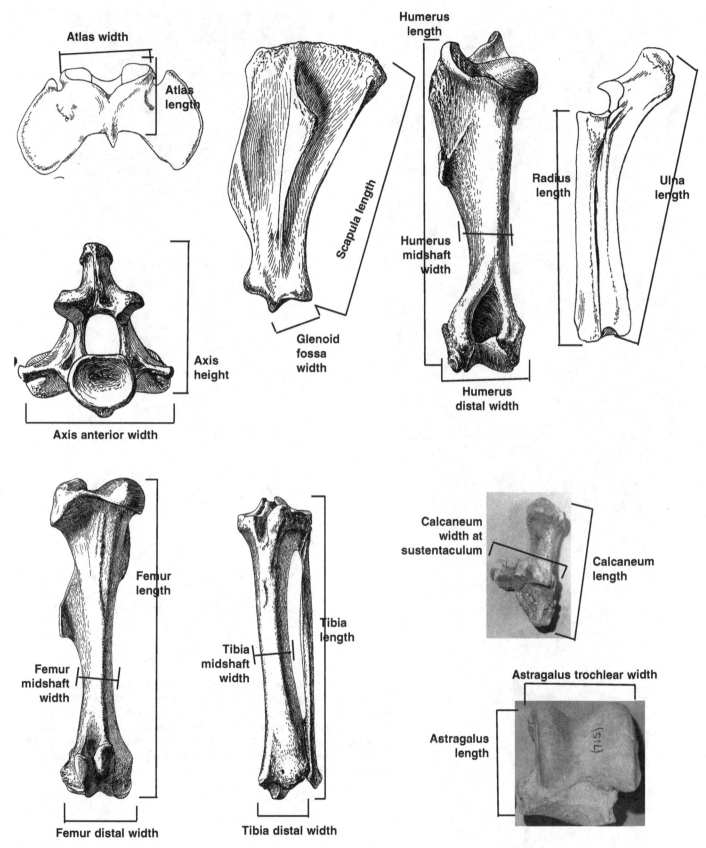

Figure 5.1. Standard dimensions and measuring landmarks of postcranial elements measured and discussed in this chapter (see Tables 5.1–5.9). (Bones of *Menoceras arikarense* after Peterson, 1920).

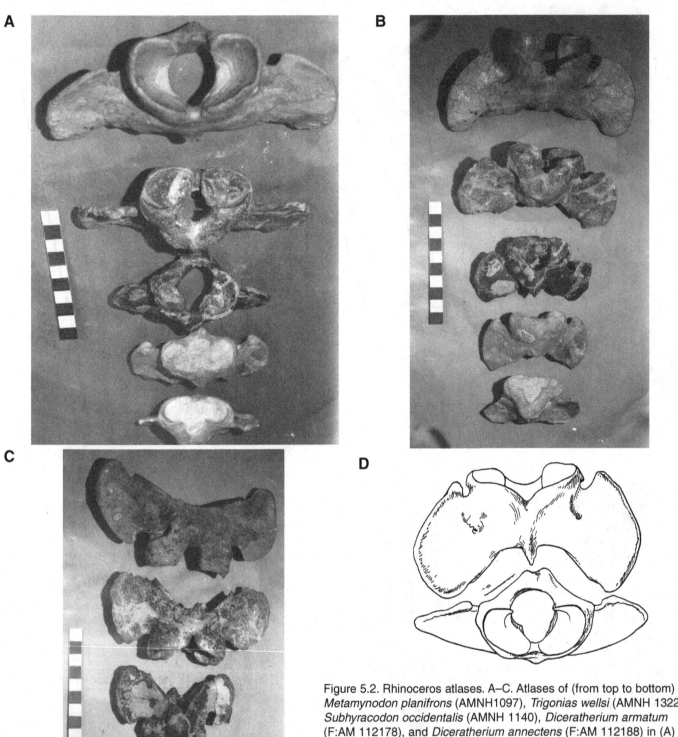

A

B

C

D

Figure 5.2. Rhinoceros atlases. A–C. Atlases of (from top to bottom) *Metamynodon planifrons* (AMNH1097), *Trigonias wellsi* (AMNH 13226), *Subhyracodon occidentalis* (AMNH 1140), *Diceratherium armatum* (F:AM 112178), and *Diceratherium annectens* (F:AM 112188) in (A) anterior, (B) dorsal, and (C) ventral views. Scale bar in 2 cm increments. D. Atlas of *Menoceras arikarense* in anterior and dorsal views (after Peterson, 1920).

faces, is fused over in all rhinos. Although *Metamynodon* still shows the anterior end of this canal as a shallow closed pocket, it is no more than a shallow indentation in all higher rhinos. Although many other shallow pits are present where foramina might once have been, they are fused over in all rhinos.

Most rhinocerotid atlases are very difficult to tell apart except by size. But post-Barstovian *Teleoceras* are marked by having very widely expanded, hatchet-shaped transverse processes, whereas even the most derived *Aphelops mutilus* retains the primitive triangular shape of the transverse processes. *T. proterum* is particularly distinct in having exceptionally long, slender transverse processes. Correlated with this change is the shallowing of

the atlantodiapophyseal notch in *Teleoceras*. In *A. mutilus,* on the other hand, this notch becomes even more deeply incised into the transverse processes.

Axis (Fig. 5.3): Rhinoceros axes all tend to have blunt cylindrical odontoid processes and broad, flaring dorsoventrally compressed anterior articular surfaces. The latter reflects the dorsoventral compression of the posterior articular surface of the atlas. The neural arch and spinous process are usually broken in rhinos, but the dorsal margin tends to slope strongly anteriorly. In most ceratomorphs except *Trigonias* and *Metamynodon*, the dorsal margin of the spinous process is straight. The transverse processes in rhinos are generally quite short, with a foramen trans-

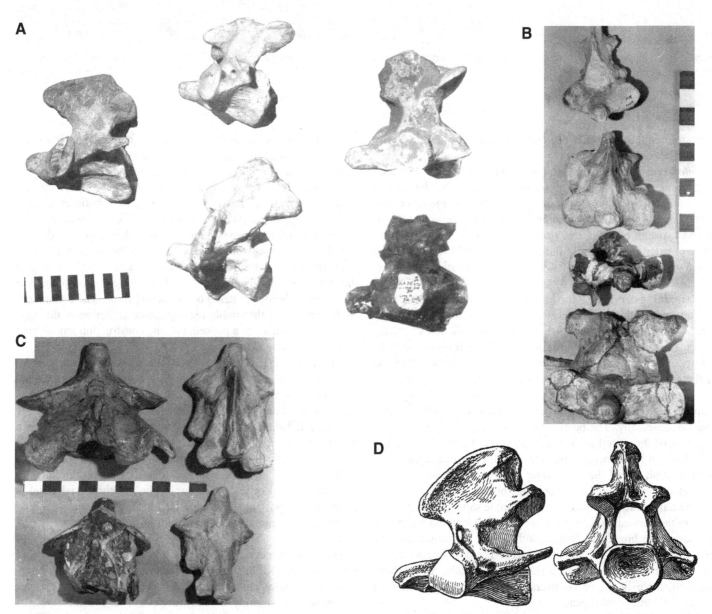

Figure 5.3. Rhinoceros axes. A. Left lateral view of *Aphelops mutilus* (F:AM 114900) (top right), *Teleoceras fossiger* (F:AM 104051) (top center), *T. medicornutum* (F:AM 115297)(bottom center), *T. hicksi* (F:AM 116094), and *T. proterum* (F:AM 116095) (bottom right). B. Anterior view. C. Dorsal view. D. Axis of *Menoceras arikarense* (after Peterson, 1920). Scale bar in A in cm; B and C in 2 cm increments.

Figure 5.4. Third cervical vertebrae of *Menoceras barbouri* from Bridgeport Quarries (unnumbered UNSM specimens) in three views. Scale bar in cm.

versarium, although this is difficult to interpret because they are usually broken.

Teleoceras and *Aphelops* axes differ in slight details. The transverse processes of the axis of *Teleoceras* tend to be shorter and more pointed than those in *Aphelops*. The anterior process of the neural arch is usually reduced and occasionally bifurcate. More derived species of *Teleoceras* (especially *T. fossiger* and *T. hicksi*) have broader, more laterally flaring postzygapophyses, unlike the narrower condition seen in *Aphelops, Peraceras*, and *T. medicornutum*.

Remaining cervicals: The **third cervical** (Fig. 5.4) is very similar to the fourth and fifth cervical in most features. There is a broad centrum with a prominent ventral keel, a convex anterior articulation and a concave posterior articulation (opisthocoelous). The neural canal is bridged by a broad, low neural arch, which has a flattened dorsal roof. Through the lateral base of the pedicel of the neural arch passes the vertebrarterial canal. The robust prezygapophyses flare widely laterally, with their articular surfaces facing dorsomedially. The postzygapophyses are more in line with the roof of the neural arch and do not have a wide lateral flare. They extend posteriorly beyond the centrum to articulate with the prezygapophyses of the succeeding vertebra. Their articular surfaces face ventrolaterally, and are quite broad.

The transverse processes and neural spine show the most variation in the cervical vertebrae. In the third cervical, the transverse processes are broad and flaring, with a long finger-like anterior process that extends halfway up the axis. The posterior branch of the transverse process is much shorter and broader. In the third cervical, there is almost no neural spine. A small vascularized ridge is present in most genera.

The **fourth cervical** (Fig. 5.5A–B) has the same long finger-like anterior branch of the transverse process as the third cervical, but the posterior branch is much longer and more posteriorly extended. The roof of the neural canal is slightly shorter, with part of the anteroventral centrum exposed. The neural spine is usually no more than a strong ridge or occasionally a prominent knob.

By the **fifth cervical** (Fig. 5.5C), the transverse process is one broad ventrally deflected flange, with no separate anterior and posterior branches. The neural spine is now a noticeable ridge with a triangular profile sloping anteriorly. As a consequence, the dorsal roof of the neural arch is subdivided into two distinct trough-like areas on either side, rather than the flat surfaces seen in the third or fourth cervical. There is also a further shortening of the anteroposterior length of the roof, resulting in a deep notch in the posterior border centered on the neural spine. The neural arch is changing from the simple rectangular construction of the third and fourth cervicals to a peaked, curving construction reminiscent of an oriental pagoda.

The **sixth cervical** (Fig. 5.6A–B) has an even more prominent neural spine than the fifth, although not yet a long process as shown by Peterson (1920, fig. 20). It is particularly distinctive in having both a lateral and ventral lamella on the transverse process. The ventral lamella of the transverse process (neural arch) is a broad flange that is broken in most specimens. Where complete, it is usually arcuate or hatchet-like in shape. There is also a separate knob-like lateral transverse process that extends posterolaterally form the pedicel of the neural arch. The vertebrarterial canal is much larger, and laterally compressed. It lies immediately lateral to the centrum, rather than dorsolaterally, as in more anterior cervicals.

The **seventh cervical** (Fig. 5.6C) has an even longer neural spine, and the roof of the neural arch is much shorter anterioposteriorly and more saddle-shaped. As a consequence, the prezygapophyses are strongly dorsally deflected, as are the postzygapophyses, which merge with the base of the neural spine. There is no ventral lamella of the transverse process. The lateral lamella is short and dorsoventrally flattened. There is no vertebrarterial canal, unlike the condition in other cervicals.

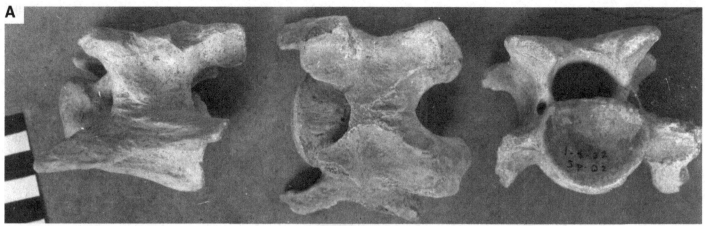

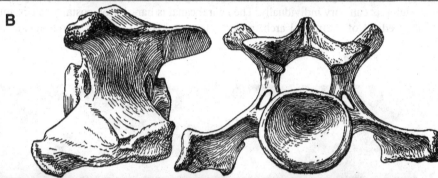

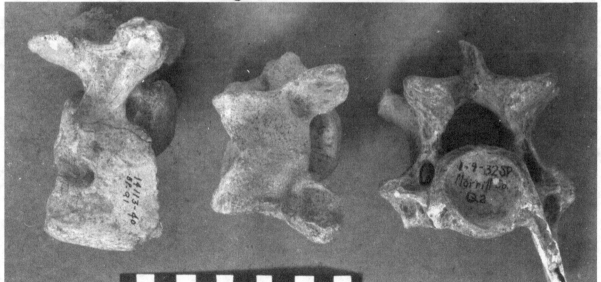

Figure 5.5. A. Fourth cervical of *Menoceras barbouri* (unnumbered UNSM specimens from Bridgeport Quarries). B. Fourth cervical of *M. arikarense* (after Peterson, 1920). C. Fifth cervical of *Menoceras barbouri* (unnumbered UNSM specimens from Bridgeport Quarries). Scale bar in cm.

Thoracic vertebrae: The number of thoracic vertebrae has been a point of uncertainty. Peterson (1920) thought that *M. arikarense* had 19, since the articulated type specimen of *Diceratherium tridactylum* (Fig. 4.15A) has 19. However, the Bridgeport *M. barbouri* sample contains only 18 identifiable elements, although the eighteenth and nineteenth thoracics are so similar that they cannot be distinguished. In articulated skeletons of higher rhinoceroses (such as *Teleoceras*), 19 thoracic vertebrae are known.

The **first thoracic** (Fig. 5.7A–B) has a very long neural spine which is prominent in any mounted or articulated specimen, because it is so much taller than those of the cervical vertebrae (to form an attachment point for the nuchal ligament and the main neck retractor muscles). The neural arch is consequently much modified. It has a very short, thick pedicel, which is emarginated posteriorly, so that the posterior roof of the neural arch overhangs this centrum. The postzygapophyses are no longer separate, but instead are flaring facets at the base of the neural spine. The prezygapophyses, however, remain as distinct knob-like protrusions that flare dorsolaterally. The transverse processes are very short, thick, and blunt. There is no vertebrarterial canal. The centrum is distinctive in that it is very dorsoventrally flattened, with only a weak ventral keel. There is a small concave costal facet just lateral to the posterior concavity of the centrum, and a large facet lateral to the anterior face of the centrum.

The **second thoracic** (Fig. 5.7C) is like the first in most details. The neural spine is slightly longer and more posteriorly curved. The prezygapophyses are now less prominent knobs, and the postzygapophyses are reduced to broad facets at the base of the neural arch. The transverse processes are even more knob-like, with a distinctive ventral costal facet. The posterior costal facet lateral to the centrum is also more distinct and concave. The anterior costal facet is large and prominent.

The **third thoracic** (Fig. 5.8A) usually has the longest neural spine, although this can vary individually. The neural canal is narrower as the pedicel of the neural arch becomes more massive.

The prezygapophyses are even less distinct from the rest of the neural arch and the postzygapophyseal facets, and rotated posteriorly. They form a notch in the posterior base of the neural spine as seen in lateral view. The anterior costal facet is deeper and more laterally directed. The posterior costal facets face almost directly posteriorly and are directly lateral to the flattened centrum. The costal facets at the base of the short stumpy transverse processes are weaker than in the first and second thoracics. The **fourth thoracic** is practically indistinguishable from the third, except that the neural spine is usually shorter.

The **fifth thoracic** (Fig. 5.8B) has an even shorter, more posteriorly deflected neural spine. As a consequence, the postzygapophyseal facet is much more ventrally deflected, forming broad, laterally flared facets that overhang the centrum. The anterior costal facet is much larger, as is the posterior costal facet. There is no longer a costal facet on the ventral surface of the trans-

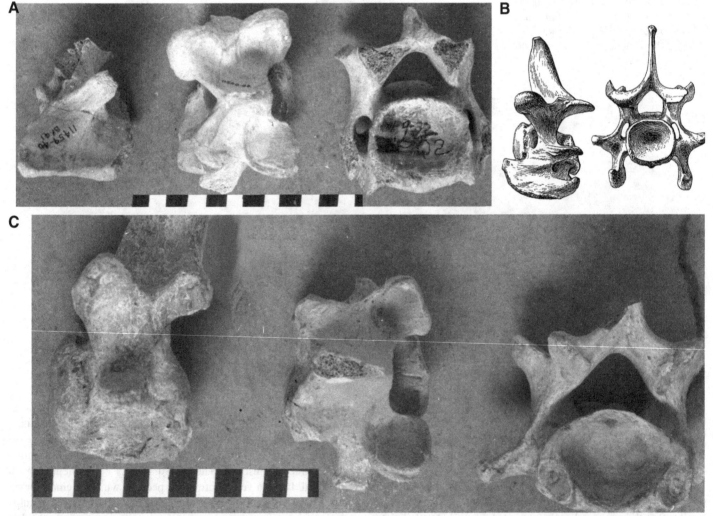

Figure 5.6. A-B. Sixth cervical vertebrae. A. *Menoceras barbouri* (unnumbered UNSM specimens from Bridgeport Quarries). B. *Menoceras arikarense* (from Peterson, 1920). C. Seventh cervical vertebra of *Menoceras barbouri* (unnumbered UNSM specimens from Bridgeport Quarries).Scale bar in cm.

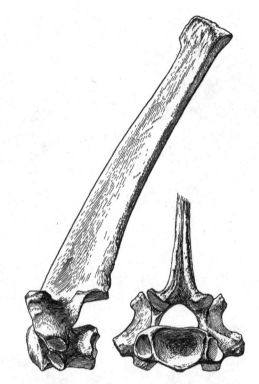

Figure 5.7. A. First thoracic of *Menoceras barbouri* (unnumbered UNSM specimens from Bridgeport Quarries). B. First thoracic of *M. arikarense* (after Peterson, 1920). C. Second thoracic of *Menoceras barbouri* (unnumbered UNSM specimens from Bridgeport Quarries). Scale bar in cm.

verse processes. A distinct intervertebral notch is present on the dorsolateral side of the centrum, posterior to the neural arch. The centrum is slightly more flattened than in the third or fourth thoracics.

The **sixth thoracic** has an even shorter and more posteriorly directed neural spine. As a consequence, the intervertebral notch is very distinct, forming a sulcus that runs along the dorsolateral surface of the posterior end of the centrum. The most striking feature is that the prezygapophyses have become pocketed into the base of the neural spine, with a ridge dividing them.

The **seventh thoracic** (Fig. 5.9A) continues the trend of shorter, more posteriorly directed neural spines. The prezygapophyses have almost disappeared into the transverse processes, and there are broad prezygapophyseal facets in the anterior base of the neural spine. The anterior and posterior costal facets are larger and deeper, as is the costal facet on the transverse process. The intervertebral canal is very sharp and distinct.

The **eighth thoracic** is very similar to the seventh in most features. It has a slightly shorter, more posteriorly deflected neural spine. It carries to an extreme the dorsoventral compression of the

centrum and neural canal. Further posteriorly, the centrum and neural canal begin to resume a more rounded shape.

In the **ninth thoracic** (Fig. 5.9B), the most striking difference is that the transverse process has shifted dorsally. It bears facets not only for the costae, but also has a deep anterodorsal sulcus. The transverse processes are also larger and more massive than in any of the more anterior vertebrae. The neural spine is shorter, and the prezygapophyses and postzygapophyses are completely incorporated into its base. The intervertebral canal is now a shallow notch in the posterior face of the pedicel of the neural arch. The anterior costal facet is a large, shallow, oval-shaped depression extending dorsally into the base of the transverse process. The posterior costal facet is smaller and shallower. From the ninth thoracic on, the neural canal and centrum become more rounded and less dorsoventrally flattened.

The **tenth thoracic** (Fig. 5.9C) differs from the ninth in that the neural spine shows a slight anterior curvature near the tip. There is a deep sulcus on the posterior margin of the spine, giving it a V-shaped cross-section. The prezygapophyseal and postzygapophyseal facets are almost invisible. The transverse processes continue

A

Figure 5.8. A. Third thoracic vertebra of *Menoceras barbouri* (unnumbered UNSM specimens from Bridgeport Quarries). B. Fifth thoracic of the same rhino. Scale bar in cm.

B

to be massive and dorsally inflected, with a distinct anterodorsal sulcus. The coastal facets of the transverse processes are laterally facing, but the anterior and posterior costal facets are very distinct and flare laterally from the centrum.

The **eleventh thoracic** (Fig. 5.9D) is the last that has a long, thin, posteriorly deflected neural spine. The spine is shorter than in any preceding vertebra. The transverse processes have become massive knobs. The prezygapophyseal and postzygapophyseal facets are still barely distinguishable at the base of the neural spine. The anterior and posterior costal facets are reduced in size compared to more anterior vertebrae. The articular surfaces of the centrum of preceding vertebrae become progressively less concavoconvex, and by the eleventh thoracic the articular surface of the centrum is practically flat.

The **twelfth thoracic** (Fig. 5.10A) has a highly variable neural spine. Some are still thin and posteriorly deflected, like those of the vertebrae anterior to it. Others are rectangular with no posterior deflection, as in the more posterior vertebrae. In most other regards, the twelfth thoracic resembles the eleventh thoracic: massive, knobby transverse processes; faint prezygapophyseal and postzygapophyseal facets; deep intervertebral canal; circular neural canal and centrum; flat anterior and posterior central articular surface.

The **thirteenth thoracic** (Fig. 5.10B) shows similar variations in the neural spine seen in the twelfth thoracic. The transverse processes are even more massive, lumpy, and dorsally deflected, with only a faint costal facet. The anterior and posterior costal facets are also becoming fainter. The intervertebral canal is very sharply demarcated along the posterior margin of the pedicel. The body of the centrum is small relative to the size of the neural arch.

The **fourteenth thoracic** (Fig. 5.10C) shows only a slight posterior deflection of the neural spine, and in many individuals it points vertically. The intervertebral canal has become a very deep groove that is almost closed by the overhang of the posterior rim of the neural arch. The transverse processes are massive, faceted knobs. The costal articulations are further reduced in size. The prezygapophyseal and postzygapophyseal facets are distinct but very small along the base of the neural spine.

The **fifteenth thoracic** is very similar to the fourteenth. The neural spine tends to be vertical in most examples, but there are still some specimens in which it is slightly deflected posteriorly. The transverse processes tend to be a little less massive than they are in the fourteenth thoracic.

By the **sixteenth thoracic** (Fig. 5.10D), the neural spine is a short, hatchet-shaped vertical rectangle. The postzygapophyseal facets are raised higher and pinched. This is reflected by a small anterodorsal knob on the transverse process which lies just lateral to the prezygapophyseal facet. The transverse processes are less massive, and the costal facets are smaller. The anterior and posterior costal facets on the centrum are very small, because these articulate with a "floating" rib. A small vertebrarterial foramen appears near the anterolateral base of the neural spine. The intervertebral notch forms a deep sulcus on the posterior margin of the pedicel, and in some individuals it closes to form a foramen.

The **seventeenth** thoracic is distinguished by having a neural spine that curves slightly anteriorly at the tip. The prezygapophyses and postzygapophyses are saddle-shaped and anteroposteriorly elongated. The transverse processes are reduced, and their costal articulations face slightly ventrally.

The **eighteenth thoracic** (and the nineteenth, if there are indeed

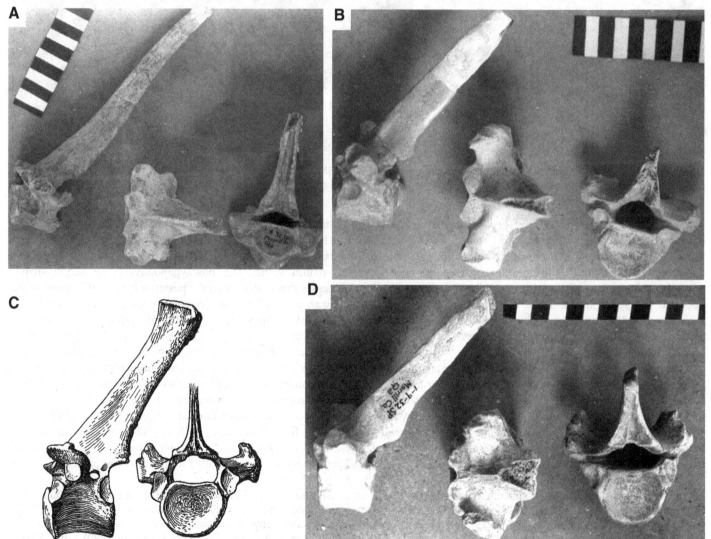

Figure 5.9. A. Seventh thoracic of *Menoceras barbouri* (unnumbered UNSM specimens from Bridgeport Quarries). B. Ninth thoracic of same taxon. C. Tenth thoracic of *M. arikarense* (after Peterson, 1920). D. Eleventh thoracic of *Menoceras barbouri* (unnumbered UNSM specimens from Bridgeport Quarries). Scale bar in cm.

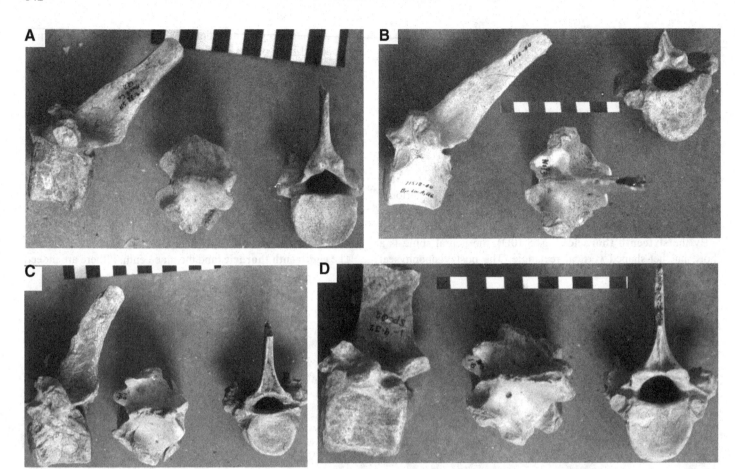

Figure 5.10. A. Twelfth thoracic of *Menoceras barbouri* (unnumbered UNSM specimens from Bridgeport Quarries). B. Thirteenth thoracic of same taxon. C. Fourteenth thoracic of same taxon. D. Sixteenth thoracic of same taxon. Scale bar in cm.

that many) has a neural spine (Fig. 5.11A, D) that is vertical and hatchet-shaped. In many individuals, it is topped by a ridge-like rugose knob of bone. This makes the neural spine resemble an upside-down keel with a stabilizing weight, as seen on a boat. The postzygapophyses form distinct paired saddles that fit into deeply pocketed prezygapophyses. There is a small knob-like process anterolateral to the prezygapophyseal sulcus. The transverse processes are reduced to a knob that supports the last rib. The costal facets on the centrum are very faint.

Ribs (Fig. 5.11B, C): The ribs of most rhinos are unremarkable. They taper gradually backward toward the rear. The ribs of some taxa (like *Teleoceras*) are much broader, heavier, and flatter than those of more gracile taxa, but they are still not true pachyostotic ribs of aquatic animals like sirenians. Some taxa, like *Trigonias*, have unusually long posterior ribs, giving them a "pot-bellied" appearance in the mounted skeleton (Fig. 4.7A).

Lumbar vertebrae: Most mammals, including rhinos, have five lumbar vertebrae. The **first lumbar** (Fig. 5.11E) is very much like the last thoracic, with a vertical (or slightly anteriorly deflected) neural spine. The postzygapophyses are saddle-shaped processes that project from the posterior base of the neural spine. They are matched by deep prezygapophyseal facets on the anterior base of the spine. Lateral to the prezygapophyseal facets are short stubby processes. There are no costal articulations, of course, but there are broad, flat, wing-like transverse processes. In most specimens, however, these are broken off. The centrum is very symmetrical and spool-like with a slightly concave anterior articulation and convex posterior articulation.

The **second and third lumbars** (Figs. 5.11F, G) are very similar to the first lumbar. They all have stubby, vertical neural arches, with well-marked prezygapophyses and postzygapophyses. They have wing-like, laterally flaring transverse processes that are

Figure 5.11. (opposite page) A. Eighteenth thoracic of *Menoceras barbouri* (unnumbered UNSM specimens from Bridgeport Quarries). Scale bar in cm. B. Detail of mounted skeleton of *Teleoceras fossiger* from Long Island Rhino Quarry, Kansas, in the AMNH, showing the configuration of the vertebrae, ribs, and sternum. C. Detail of the mounted skeleton of *Peraceras superciliosum* from UNSM locality Bn-10, showing details of the vertebrae, ribs, and sternum. D. Eighteenth thoracic of *M. arikarense* (after Peterson, 1920). Lumbar vertebrae. E. First lumbar of *Menoceras barbouri* (unnumbered UNSM specimens from Bridgeport Quarries). F. Second lumbar of *M. arikarense* (after Peterson, 1920). G. Third and fourth lumbars of *M. barbouri* (unnumbered UNSM specimens from Bridgeport Quarries). H. Fifth lumbars of *M. barbouri*. Scale bar in cm.

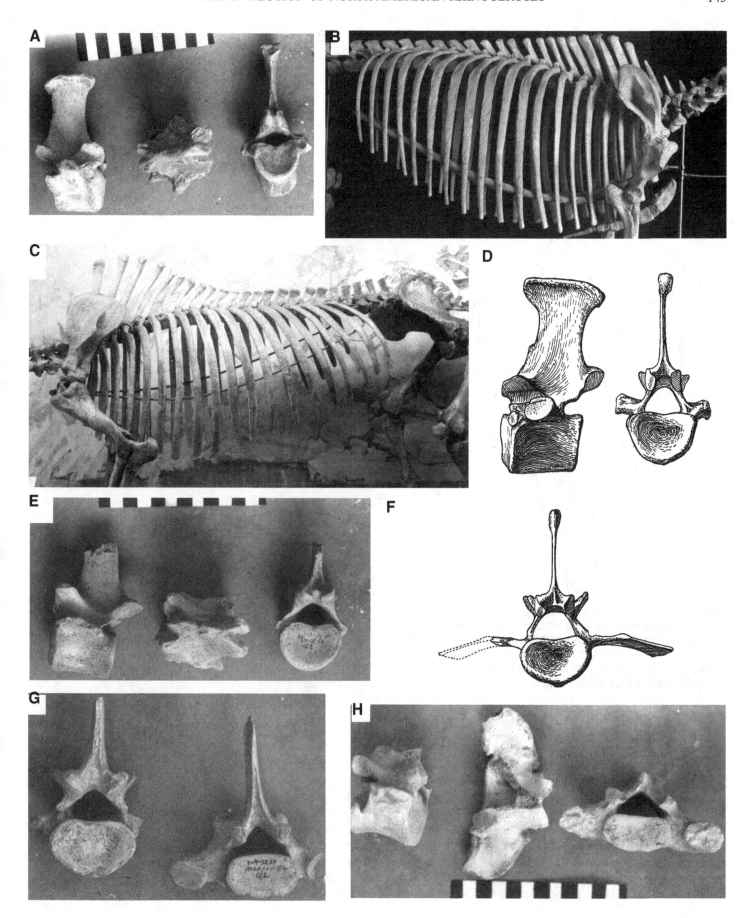

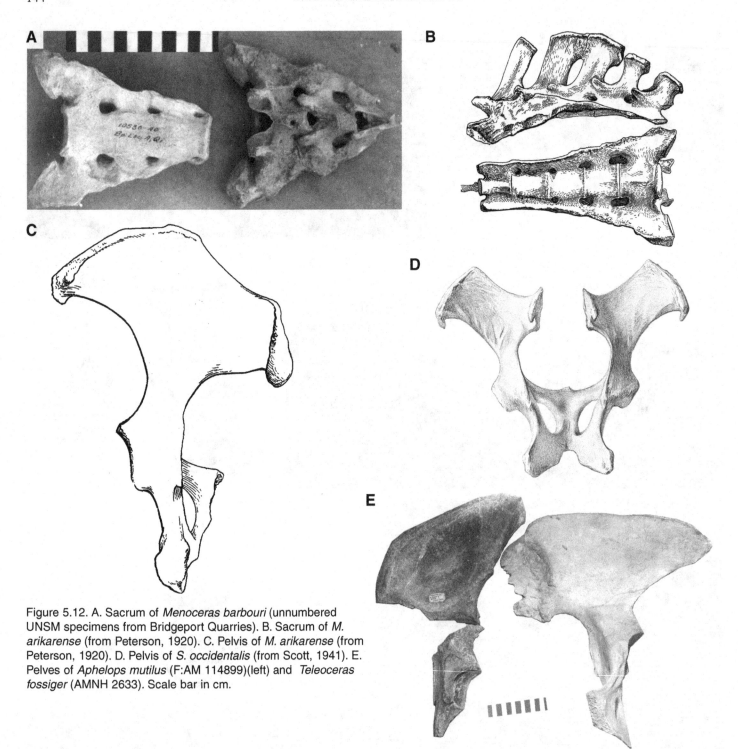

Figure 5.12. A. Sacrum of *Menoceras barbouri* (unnumbered UNSM specimens from Bridgeport Quarries). B. Sacrum of *M. arikarense* (from Peterson, 1920). C. Pelvis of *M. arikarense* (from Peterson, 1920). D. Pelvis of *S. occidentalis* (from Scott, 1941). E. Pelves of *Aphelops mutilus* (F:AM 114899)(left) and *Teleoceras fossiger* (AMNH 2633). Scale bar in cm.

usually broken off. The third lumbar is distinct in that its centrum is more dorsoventrally flattened. The intervertebral canal is very shallow.

The fourth lumbar (Fig. 5.11G) is very much like the third lumbar, except that the transverse processes are modified into a buttress that articulates with the fifth lumbar. The centrum is even more dorsoventrally flattened. The neural canal is no longer a flat-bottomed circle, but is actually slightly C-shaped, with the open end of the "C" ventral. The prezygapophyses and postzygapophy-

ses are much weaker, as is the process lateral to the prezygapophyses.

The fifth lumbar (Fig. 5.11H) is the most dorsoventrally flattened of the series, with a very oval-shaped centrum. The neural canal is nearly triangular in posterior view. The neural spine is a short vertical blade, and the prezygapophyses and postzygapophyses are very weak. The most striking feature is the robust, posteriorly curved transverse processes with extra articulations for the sacrum. On the anterior face of the transverse processes are facets

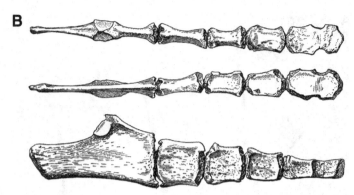

Figure 5.13. Caudals. A. *Menoceras barbouri* (unnumbered UNSM specimens from Bridgeport Quarries). Scale bar in cm. B. *M. arikarense* (from Peterson, 1920).

for the posterior transverse process facets of the fourth lumbar.

Sacrum and pelvis (Fig. 5.12): The pelvic material of most rhinos known from skeletons is usually badly crushed, so that interpretation of this part of the skeleton is difficult. The sacra of most rhinos have broad, anteriorly flaring iliac blades with a semicircular anterior margin. In *Subhyracodon* and *Diceratherium*, these do not flare as widely, and the anterior margin of the ilium is embayed between the lateral and medial edges. The ischial portion is much broader and more flaring in rhinos than in most other mammals. The iliac blade in *Teleoceras* is much broader than that of *Aphelops* or any more primitive rhino. The acetabulum is deeply concave, with a ventral cotyloid notch that leads to the obturator foramen.

Caudals (Fig. 5.13): Scott (1941) reports 23 caudal vertebrae in

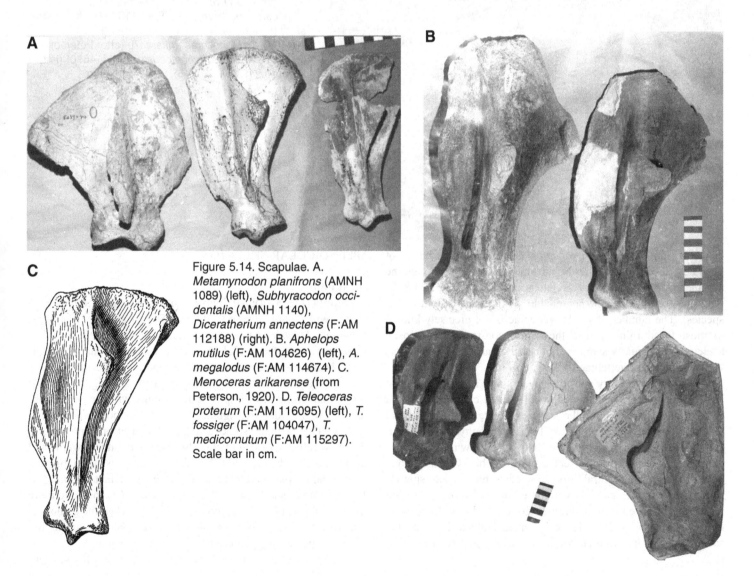

Figure 5.14. Scapulae. A. *Metamynodon planifrons* (AMNH 1089) (left), *Subhyracodon occidentalis* (AMNH 1140), *Diceratherium annectens* (F:AM 112188) (right). B. *Aphelops mutilus* (F:AM 104626) (left), *A. megalodus* (F:AM 114674). C. *Menoceras arikarense* (from Peterson, 1920). D. *Teleoceras proterum* (F:AM 116095) (left), *T. fossiger* (F:AM 104047), *T. medicornutum* (F:AM 115297). Scale bar in cm.

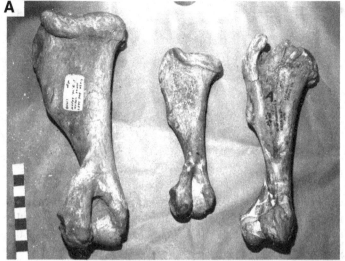

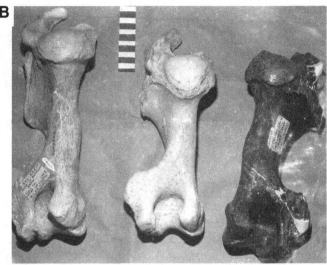

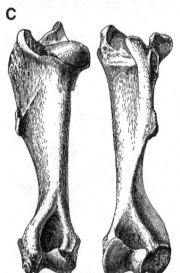

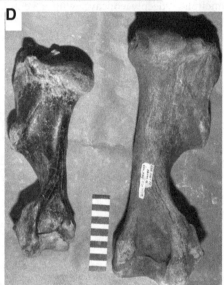

Figure 5.15. Humeri. A. Giant dicerathere (F:AM 111846) (left), *Diceratherium annectens* (F:AM 112188), and *D. armatum* (F:AM 112178). B. *Teleoceras medicornutum* (F:AM115297)(left), *T. fossiger* (F:AM 104047), and *T. proterum* (F:AM 104197). C. *Menoceras arikarense* (after Peterson, 1920). D. *Aphelops megalodus* (F:AM 114670) (left), *A. mutilus* (F:AM 114845). Scale bar in cm.

Trigonias, and 21 in *Subhyracodon*; the exact counts have not been described for any other North American rhino (although the articulated *Teleoceras major* skeletons at Ashfall Fossil Bed State Park in Nebraska have the potential to yield a caudal count in this species). The number of caudal vertebrae is not precisely known in most fossil rhinos, since they are reconstructed composite skeletons from quarry samples, not based on articulated skeletons with complete tail vertebrae.

Caudal vertebrae vary considerably in shape and size, but most have a dorsoventrally flattened spool-shaped centrum with flat anterior and posterior articular surfaces. There is a small neural arch and tiny prezygapophyses and postzygapophyses. The neural canal is reduced to a narrow channel only a few millimeters in diameter. Two broad ridges representing the bases of the hemal arch run along the ventral surface of the centrum. The spinous processes of the caudal vertebrae in *Subhyracodon* and *Diceratherium* are longer and more robust than those of *Trigonias* or more primitive rhinos. There is considerable fusion with the sacrals in the spinous processes, as well as on the lateral processes.

APPENDICULAR SKELETON

Front limb

Scapula (Fig. 5.14): All rhino scapulae are roughly triangular in shape, with more cursorial rhinos (*Subhyracodon, Trigonias, Aphelops, Menoceras*) having slender, parallel-sided scapulae. More graviportal forms (*Teleoceras*) have scapulae that are shorter and broader and very triangular in shape. The anterior border is gently convex in shape, and a the scapula has a straight posterior border that terminates abruptly at the juncture with the gently convex vertebral border. In *Trigonias*, the vertebral border is strongly convex. The scapular spine generally has a strong, posteriorly deflected tuber spinae, but this diminishes rapidly toward, and does not reach, the glenoid. The tuber spinae is usually large and bulbous and separated from the glenoid by a distinct channel. All rhino scapulae lack a distinct coracoid process. Compared to more primitive forms like *Metamynodon* and *Hyracodon*, the glenoid is relatively small in *Subhyracodon* and *Diceratherium* and higher rhinos. None has any glenoid notch, as seen in equids.

Aphelops scapulae have a long, high spine with a thick, triangular knob-like tuber spinae that is deflected posteriorly. The infraspinous fossa is slightly larger than the supraspinous fossa, and the posterior margin of the scapula has a rugose, thickened edge (particularly well developed in *A. mutilus*). The coracoid process is round and knob-like, and the glenoid fossa is ovoid in shape and shallowly concave.

By contrast, the scapula in *Teleoceras* is so short that it almost resembles an equilateral triangle. The scapula also gets significantly smaller in more derived species of *Teleoceras* as their limbs become shorter and stumpier. The scapular spine is much shorter and lower than in *Aphelops*, but the tuber spinae is a massive, rugose triangular knob of bone, sharply reflected posteriorly. The infraspinous fossa is significantly larger than the supraspinous fossa. There is no thickening in the posterior margin of the infraspinous fossa. A very faint ridge runs along the coracoid border of the supraspinous fossa. The coracoid process is angular and rugose, rather than the smooth ball-shaped feature of *A. mutilus*. The glenoid fossa resembles the condition seen in *Aphelops*.

Primitive ceratomorphs like *Heptodon* (Radinsky, 1965) had an acromion, but *Tapirus* and all more advanced perissodactyls have

lost it. This loss appears to be related to the loss of the clavicle (Radinsky, 1965, p. 84).

Humerus (Fig. 5.15): Rhinoceros humeri tend to be quite robust, with heavy articulations and processes, as would be expected of large-bodied mammals. The small cursorial forms such as *Menoceras*, *Penetrigonias,* and *Teletaceras* are exceptions, with relatively gracile humeri (where known). The deltopectoral crest is usually quite heavy and extends about a third of the way down the shaft (about halfway in graviportal forms like *Teleoceras*). The supinator crest flares out to the ectocondyle, which is broad and robust. The deltoid tuberosity is distinctly raised and reflected posteriorly. The lateral tuberosity is a long anteriomedially inflected flange, extending well above the head of the humerus. The posterior end of the lateral tuberosity is also a broad, thick flange, raised above the head. The medial tuberosity is also distinct from, and raised above, the head of the humerus. It is separated from the lateral tuberosity by a deep bicipital groove that passes down about a quarter of the shaft before disappearing.

The distal end of the humerus in rhinos frequently had an asymmetrical trochlea, with the medial condyle much stronger than the

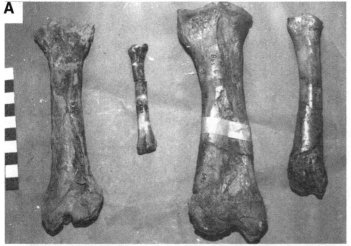

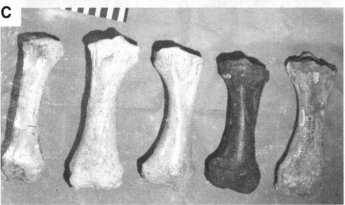

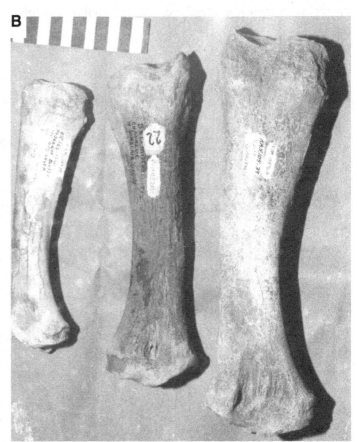

Figure 5.16. Radii. A. (From left to right) *Metamynodon planifrons* (AMNH 554), *Hyracodon nebraskensis* (AMNH 9789), *Amphicaenopus platycephalus* (AMNH 548), and *Subhyracodon occidentalis* (AMNH 1132). B. (From left to right) *Peraceras profectum* (F:AM 114970), *Aphelops megalodus* (F:AM 115203C), and *A. mutilus* (F:AM 114896A). C. (From left to right) *Teleoceras americanum* (AMNH 20461), *T. medicornutum* (F:AM 115297), *T. fossiger* (AMNH 2650), *T. proterum* (F:AM 104199), and *T. guymonense* (F:AM 116092). Scale bar in cm.

lateral condyle. The lateral condylar crest and lateral epicondyle are correspondingly stronger. This asymmetry is very marked in *Diceratherium* and higher rhinos, and less so in small gracile forms like *Menoceras*. The coronoid (olecranon) fossa is deep and long, with a strong medial tuberosity overhanging it slightly. The anconeal fossa is quite deep, and adjoins the strongly asymmetrical trochlea.

The humeri of *Teleoceras* are much shorter and more robust than those of any other North American rhino, with a wider proximal end and a broader ectocondyle, which becomes even broader in later species. The bicipital tubercle in *Teleoceras* is also much longer and more recurved, and surrounds a deep bicipital groove.

Radius (Fig. 5.16): The radius generally shows little variation among mammals other than in proportions. Graviportal animals tend to have much more robust radii than mediportal or cursorial mammals. Among the rhinos, this is especially true. *Teleoceras* has a very short, stout radius, with those of *T. proterum, T. hicksi,* and *T. guymonense* being the shortest and stoutest. *Trigonias* and *Subhyracodon* have more moderate proportions, while *Menoceras* has a slender radius. Surprisingly, the radius (and ulna) of *Diceratherium* is much longer and more slender for an animal of its size than was previously realized. It is considerably more long-

limbed (and presumably more cursorial) than *Subhyracodon, Menoceras,* or even the "long-limbed running rhinoceros," *Aphelops*.

Ulna (Fig. 5.17): Like the radius, the proportions of the ulna also reflect the degree of cursoriality or graviportality of the limb. Thus, the ulna of *Teleoceras* is stout, with a thick triangular cross-section and a broad distal end. *Trigonias, Subhyracodon,* and most higher rhinos tend to be more generalized in proportions. Cursorial *Diceratherium* and *Menoceras* have very slender ulnae, with a very small distal articulation and a tendency toward radial-ulnar fusion (seen in fully cursorial mammals). All rhinos have fairly massive, rugose olecranons, with a rectangular outline in lateral view. The semilunar notch is triangular in proximal view, with a prominent overhanging anconeal process. The proximolateral and proximomedial articulations for the radius flare widely, with the lateral side being the larger of the two. There is a distinct pit-like radial notch between them.

Manus

The manus is now known for most of the genera of North American rhinocerotoids, but except for the descriptions of *Metamynodon, Hyracodon, Trigonias,* and *Subhyracodon* by Scott

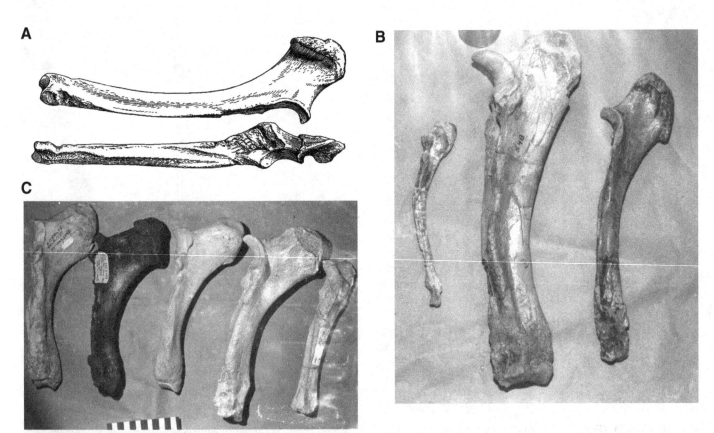

Figure 5.17. Ulnae. A. *Menoceras arikarense* in lateral and anterior views (after Peterson, 1920). B. (From left to right) *Peraceras profectum* (F:AM 114970), *Aphelops mutilus* (F:AM 114897A), *A. megalodus* (F:AM 115204C). C. (From left to right) *Teleoceras guymonense* (F:AM 116003), *T. proterum* (F:AM 104200), *T. fossiger* (F:AM 104504), *T. medicornutum* (F:AM 115297), *T. americanum* (F:AM 116096). Scale bar in cm.

(1941) and of *Menoceras arikarense* by Peterson (1920), there are few descriptions and even fewer useful illustrations of the elements of the manus and pes in the literature. Yet a survey of the fossil rhino collections in many museums has shown that there are often hundreds of rhino foot bones in the collections, which are usually not identified because no one is familiar with their shapes and systematics. The following descriptions and figures should make it possible for anyone to identify any rhino foot bone in any collection, and so add to the faunal lists of many localities. The terminology follows that of Peterson (1920), Scott (1941), Radinsky (1965), Yalden (1971), and Klaits (1972, 1973), except that I prefer to use the terms "medial" and "lateral" rather than "ectal" and "ental" for clarity, and "dorsal" (front surface of the hand) vs. "plantar" (palmar surface of the hand) rather than "anterior" and "posterior" to avoid ambiguity.

Figure 5.18 shows a series of articulated hands of different rhinoceros taxa. As can be seen, they are all fundamentally similar in the size and arrangement of all the bones, but differ largely in proportions. Most rhino hands are mediportal in proportions (*Trigonias, Subhyracodon, Diceratherium*), a few are gracile (*Menoceras*), others have tendencies toward graviportal proportions (*Metamynodon, Amphicaenopus*), and *Teleoceras* is extremely graviportal so that most elements are proximodistally compressed. Thus, the descriptions below will generalize about the typical shape of a given element within the Rhinocerotidae, and if possible, describe exceptions to the general rhinocerotoid morphology observed in some genera.

Scaphoid: The first row of carpals begins with the scaphoid bone at the radial (medial) end of the row, with the lunar bone in the middle of the row, and the cuneiform on the ulnar (lateral) end of the row. The scaphoid bone (Fig. 5.19) is easily recognized by its broad, massive roughly rectangular or trapezoidal shape in medial view, and its rugose surface on the medial exterior face that has no contact with the other carpal bones. It is also distinctive in having a deeply saddle-shaped triangular proximal facet for one of the trochleae of the radius. At the distal end, it has another smaller saddle-shaped facet for the trapezoid, adjacent to a smaller semicircular concave facet for the magnum toward the distal lateral side. Next to this is a small facet for the trapezium. On the lateral plantar face of the bone are three facets for articulation with the lunar. One is a broad dorsoplantar rectangular facet that adjoins the radial facet on the proximal surface. The second is a small raised process with a faceted tip for the plantar hook of the lunar (in most rhino taxa). The third is a broad, curved triangular facet on the plantar side of the mediodistal spur of the scaphoid; this facet is confluent with the magnum facet on the distal side of the bone.

As can be seen in Figure 5.19, the scaphoid is very consistent in overall shape, but the proportions can vary considerably. In most mediportal taxa (*Trigonias, Subhyracodon, Diceratherium*), the bone is massive and trapezoidal in shape in medial view, with a broader distal end due to the mediodistal expansion of the process that contacts the lunar and magnum (visible on the dorsal surface), and thick knob of bone that extends in the medioplantar direction

of inside back part of the wrist, where there is no constraint of other wrist bones. However, in more gracile taxa like *Menoceras*, and in the tapirs (Radinsky, 1965, fig. 9), the bone is almost rectangular in medial view, and the two distal spurs of bone are less well developed. By contrast, the scaphoids in *Metamynodon* and *Teleoceras* are very proximodistally compressed, so that the bone has a very short proximodistal axis and a relatively long mediolateral axis.

Lunar (Fig. 5.20): In the middle of the proximal row of carpals is the lunar, which is easily recognized because of its half-moon (hence the name) or crescentic shape in lateral or medial view. The proximal surface of the lunar has a broad, convex facet for articulation with the distal end of the radius. In proximal view, this facet may be narrow and oval-shaped (as in *Menoceras*), or have a broader polygonal shape (as in *Trigonias* and *Metamynodon*). The lateral and medial surfaces on the dorsal side of the radial facet have small facets for the adjoining scaphoid and cuneiform. Extending in the dorsal and distal direction from this radial facet is the rugose dorsal exposure of the lunar, which is trapezoidal or triangular in shape, tapering in the distal direction away from the radial facet. The distal end of any lunar is very distinctive in having two broad, deeply concave facets which meet with a sharp ridge between them, somewhat resembling an arête separating two cirques in glacial geology. The mediodistal facet articulates with the proximal facet of the magnum; the larger mediolateral facet articulates with the proximal facet of the unciform. The most distinctive feature, however, is the long "hook" or volar process that extends in a plantar and distal direction from the body of the lunar. In most rhinos, it is thick but long, but in more graviportal forms like *Metamynodon* and *Teleoceras*, it is short and stubby. By comparison, it does not extend out much at all in tapiroids (Radinsky, 1965, fig. 10), so this is a striking contrast between rhinos and tapirs.

Cuneiform (Fig. 5.21): The lateral bone in the proximal row of carpals is the cuneiform (also called the pyramidal by some authors, such as Klaits, 1972, 1973). This bone can vary considerably in shape. At the proximal end is a broad, concave saddle-shaped facet that contacts the distal end of the ulna. This facet may be slightly triangular in proximal view, since it has a long extension on the plantar surface for the continuation of the ulnar contact. However, in graviportal rhinos like *Metamynodon* and *Teleoceras*, there is no plantar extension of the facet, and it is just a simple saddle in shape. At the distal end is another large concave or saddle-shaped facet that contacts the proximal facet of the unciform. There may also be a small facet adjoining this one on the medial side for articulation with the lunar. In dorsal view, the rugose exposure of the cuneiform is roughly triangular in shape in many rhinos (e.g., *Trigonias, Menoceras, Subhyracodon*), with the proximal "neck" of bone contacting the ulna, and the broad "base" of bone contacting the unciform. However, this same dorsal surface can be compressed into a narrow rectangle that is shorter in the proximodistal direction than it is in the mediolateral direction in graviportal taxa such as *Metamynodon* or *Teleoceras*. In many cuneiforms, there may also be a rugose knob-like process which

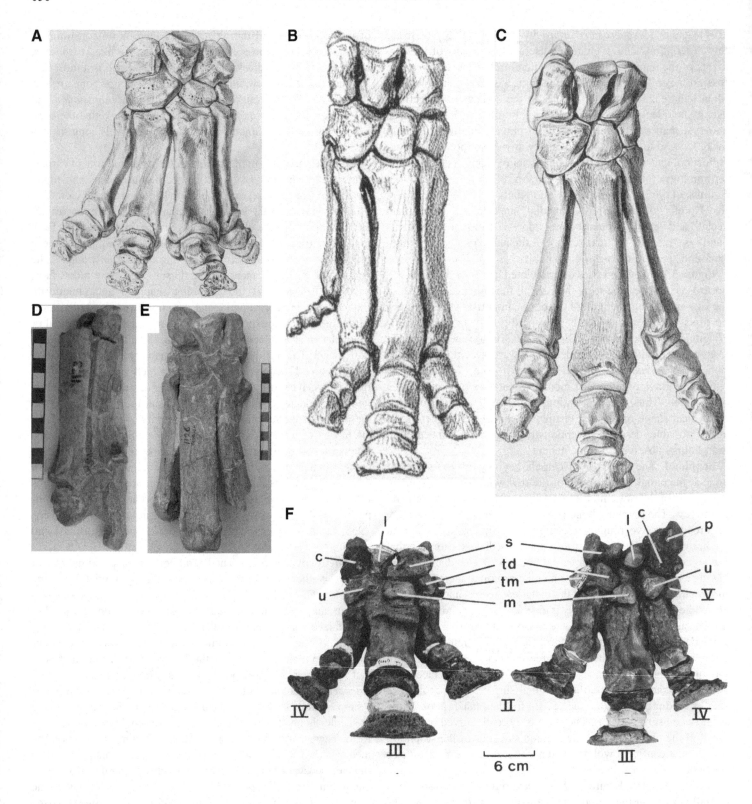

Figure 5.18. The manus in different genera of rhinocerotoids. A. *Metamynodon planifrons* (After Scott, 1941). B. *Trigonias osborni* (After Scott, 1941). C. *Subhyracodon occidentalis* (After Scott, 1941). D. Partial manus of *Penetrigonias dakotensis* (AMNH 1173). E. Partial manus of *Diceratherium tridactylum* (AMNH 1126). F. Dorsal (left) and plantar(right) views of articulated manus of *Teleoceras proterum*. The proximal carpals are the scaphoid (s), lunar (l), and cuneiform (c); the distal series is the trapezium (tm), trapezoid (td), magnum (m), and unciform (u). Metacarpals and phalanges II, III, and IV are numbered. (After Harrison and Manning, 1983, figure 1).

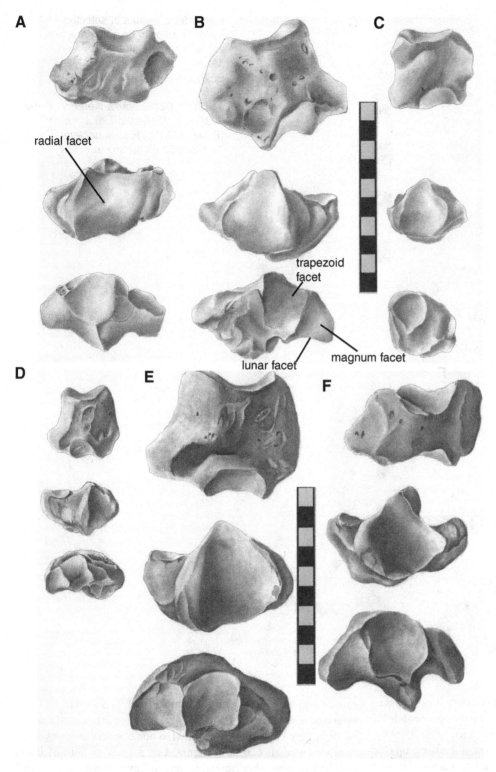

Figure 5.19. Scaphoids of selected rhinocerotoids in medial, proximal and distal views (top to bottom). A. *Metamynodon planifrons* (AMNH 1089). B. *Trigonias wellsi* (AMNH 13226). C. *Subhyracodon occidentalis* (AMNH 1132). D. *Menoceras arikarense* (F:AM 99260). E. *Diceratherium armatum* (F:AM) 112178. F. *D. annectens* (F:AM 112188). Scale bar in cm.

radial facet

trapezoid facet

lunar facet

magnum facet

extends in the lateral and plantar direction, unconstrained by articulation with other bones. It is particularly well marked in *Metamynodon, Trigonias,* or *Teleoceras,* but it is just a small ridge of bone in gracile forms like *Menoceras.* By contrast, the cuneiform of tapirs (Radinsky, 1965, fig. 11) is more trapezoidal in shape, with no "neck" on the dorsal exposure for the ulnar end of the bone, and a broad knob of bone on the lateral end.

Pisiform (Fig. 5.22): Although not preserved in many taxa, the complete carpus includes a distinctive teardrop-shaped bone known as the pisiform. It extends from its articulation with the plantar surfaces of the ulna and cuneiform, and then curves in a distal-plantar direction, comparable to the distal curvature of the plantar volar process on the lunar, magnum, and unciform. This is most apparent in the plantar or lateral view of an articulated car-

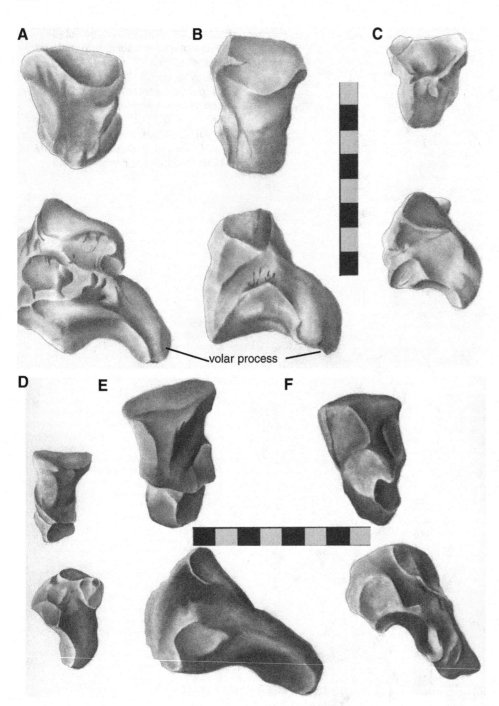

volar process

Figure 5.20. Lunars of selected rhinocerotoids in dorsal and lateral views (top and bottom). A. *Metamynodon planifrons* (AMNH 1089). B. *Trigonias wellsi* (AMNH 13226). C. *Subhyracodon occidentalis* (AMNH 1132). D. *Menoceras arikarense* (F:AM 99260). E. *Diceratherium armatum* (F:AM) 112178. F. *D. annectens* (F:AM 112188). Scale bar in cm.

pus (Fig. 5.23). In isolation (Fig. 5.22), the pisiform is easy to recognize. It looks like a flattened teardrop (flat in the proximodistal direction, broad in the mediolateral direction), with knobby, rugose surfaces on all sides except at the point of the teardrop (the dorsal end), where there are small triangular facets for the ulna and cuneiform. There is a small neck of bone separating the ulnar and cuneiformal facets (the "point" of the teardrop) from the broad flat body of the pisiform. The pisiform is not completely symmetrical, but has a slight tubercle on the medial side, presumably for attachment of ligaments that bound the pisiform to the rest of the carpus.

Trapezium (Fig. 5.23): The medial side of the second (distal)

row of carpals begins with the trapezium. It is so small that it is rarely preserved except on unusually complete articulated manuses, or in very large collections of wrist bones where every element is represented. It can be recognized as a small nubbin of bone, smaller than any other carpal element in a given taxon. It is shaped like an inflated triangle, which sits on the medial side of the trapezoid. On the lateral face of the trapezium is a small facet for articulation with the trapezoid.

Trapezoid (Fig. 5.24): Moving laterally, the next bone in the second row of carpals is the trapezoid. It has a very distinctive shape like an hourglass in medial view, since it has a large concave facet on the proximal surface for articulation with the scaphoid, and a similar large concave facet on the distal surface for articulation with Mc2. On the medial side there is a small flat facet for the trapezium, and on the lateral side is a much larger flat facet for the magnum. In most ceratomorphs (such as tapirs—Radinsky, 1965, Fig. 12) and graviportal forms such as *Teleoceras* (Fig. 5.18), the trapezoid is almost rectangular in shape in medial view. However, in many rhinocerotids, such as *Trigonias* and *Menoceras*, it is very strongly hourglass-shaped, with a dramatic constriction of the middle of the body of the bone between the proximal scaphoid facet and the distal Mc2 facet, with the dorsal and plantar surfaces bulging out at each end.

Magnum (Fig. 5.25): The middle bone in the second (distal) row of carpals is the magnum. Like the lunar, it tends to have a distinctive crescentic shape in medial or lateral view with a hook-like volar process that curves distally from the plantar side of the bone. On the distal side it has two large rectangular or trapezoid-shaped concave facets for the Mc2 and Mc3, separated by a sharp ridge. The Mc3 facet is much larger, since the magnum is the main weight-bearing element articulating with the Mc3, which is the largest of the metacarpals. The proximal side of the magnum is marked by a large knob of bone with convex curved facets along the top for the scaphoid and lunar, again separated by a ridge.

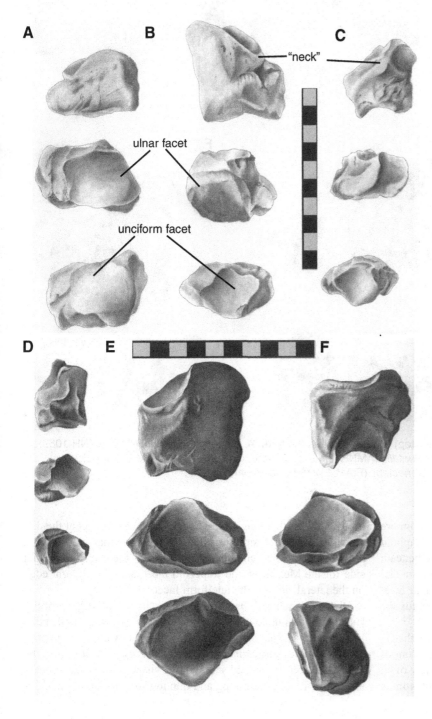

Figure 5.21. Cuneiforms of selected rhinocerotoids in dorsal, proximal, and distal views (from top to bottom). A. *Metamynodon planifrons* (AMNH 1089). B. *Trigonias wellsi* (AMNH 13226). C. *Subhyracodon occidentalis* (AMNH 1132). D. *Menoceras arikarense* (F:AM 99260). E. *Diceratherium armatum* (F:AM 112178). F. *D. annectens* (F:AM 112188). Scale bar in cm.

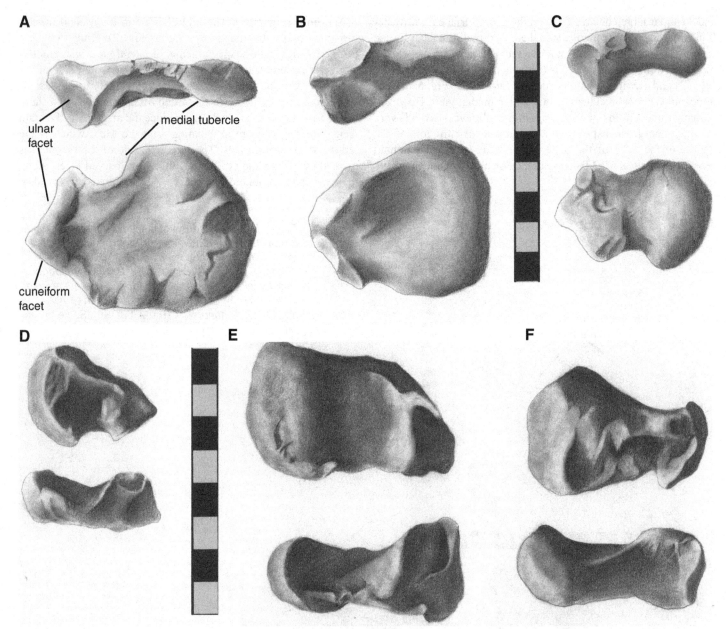

Figure 5.22. Pisiforms of selected rhinocerotoids in medial (top) and distal views (bottom). A. *Metamynodon planifrons* (AMNH 1089). B. *Trigonias wellsi* (AMNH 13226). C. *Subhyracodon occidentalis* (AMNH 1132). D. *Menoceras arikarense* (F:AM 99260). E. *Diceratherium armatum* (F:AM) 112178. F. *Diceratherium annectens* (F:AM 112188). Scale bar = 5 cm.

Unlike the proximal facets on the lunar, those of the magnum are on a distinct raised ridge of bone, and the edges of the scaphoid and lunar facets overhang the supporting buttress of bone beneath them. Between the scaphoid facet and Mc2 facet on the medial side is a small flat facet for the trapezoid. Between the lunar facet and Mc3 facet on the lateral side is a much larger facet for the unciform.

On the dorsal side of the magnum is the rugose area that is exposed between the surrounding wrist bones, facing the front of the wrist (and the only part that is visible on the dorsal view of an articulated carpus). It is typically an asymmetrical pentagon in dorsal view, with a long straight edge for the Mc3 facet at the distal side, a parallel straight edge on the proximal side for the scaphoid facet, two smaller edges between them on the medial side for the Mc2 facet and trapezoid facet, and one straight edge on the lateral side for the unciform facet.

On the plantar side is the knob-like "hook" or volar process. Usually it is a simple knob or cylinder that flexes slightly distally. However, as Harrison and Manning (1983) showed, in large population samples of magna (such as *Teleoceras proterum* from the F:AM Mixson's bone bed sample), the shape of the volar process is highly variable, from long and thin to short and stubby to com-

Figure 5.23. Articulated carpus of *Amphicaenopus platycephalus* (AMNH 12453) in (A) medial; and (B) dorsal views. Location of pisiform on plantar side of cuneiform is shown.

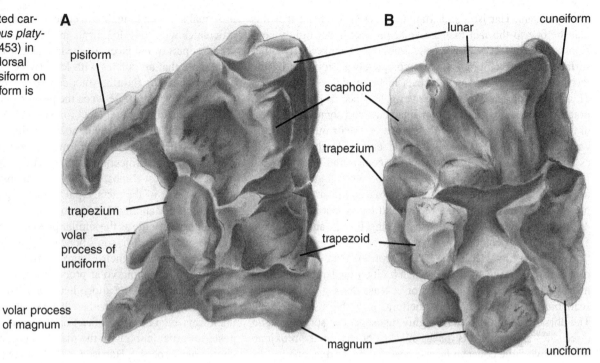

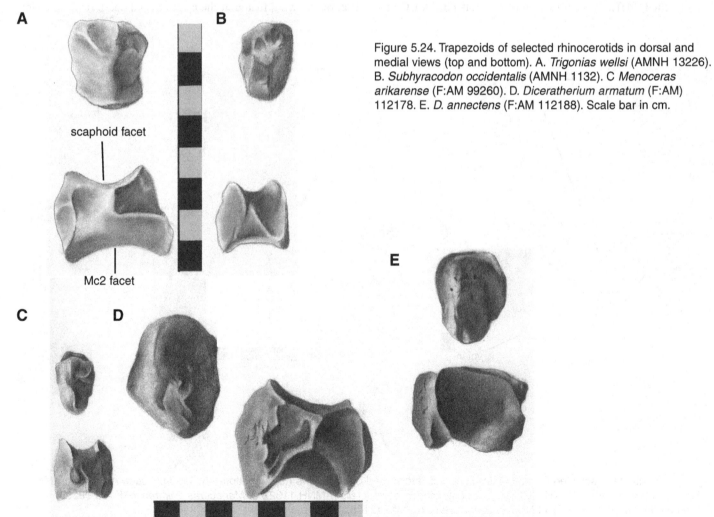

Figure 5.24. Trapezoids of selected rhinocerotids in dorsal and medial views (top and bottom). A. *Trigonias wellsi* (AMNH 13226). B. *Subhyracodon occidentalis* (AMNH 1132). C *Menoceras arikarense* (F:AM 99260). D. *Diceratherium armatum* (F:AM) 112178. E. *D. annectens* (F:AM 112188). Scale bar in cm.

pletely absent. Harrison and Manning (1983) point out that this may be due to the fact that the volar process is not tightly constrained by contact with adjacent foot bones (unlike every other surface of the magnum, which is completely surrounded by the rest of the carpals and metacarpals in the middle of the wrist). Thus, it is apparently much more variable, since its primary function is to connect with the flexor ligaments and some of the short muscles of the toes, but not for the major flexor muscles of the wrist (Yalden, 1971). According to Yalden (1971), these processes and their attached ligaments help to control the flexion of the manus, stabilizing it during hyperextension when the foot is pushing off at the most posterior point of its backward stroke. Similar processes are found in living hippos and tapirs, but are absent in most other ungulates, including sheep, pigs, horses, and elephants, and in giant indricothere rhinocerotoids.

Unciform (Fig. 5.26): The most lateral element of the second row of carpals is the unciform, which is either the largest and most massive of the carpal elements, or at least the same size as the scaphoid. In dorsal view, the unciform is highly asymmetrical. The dorsal face of the bone is highly rugose, and is surrounded on three sides by large facets for the adjacent bones. On the proximal side are two large concave facets separated by a ridge, one for the lunar (medial side) and one for the cuneiform (lateral side). On the

proximal side is a long, broad concavoconvexly curving facet that articulates with the Mc3 on its medial side, and with Mc4 in the broadest part of the facet. To the lateral side of this facet there is a small area that articulates with Mc5 (which is usually reduced to a nubbin in most North American rhinos except *Trigonias* and some aceratheriines). Between the proximal facet for the lunar and cuneiform, and the distal facet for Mc3–Mc4–Mc5 is a large facet on the medial side for articulation with the magnum.

Like the magnum, the unciform also has a plantar knob of bone, or volar process, projecting slighly distally from the main body of the bone. In most rhinos, it is a short, stubby rugose knob, which may have part of the metacarpal facet along its distal surface. As Harrison and Manning (1983) showed, however, this volar process is as highly variable as the similar process on the magnum of some rhinos, such as the Mixson's bone bed *Teleoceras proterum*. Some specimens shown in their Figure 6 have relatively long and slender or even branching volar processes, while others are short and stubby, or even lacking altogether. The most interesting feature of this carpal variability is the fact that a small number of *T. proterum* unciforms even develop a facet on the volar process for a *de novo* posterior articulation with the magnum (which has a similar facet on its volar process). Harrison and Manning (1983) argue that this is a trend toward fusion of the magnum and unciform, a phenom-

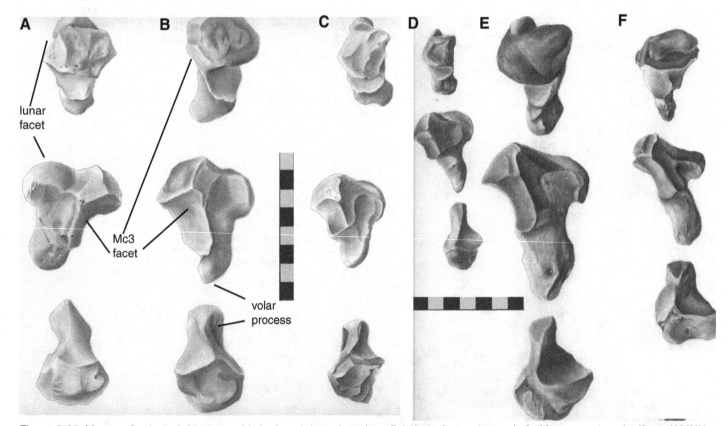

Figure 5.25. Magna of selected rhinocerotoids in dorsal, lateral, and medial views (top to bottom). A. *Metamynodon planifrons* (AMNH 1089). B. *Trigonias wellsi* (AMNH 13226). C. *Subhyracodon occidentalis* (AMNH 1132). D. *Menoceras arikarense* (F:AM 99260). E. *Diceratherium armatum* (F:AM) 112178. F. *D. annectens* (F:AM 112188). Scale bar in cm.

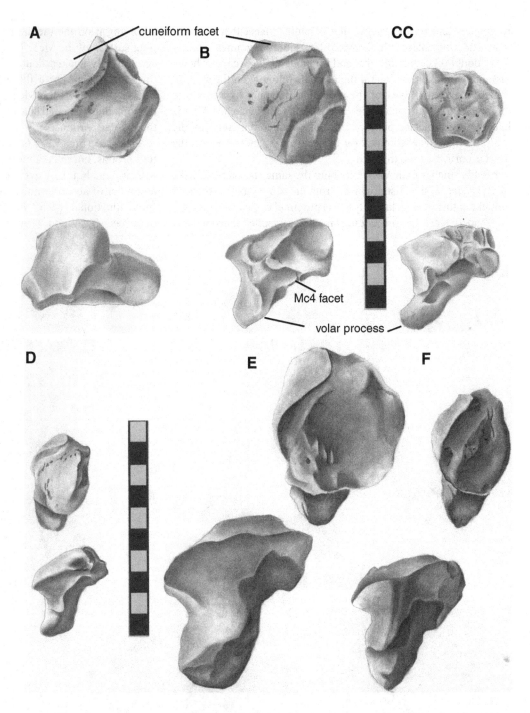

Figure 5.26. Unciforms of selected rhinocerotoids in dorsal and lateral views (top and bottom). A. *Metamynodon planifrons* (AMNH 1089). B. *Trigonias wellsi* (AMNH 13226). C. *Subhyracodon occidentalis* (AMNH 1132). D. *Menoceras arikarense* (F:AM 99260). E. *Diceratherium armatum* (F:AM 112178). F. *D. annectens* (F:AM 112188). Scale bar in cm.

enon that happens in many mammals with highly reduced and fused wrist elements.

Metacarpals (Fig. 5.27): All perissodactyls have lost Mc1, so the first metacarpal on the medial side is Mc2. Like all metacarpals, it has a large flat articular surface on the proximal end for the carpus, a long shaft which is flattened in the dorsoplantar direction and widest mediolaterally, with a distinct raised trochlea on the distal end which curves from the dorsal to the plantar side, and has a keel in the middle on that plantar side. **Mc2** can be recognized by its distinctive triangular proximal facet that articulates with the trapezoid and magnum. There is a small notch

on the plantar-lateral side of this triangular facet that allows the articulation of a small process on the adjacent Mc3. The other distinctive feature is the fact that the shaft of Mc2 tends to curve slightly medially at the distal end, so that the toes flare away from the central axis of the hand.

By contrast, the **Mc3** is the central, weight-bearing element of the hand, so its shaft is straight and more massive than that of the other metacarpals. Its large proximal facet is triangular but saddle-shaped (to articulate with the surface of the magnum), and adjoins another smaller raised facet on the proximolateral side that articulates with the unciform. Beneath this raised platform for the unciform facet and on the lateral side is another facet that overlaps with, and articulates with, the Mc4. This highly asymmetrical elevated double-faceted proximal end is unique to the Mc3, and helps distinguish it from the lateral metacarpals (along with the straight shaft). The medial apex of the triangular magnum facet fits into the notch on the articular facet of the Mc2. Otherwise, the Mc3 is like the other metacarpals, with a dorsoplantar flattening of the shaft, a broad distal trochlea for the phalanges, and a keel on the plantar portion of the trochlea.

Mc4 is smaller than Mc3, but about the same size as Mc2. Like Mc2, its shaft flares distally away from the central axis of the foot (which, in this case, is laterally). The proximal end has a large saddle-shaped facet for the unciform, with a smaller narrow facet adjoining it on the medial side for articulation with the overlapping surface of the Mc3. This distiguishes it from the Mc3 (which has a highly peaked double-faceted proximal surface), and from the Mc2 (which has a flat articular surface with a notch for the Mc3). The rest of the Mc4 is like the other metacarpals, with a dorsoplantar-flattened shaft, a broad distal trochlea with a keel on the plantar portion.

The **Mc5** (Fig. 5.28) is primitively present in many rhinocerotoid groups, but in taxa more derived than *Trigonias*, it is reduced to a tiny nubbin. However, as reviewed by Prothero *et al.* (1986), a number of aceratheriines appear to have secondarily regained a fully functional Mc5. The Mc5 of *Floridaceras whitei* was described by Wood (1964) and that of *Peraceras hessei* by

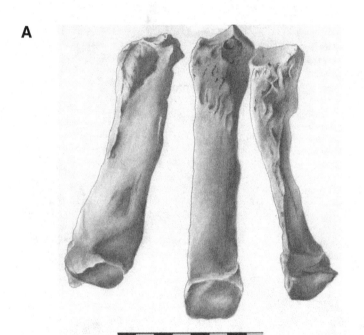

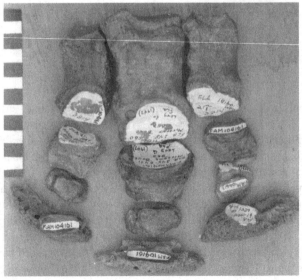

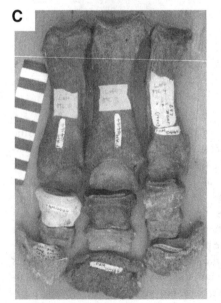

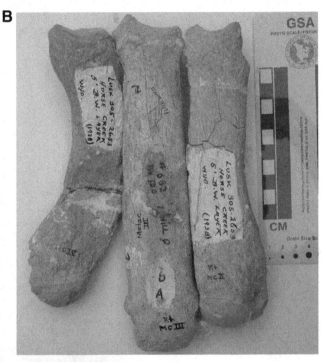

Figure 5.27. Metacarpals 2-4. A. *Amphicaenopus platy-cephalus* (AMNH 12453). B. *Diceratherium annectens* (F:AM 112188). C-D. Metacarpals and phalanges. C. *Aphelops malacorhinus* (F:AM 104164). D. *Teleoceras proterum* (F:AM 104163). Scale bar in cm.

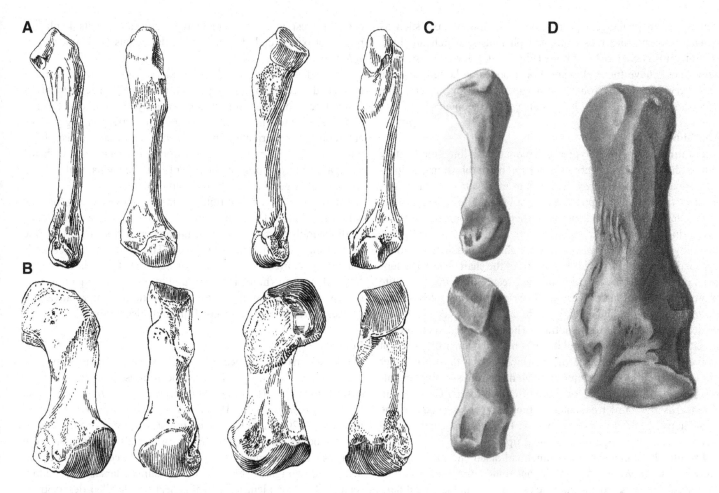

Figure 5.28. Fifth metacarpals. A. *Trigonias osborni* and B. *Floridaceras whitei* (after Wood, 1964). C. *Peraceras hessei* (after Prothero and Manning, 1987). D. *Aphelops megalodus*.

Prothero and Manning (1987). An Mc5 of *Aphelops* is shown in Figure 5.28D. In these taxa, the Mc5 tends to be much smaller than the remaining metacarpals, with a thinner shaft and small triangular proximal articular facet. It also tends to flare out sharply at its distal end from the axis of the hand.

Beyond these generalities, the individual genera vary largely in their metacarpal proportions. Those of graviportal forms such as *Teleoceras* are remarkably short, stubby, and broad, while those of cursorial and mediportal forms such as *Menoceras, Subhyracodon,* and *Diceratherium* are much longer and more slender. These size and ratios of shape are best demonstrated by the measurements in the tables at the end of this chapter.

Phalanges: Each finger in the rhinoceros manus has three phalanges (Fig. 5.27C, D). The **proximal phalanx** is the largest, with a deep concave facet on its proximal end to articulate with the trochlea of the metacarpal. It is cylindrical in shape, tapering to the distal end, with a broad flat distal articular surface for the second phalanx. On the plantar side of the proximal end are raised areas for attachment of the collateral ligaments. The sesamoidean ligaments probably inserted in the areas between them. There are tuberosities on the medial and lateral side of the plantar face of the distal end, which mark the insertion of the superficial digital flex-or tendon. On each side of the distal articular facet are pits surmounted by a low tuberosity that indicate the attachment of the collateral ligament and suspensory ligament of the navicular sesamoid.

The **second phalanx** is much smaller than the first, and consists of a small cylinder of bone with broad proximal and distal facets for articulation with the first and third phalanges. Low ridges at the proximal end of the plantar face serve for insertion of the tendo perforatus and the collateral ligaments on the lateral side. Shallow pits on each side of the distal condyle are for attachment of the collateral ligaments of the third phalanx.

The **third**, or **ungual**, **phalanx**, is a broad, flattened hoof-like structure that supports the keratinous hoof sheath of the hand. In most rhinos, it has a deep concave facet for articulation with the second phalanx, which tapers to a broad flattened tip that flares out from the body of the phalanx. In most rhinos, the central anterior edge of the phalanx has a fissure that helps anchor the keratinous hoof sheath, and in many rhinos, the leading edge of the ungual phalanx is jagged as well. The third phalanx of the middle digit (3) is nearly symmetrical, but those of the side digits (2 and 4) are asymmetrical away from the central axis of the hand.

As with the metacarpals, the principal differences between the genera are in phalangeal proportions. Graviportal taxa such as *Teleoceras* have the most flattened phalanges, which are little more than flattened disks of bone (Fig. 5.27C). By contrast, most rhino hands have more elongate finger bones, although they are still very broad and robust. But gracile running rhinos like *Menoceras* have relatively long phalanges in the manus.

Hindlimb

Femur (Fig. 5.29): Rhinoceros femora have a long, moderately robust shaft that is noticeably shorter and more robust in graviportal animals like *Teleoceras* than it is in more cursorial taxa. The head of the femur is supported by a robust buttress. The greater trochanter is broad, with a narrow posteromedial portion. In primitive rhinos, the surface of the greater trochanter is inclined anteromedially, but in *Subhyracodon* and higher rhinos it tends to be inclined more at right angles to the axis of the shaft. The ridge supporting the anteromedial face of the greater trochanter is quite prominent. Rhinos have a very weak lesser trochanter; it is a weak ridge running along the medial side of the shaft, and does not protrude much from the shaft outline. The third trochanter, on the other hand, is very strong, and its tip is reflected anteriorly (except in the Haughton astrobleme rhino, which lacks a third trochanter entirely). The third trochanter is much more robust in higher rhinos than it is in *Trigonias* or more primitive forms. However, in *Teleoceras* the third trochanter is progressively reduced in the more derived and graviportal species. In *T. fossiger* and *T. proterum*, it is a faint ridge with a rugose tip.

Primitively, rhinos have a symmetrical distal end of the femur. However, in *Subhyracodon* and higher rhinos, the medial half of the anterior patellar trochlea is much stronger than the lateral half. This seems correlated to the asymmetry seen in the distal humerus of rhinos. The posterior portions of the condyle show relatively little variation or asymmetry. The epicondyles are weak rugose bumps in all rhinos, which barely project beyond the profile of the shaft as seen in anterior view.

Patella (Fig. 5.30): The patella in rhinos has a highly rugose anterior surface, and an internal surface divided into a medial and lateral facet by a sharp vertical ridge. The triangular distal process is quite long in *Trigonias* and most higher rhinos, but less so in more primitive forms. All rhinocerotid patellae have a distinct medial groove that follows the contour of the two internal surfaces. The patellae of *Teleoceras* are distinctive in that they are more massive, transversely wider, and proximodistally shorter than those of *Aphelops* or other more cursorial rhinos.

Tibia (Fig. 5.31): The tibia, like the femur, is usually quite robust in rhinos, and noticeably even more robust in graviportal forms. The tibial trochleae are very symmetrical, except that the lateral trochlear spine is higher than the medial spine in all rhinocerotoids. The cnemial crest in rhinos has a broad, lumpy proximal termination that is inflected medially. The crest itself is not very distinct in shape and merges with the shaft at about a third of the way distally. However, this is a derived condition. In more primitive rhinocerotoids (*Hyrachyus, Hyracodon*), the cnemial crest was much more sharp and prominent. In *Teleoceras*, the

cnemial crest is robust and heavy, especially in its knob-like rugose proximal end. The intermuscular sulcus is also deeper in *Teleoceras*.

On the distal end of the tibia, the lateral and medial malleoli are broad and rugose, but do not project transversely. The distal articular surfaces in rhinocerotoids are generally less anteroposteriorly compressed; their intermediate ridges are correspondingly more anteroposteriorly oriented. In *Teleoceras*, the distal tibia is robust, with a slightly more symmetrical trochlear articulation with the astragalus. In *Aphelops*, on the other hand, the medial malleolus is deeper and larger than the lateral malleolus.

Fibula (Fig. 5.31): The fibula in most rhinos is a long, narrow split with flattened proximal and distal ends. It is also flattened mediolaterally as well. There is no fusion of the fibula with the tibia in most North American rhinos. The fibula of *Teleoceras* is noticeably shorter and more robust than that found in any other North American rhino. It is so durable that it is frequently preserved, and it frequently fuses with the tibia in more advanced species of *Teleoceras* and most Eurasian teleoceratines.

Pes

Like the manus, the pes (Fig. 5.32) is now known for most of the genera of North American rhinocerotoids, but except for the descriptions of *Metamynodon, Hyracodon, Trigonias,* and *Subhyracodon* by Scott (1941) and of *Menoceras arikarense* by Peterson (1920), there have been few descriptions of the individual foot bones. Also like the manus, the tarsals of rhinos are very common in many museum collections, but except for the calcaneum and astragalus, most of the remaining tarsals are seldom separated out or identified. In this section, detailed descriptions and illustrations of all the tarsals, metatarsals, and phalanges are provided to allow easy identification of any rhino foot bone in any collection.

Tarsals: Like the carpals, the tarsals of most rhinos are very similar and stereotyped except for their proportions. Most rhinocerotoids have relatively robust but slender metatarsals, but the graviportal forms like *Teleoceras* have extremely short and stubby metatarsals and compressed tarsals (Fig. 5.32J), and the gracile taxa like *Menoceras* have relatively long, slender metapodials.

Calcaneum (Fig. 5.33): The calcaneum is a very distinctive element, with a thick, stubby calcaneal tuberosity for the attachment of the Achilles tendon, and two large facets and one small facet on the dorsal distal surface for the three facets of the astragalus. The medial of these two facets is on the dorsal side of a distinctive medial process that emerges at right angles to the axis of the calcaneal tuberosity. At the very distal end, the calcaneum has a somewhat saddle-shaped facet for the articulation of the lateral proximal cuboid facet.

Most rhinocerotoids have very similar proportions in their calcanea except for the highly graviportal taxa such as *Teleoceras* (Fig. 5.33C), which have very short, robust calcanea with a short calcaneal tuberosity.

Astragalus (Fig. 5.34): Rhinoceros astragali are also very dis-

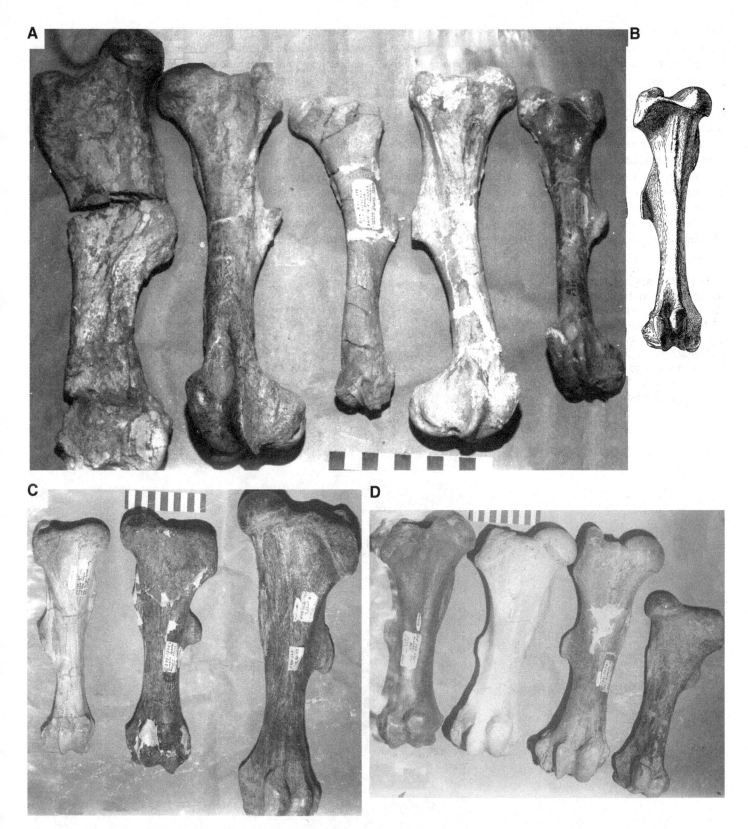

Figure 5.29. Rhinocerotoid femora. A. (from left to right) *Metamynodon planiforns* (AMNH 1088), *Trigonias wellsi* (AMNH 13226), *Subhyracodon occidentalis* (AMNH1132), *Diceratherium armatum* (F:AM 112178), *D. annectens* (F:AM 112188). B. *Menoceras arikarense* (after Peterson, 1920). C. *Peraceras profectum* (F:AM 114970), *Aphelops megalodus* (F:AM 114680), *A. mutilus* (F:AM 114980). D. *Teleoceras americanum* (F:AM 115028), *T. medicornutum* (F:AM 115134), *T. fossiger* (F:AM 104055), *T. proterum* (F:AM 104155). Scale bar in cm.

tinctive and easily recognized. As in all perissodactyls, in rhinos the neck of the astragalus is very short, with a nearly flat or gently saddle-shaped distal articular facet for articulation with the proximal facets of the cuboid and navicular. The astragalar trochleae is large and well defined, with sharp ridges on its dorsal side that fit into the grooves at the distal end of the tibia. The trochleae are nearly rectangular in dorsal view, with its axis aligned with the axis of the calcaneum. On the plantar surface is a large sustentacular facet for articulation with medial facet of the calcaneum, and a smaller ectal facet for articulation with the lateral process of the calcaneum. There is relatively little variation in the shape of astragalus, except for the graviportal forms like *Teleoceras*. In these taxa, the astragalus is highly compressed proximodistally so that it almost appears flattened (Fig. 5.32C).

Navicular (Fig. 5.35): The medial bone in the distal row of tarsal bones is the navicular. It is a very distinctive and easily recognized bone, since its shape is completely unlike any other carpal or tarsal. In rhinos, it is somewhat like a flat, compressed disk squashed between the main distal facet on the astragalus and the proximal facet of the ectocuneiform. However, the proximal facet (which articulates with the astragalus) is not flat, but concave and saddle-shaped. The distal facet (which articulates with the ectocuneiform) is slighly convexly curved. In proximal view (Fig. 5.36 bottom row), it is not circular in shape, but more like an irregular capital D, with the flat surface facing in the lateral direction and bearing a facet for articulation with the cuboid, and the convex surface facing medially. The dorsal surface on the exposures between the flattened facets is rugose, as is the exposed medial

Figure 5.30. Patellae of selected rhinocerotoids in dorsal (A) and plantar (B) view. Top row (from left to right): *Metamynodon planifrons* (AMNH 1088); *Trigonias wellsi* (AMNH 13226); *Subhyracodon occidentalis* (AMNH 1132). Bottom row (left to right): *Diceratherium armatum* (F:AM 112178); *D. annectens* (F:AM 112188). Scale bar in cm.

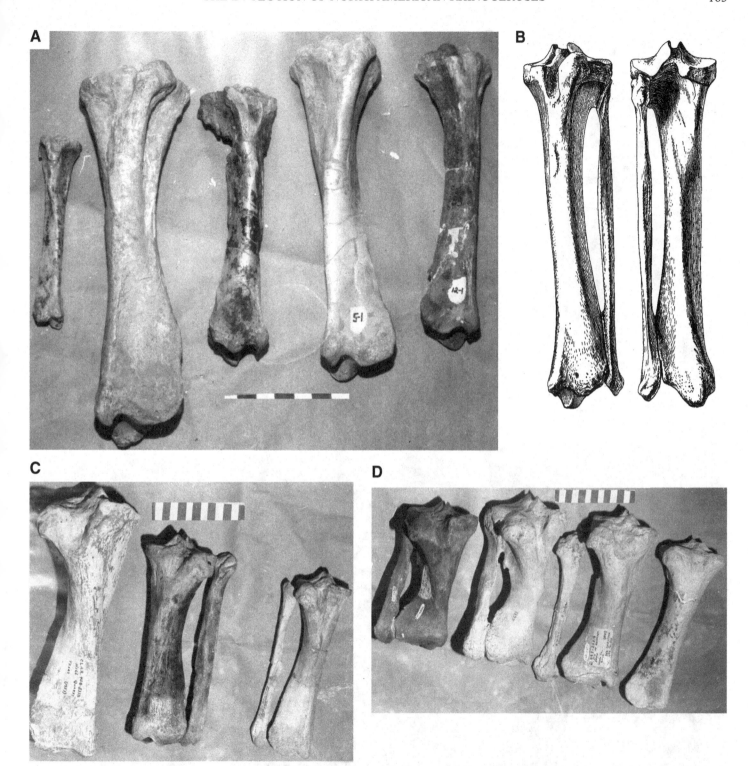

Figure 5.31. Tibiae and fibulae of selected rhinocerotoids. A. (from left to right) *Hyracodon nebraskensis* (AMNH 9289), *Trigonias wellsi* (AMNH 13226), *Subhyracodon occidentalis* (AMNH 1132), *Diceratherium armatum* (F:AM 112178), *D. annectens* (F:AM 112188). B. *Menoceras arikarense* (after Peterson, 1920). C. (from left to right) *Aphelops mutilus* (F:AM 114998), *A. megalodus* (F:AM 11482), *Peraceras profectum* (F:AM 114970). D. (from left to right) *Teleoceras proterum* (F:AM 104156), *T. fossiger* (AMNH 2637), *T. medicornutum* (F:AM 115297), *T. americanum* (AMNH 8923). Scale bar in cm.

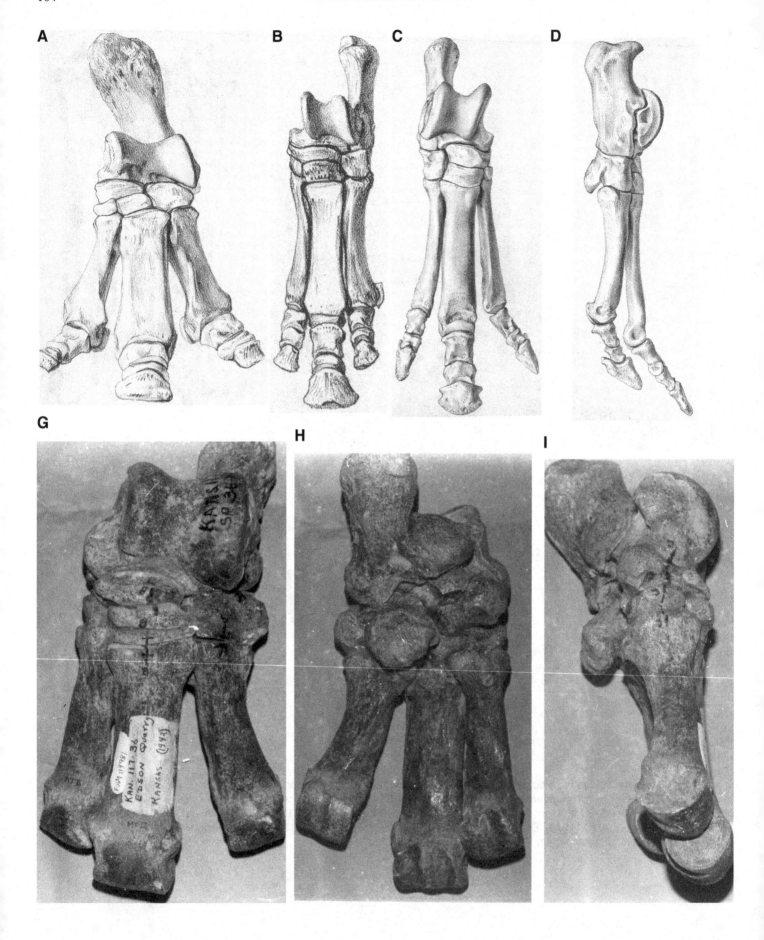

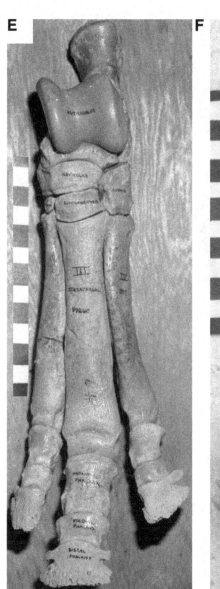

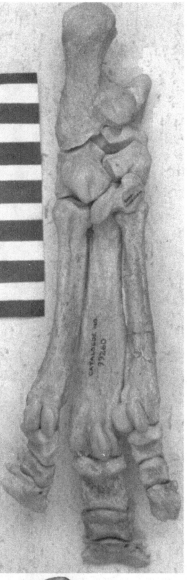

Figure 5.32. Pedes of selected rhinocerotoids. A. *Metamynodon planifrons*. B. *Trigonias osborni*. C–D. *Subhyracodon occidentalis* in dorsal (left) and lateral (right) views. (A-D after Scott, 1941). E–F. *Menoceras arikarense* (F:AM 99260) in dorsal (left) and plantar (right) views. G–I *Aphelopus mutilus* (F:AM 114981) in dorsal (left), plantar (middle), and lateral (right) views. J. Articulated pes of *Teleoceras fossiger* (AMNH 9745) and (K) *Aphelops megalodus* (AMNH 8293). Scale bar in cm.

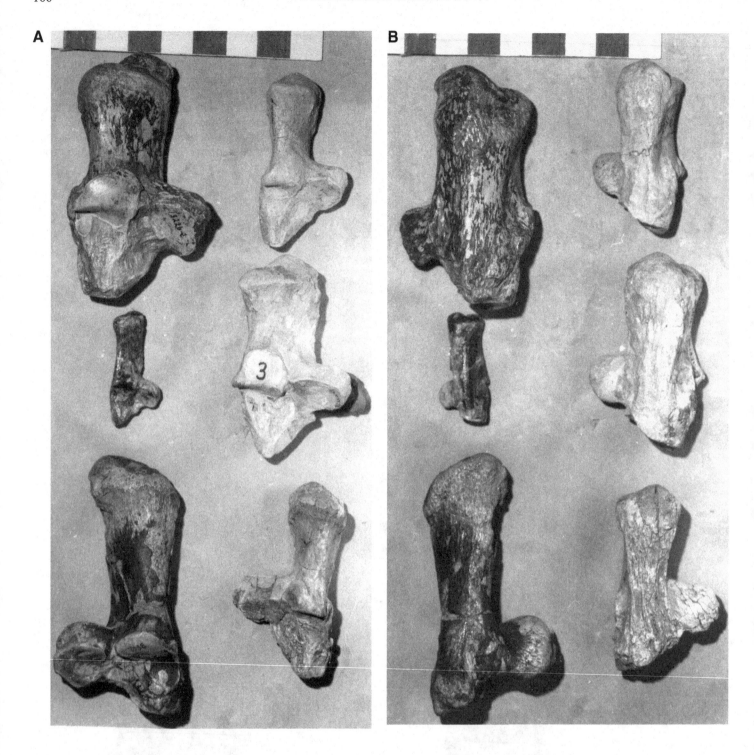

Figure 5.33. Calcanea of selected rhinocerotids in dorsal (A) and plantar (B) views. Left row, top to bottom: *Trigonias wellsi* (AMNH 13226), *Hyracodon nebraskensis* (AMNH 9289), *Metamynodon planifrons* (AMNH 1088). Right row, top to bottom: *Diceratherium annectens* (F:AM 112188), *D. armatum* (F:AM 112178), *Subhyracodon occidentalis* (F:AM 536). Scale bar in 2 cm increments.

A

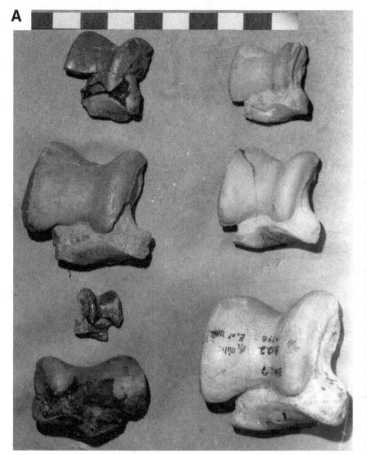

B

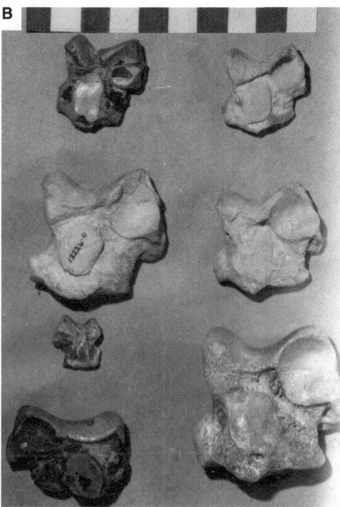

C

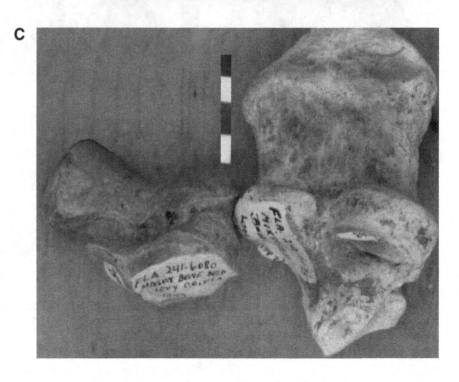

Figure 5.34. Astragali of selected rhinocerotoids in dorsal (A) and plantar (B) views. Left row, from top to bottom: *Subhyracodon occidentalis* (AMNH 536); *Trigonias wellsi* (AMNH 13226); *Hyracodon nebraskensis* (AMNH 9289); *Metamynodon planifrons* (AMNH 1089). Right row, from top to bottom: *Diceratherium annectens* (F:AM 112188); *D. armatum* (F:AM 112178); giant Arikareean astragalus (F:AM 111846). C. Astragalus and calcaneum of *Teleoceras proterum* (F:AM 104162). Scale bar in 2 cm increments.

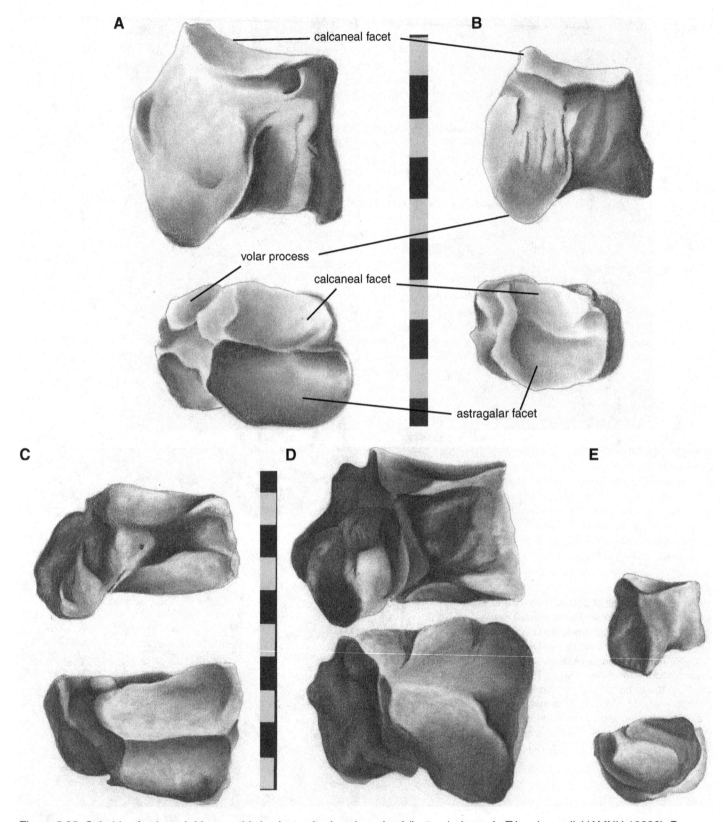

Figure 5.35. Cuboids of selected rhinocerotids in plantar (top) and proximal (bottom) views. A. *Trigonias wellsi* (AMNH 13226). B. *Subhyracodon occidentalis* (AMNH 1132). C. *Diceratherium annectens* (F:AM 112188). D. *Diceratherium armatum* (F:AM 112178). E. *Menoceras arikarense* (F:AM 99260). Scale bar in 1 cm increments.

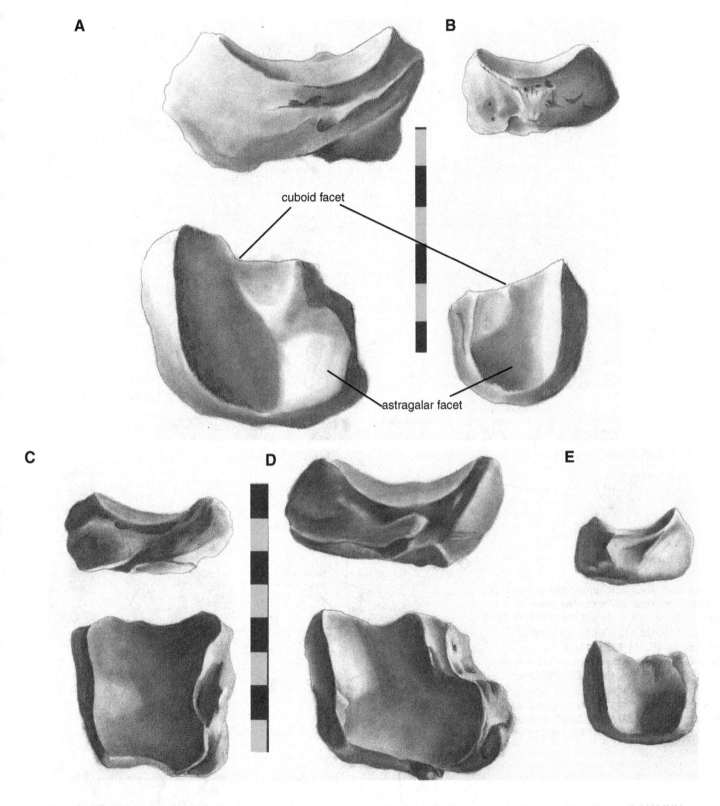

Figure 5.36. Naviculars of selected rhinocerotids in dorsal (top row) and proximal (bottom row) views. A. *Trigonias wellsi* (AMNH 13226). B. *Subhyracodon occidentalis* (AMNH 1132). C. *Diceratherium annectens* (F:AM 112188). D. *Diceratherium armatum* (F:AM 112178). E. *Menoceras arikarense* (F:AM 99260). Scale bar in 1 cm increments.

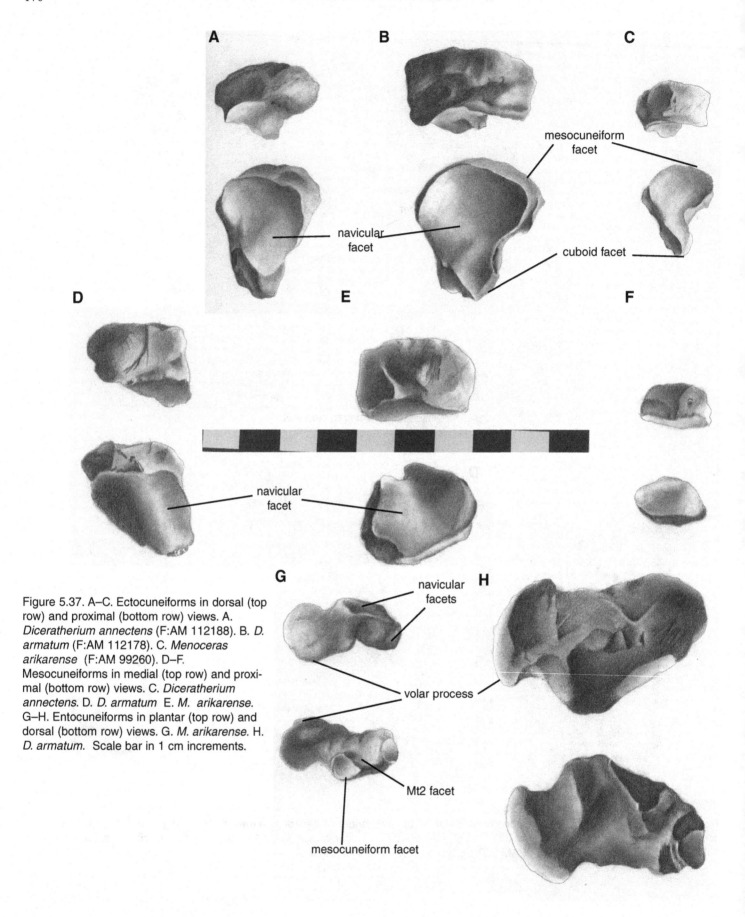

Figure 5.37. A–C. Ectocuneiforms in dorsal (top row) and proximal (bottom row) views. A. *Diceratherium annectens* (F:AM 112188). B. *D. armatum* (F:AM 112178). C. *Menoceras arikarense* (F:AM 99260). D–F. Mesocuneiforms in medial (top row) and proximal (bottom row) views. C. *Diceratherium annectens*. D. *D. armatum* E. *M. arikarense*. G–H. Entocuneiforms in plantar (top row) and dorsal (bottom row) views. G. *M. arikarense*. H. *D. armatum*. Scale bar in 1 cm increments.

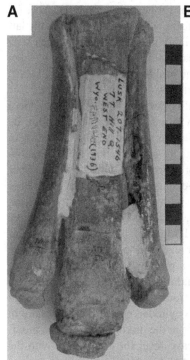

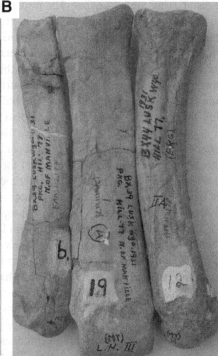

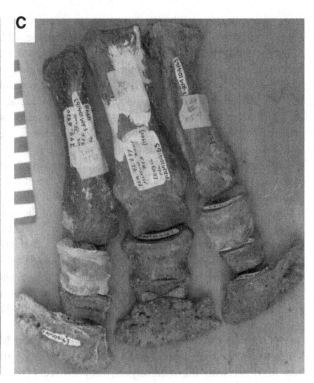

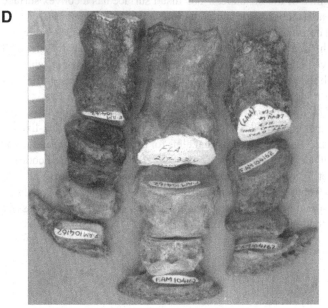

Figure 5.38.A–B. Metatarsals of (A) *Diceratherium annectens* (F:AM 112188) and (B) *D. armatum* (F:AM 112188). C–D. Metatarsals and phalanges. C. *Aphelops malacorhinus* (F:AM 104164). D. *Teleoceras proterum* (F:AM 104163). Scale bars in cm.

surface. However, the plantar surface of the bone is irregular, and in some taxa actually bears a small tuberosity that articulates with the plantar process of the cuboid.

Cuboid (Fig. 5.36): The distal row of tarsal bones starts with the cuboid on the lateral side. As the name implies, it is a cubically shaped bone in its main body, with large flat facets on the proximal end for the calcaneum and astragalus, and on the distal end for the Mt4. However it also has a massive triangular plantar process or "hook" which curves in a distal and plantar direction from the back of the main body of the cuboid. On the medial side of the bone is a small circular facet for the articulation with the navicular. Along the distal edge of the Mt4 facet is a smaller facet

for the articulation with the ectocuneiform.

The cuboid is very stereotyped in its overall shape, except that the cuboids of graviportal taxa such as *Teleoceras* are much flatter and more compressed, and less like the classic cubic shape of normal rhinocerotid cuboid bones.

Ectocuneiform (Fig. 5.37A–C): The distal surface of the navicular articulates with three small bones that form a third distal row of tarsals: the ectocuneiform, mesocuneiform, and entocuneiform. The ectocuneiform is a small flattened disk with a flat surface on the proximal side for articulation with the navicular, and a flat facet on the distal side for articulation with Mt3. In proximal or distal view, however, it is not a cylinder, but its outline is more like a crescent with rounded points in shape. On the distal medial corner is a small facet for articulation with the Mt2.

Mesocuneiform (Fig. 5.37D–F): The middle (and smallest) bone of the three in the third tarsal row is the mesocuneiform. It is a small, flat disk of bone, with flat surfaces on the proximal surface for articulation with the navicular, and on the distal surface for articulation with Mt2. It also has facets on the lateral side for articulation with the ectocuneiform, and on the medial side for articulation with the entocuneiform.

Entocuneiform (Fig. 5.37G–H): The most medial of the bones in the third row of metatarsals is the entocuneiform. It too is very distinctive, with a lateral and dorsal portion that consists of a flat disk of bone that articulates with the plantar side of the distal navicular facet, the plantar side of the mesocuneiform, and the plantar facet on the proximal end of the Mt2. Its most distinctive feature, however, is the long spur of bone (volar process) that curves in a plantar and lateral direction away from the ankle, and often articulates with the volar process of the cuboid. Presumably, these volar processes help in attaching the tendons that hold the ankle together, as in the carpus.

Metatarsals (Fig. 5.38): All perissodactyls have lost Mt1 and Mt5, so only three metatarsals are present in any rhino: Mt2, Mt3, and Mt4. As with the metacarpals, they are very similar and stereotyped in most of their features, and differ largely in size between taxa. The biggest differences are in proportion, with graviportal taxa like *Teleoceras* having very short, stubby metatarsals, while gracile taxa like *Menoceras* have relatively long, slender toes.

Mt2 is the shortest of the three metatarsals, and usually flexes slightly medially away from the midline of the foot at its distal end. It has a concave triangular-shaped facet on the proximal end for articulation with the mesocuneiform, and a small raised spur on the lateral edge of the proximal end for a small articulation with the ectocuneiform. The distal end bears a large rounded trochlea for articulation with the first phalanx, with a small medial keel on the plantar exposure of the trochlea.

Mt3 is the largest metatarsal, since it is the main weight-bear-ing toe in the foot. The proximal surface has a large flat facet for articulation with the ectocuneiform, with a raised ridge on the lateral side for a small facet (variably present) that articulates with the cuboid. As in the other metatarsals, the distal end bears a large trochlea for articulation with the middle phalanx, which is keeled as it approaches the plantar surface.

Mt4 is smaller than Mt3, but larger than Mt2. It tends to curve laterally at its distal end, away from the midline of the foot. Its proximal end has a concave facet for articulation with the main body of the cuboid, and secondary small plantar facet (variably present in different species) for articulation with the plantar spur of the cuboid. As in the other metatarsals, the distal end has a large trochlea for the fourth phalanx, with a keel on the plantar surface.

Phalanges: Like the manus, each digit of the pes bears three phalanges, which are very similar in overall topology. They differ between taxa primarily in their size, and in graviportal taxa, they are much more compressed in a proximal-distal direction that in other rhinos. The **proximal phalanges** of the foot are very much like those of the hand. They are roughly shaped like a flattened (in the anterior-posterior direction) cylinder, except that they flare in the proximal direction and taper in the distal direction. The proximal surface bears a deep concave facet for articulation with the trochlea of the metatarsal. The distal surface has a convex surface for articulation with the medial phalanx.

The **medial phalanx** is the smallest of the three phalanges on each toe. It is shaped like a flattened pillow or rectangle with curved edges. It bears a concave proximal surface for articulation with the proximal phalanx, and a convex distal surface for articulation with the distal phalanx.

The **distal**, or **ungual**, **phalanx** is the final toe bone. It is very distinct is its broad, hoof-like shape, with a rugose wedge-like tip which usually bears a medial cleft for attachment of the keratinous hoof sheath over the top. The proximal end bears a paired concave facet for articulation with the medial phalanx. There may also be small spurs of bone on the medial and lateral edges of the phalanx, presumably also for attachment surfaces of the hoof, or possibly for tendons that run up through the foot.

Table 5.1. Postcranial measurements of *Trigonias* and *Amphicaenopus* (in mm)

MEASUREMENT	T. osborni			T. wellsi			A. platycephalus		
	Mean	SD	N	Mean	SD	N	Mean	SD	N
Atlas length	104	—	1	95	5	2	—	—	—
Atlas width	199	—	1	255	12	2	—	—	—
Axis length	73	—	1	—	—	—	—	—	—
Axis anterior width	96	—	1	—	—	—	—	—	—
Axis posterior width	42	—	1	—	—	—	—	—	—
Scapula length	321	—	1	—	—	—	—	—	—
Scapula maximum width	188	—	1	—	—	—	—	—	—
Scapula neck width	69	—	1	—	—	—	—	—	—
Glenoid fossa width	56	—	1	—	—	—	—	—	—
Humerus length	280	—	1	—	—	—	454	—	1
Humerus midshaft width	98	—	1	—	—	—	124	—	1
Humerus distal width	60	—	1	—	—	—	117	—	1
Radius length	253	—	1	409	9	3	394	12	2
Radius midshaft width	61	—	1	87	3	3	101	31	2
Radius distal width	63	—	1	89	3	3	91	42	2
Ulna length	335	—	1	464	65	2	469	18	2
Ulna olecranon length	89	—	1	86	2	3	114	38	2
Ulna distal width	30	—	1	63	6	3	58	17	2
Mc2 length	108	—	1	—	—	—	166	—	1
Mc2 proximal width	27	—	1	39	1	3	59	—	1
Mc3 length	119	—	1	185	8	3	—	—	—
Mc3 proximal width	32	—	1	51	2	3	—	—	—
Mc4 length	100	—	1	165	—	1	—	—	—
Mc4 proximal width	25	—	1	34	—	1	—	—	—
Mc5 length	100	28	2	115	—	1	160	—	1
Mc5 proximal width	20	7	2	36	—	1	41	—	1
Femur length	341	—	1	494	1	2	—	—	—
Femur midshaft width	124	—	1	163	2	2	—	—	—
Femur distal width	104	7	2	130	12	2	—	—	—
Patella length	65	—	1	83	9	3	—	—	—
Patella width	67	—	1	88	4	3	—	—	—
Tibia length	254	—	1	390	6	2	—	—	—
Tibia midshaft width	87	—	1	115	14	2	—	—	—
Tibia distal width	59	—	1	78	13	2	—	—	—
Fibula length	235	—	1	—	—	—	—	—	—
Fibula midshaft width	19	—	1	—	—	—	—	—	—
Fibula distal width	13	—	1	—	—	—	—	—	—
Astragalus length	56	3	2	78	5	6	—	—	—
Astragalus trochlear width	51	1	2	76	3	6	—	—	—
Calcaneum length	105	—	1	130	5	3	—	—	—
Calcaneum width @ sustentaculum	65	—	1	77	7	3	—	—	—
Mt2 length	99	—	1	160	9	4	—	—	—
Mt2 proximal width	24	—	1	39	3	4	—	—	—
Mt3 length	129	30	2	—	—	—	—	—	—
Mt3 proximal width	35	1	2	—	—	—	—	—	—
Mt4 length	98	—	1	163	11	2	—	—	—
Mt4 proximal width	31	—	1	48	4	2	—	—	—

Table 5.2. Postcranial measurements of *Subhyracodon* and *Diceratherium* (in mm)

MEASUREMENT	S. mitis			S. occidentalis			D. tridactylum		
	Mean	SD	N	Mean	SD	N	Mean	SD	N
Atlas length	—	—	—	166	53	3	59	—	1
Atlas width	—	—	—	84	14	3	257	—	1
Axis length	170	—	1	105	17	3	100	—	1
Axis anterior width	179	—	1	101	6	3	121	—	1
Axis posterior width	138	—	1	57	4	2	72	—	1
Scapula length	—	—	—	334	18	2	352	41	2
Scapula maximum width	—	—	—	204	1	2	214	—	1
Scapula neck width	—	—	—	77	9	2	78	7	2
Glenoid fossa width	43	1	2	65	5	2	61	4	2
Humerus length	251	1	2	305	22	4	297	—	1
Humerus midshaft width	67	29	3	120	15	4	124	—	1
Humerus distal width	61	5	3	76	7	4	106	—	1
Radius length	218	34	2	227	—	1	292	6	2
Radius midshaft width	46	3	3	61	—	1	66	6	2
Radius distal width	41	1	2	56	—	1	67	4	2
Ulna length	295	—	1	366	22	2	374	—	1
Ulna olecranon length	—	—	—	99	1	2	101	—	1
Ulna distal width	—	—	—	33	8	2	42	—	1
Mc2 length	104	2	3	138	13	2	133	—	1
Mc2 proximal width	20	3	3	31	4	2	19	—	1
Mc3 length	120	1	3	152	12	3	150	—	1
Mc3 proximal width	27	2	3	44	12	3	42	—	1
Mc4 length	95	4	2	120	1	2	125	—	1
Mc4 proximal width	20	1	2	34	11	2	29	—	1
Mc5 length	20	1	2	26	—	1	—	—	—
Mc5 proximal width	—	—	—	20	—	1	—	—	—
Femur length	342	9	4	374	6	3	414	28	2
Femur midshaft width	111	7	4	127	4	3	129	30	2
Femur distal width	81	5	4	92	3	3	89	25	2
Patella length	53	8	2	61	3	5	68	13	2
Patella width	63	2	2	68	5	5	62	8	2
Tibia length	265	5	4	302	24	3	307	—	1
Tibia midshaft width	76	5	4	91	11	2	66	—	1
Tibia distal width	55	4	4	64	1	2	68	—	1
Fibula length	—	—	—	264	—	1	300	—	1
Fibula midshaft width	—	—	—	24	—	1	38	—	1
Fibula distal width	—	—	—	10	—	1	41	—	1
Astragalus length	55	5	4	59	3	6	59	1	2
Astragalus trochlear width	50	6	4	51	3	6	56	2	2
Calcaneum length	93	9	4	107	4	2	103	—	1
Calcaneum width @ sustentaculum	51	9	3	56	2	2	63	—	1
Mt2 length	116	5	3	118	—	1	106	—	1
Mt2 proximal width	29	3	3	22	—	1	—	—	—
Mt3 length	121	1	2	132	1	2	118	—	1
Mt3 proximal width	32	4	2	37	1	2	33	—	1
Mt4 length	115	—	1	110	—	1	112	—	1
Mt4 proximal width	24	—	1	22	—	1	33	—	1

Table 5.3. Postcranial measurements of *Diceratherium* (in mm)

MEASUREMENT	D. armatum			D. annectens			D. niobrarense		
	Mean	SD	N	Mean	SD	N	Mean	SD	N
Atlas length	72	12	3	65	5	3	—	—	—
Atlas width	168	9	3	154	6	2	—	—	—
Axis length	130	11	3	85	—	1	—	—	—
Axis anterior width	101	2	3	88	—	1	—	—	—
Axis posterior width	82	3	3	65	—	1	—	—	—
Scapula length	—	—	—	315	—	1	—	—	—
Scapula maximum width	—	—	—	160	—	1	—	—	—
Scapula neck width	—	—	—	73	—	1	—	—	—
Glenoid fossa width	—	—	—	53	—	1	—	—	—
Humerus length	342	16	5	293	11	7	481	—	1
Humerus midshaft width	142	5	4	106	11	7	112	—	1
Humerus distal width	97	9	5	72	4	6	57	—	1
Radius length	398	6	6	397	13	7	394	—	1
Radius midshaft width	74	3	7	84	3	7	86	—	1
Radius distal width	71	4	7	37	2	6	74	—	1
Ulna length	478	18	2	323	15	6	—	—	—
Ulna olecranon length	99	2	3	59	2	7	—	—	—
Ulna distal width	46	1	2	58	2	6	—	—	—
Mc2 length	179	4	4	149	4	6	133	—	1
Mc2 proximal width	37	6	4	27	2	5	28	—	1
Mc3 length	203	14	3	168	5	8	151	—	1
Mc3 proximal width	45	1	3	37	3	9	31	—	1
Mc4 length	171	6	3	140	7	6	—	—	—
Mc4 proximal width	31	2	3	26	4	6	—	—	—
Mc5 length	—	—	—	—	—	—	—	—	—
Mc5 proximal width	—	—	—	—	—	—	—	—	—
Femur length	459	6	4	380	10	7	465	—	1
Femur midshaft width	137	7	4	115	5	7	139	—	1
Femur distal width	106	19	3	87	6	7	103	—	1
Patella length	69	1	5	54	2	7	68	10	3
Patella width	78	2	5	64	3	7	81	6	3
Tibia length	368	10	3	328	5	5	111	9	4
Tibia midshaft width	100	2	3	83	5	5	111	9	4
Tibia distal width	79	2	3	62	3	5	87	7	4
Fibula length	335	—	1	308	6	2	346	—	1
Fibula midshaft width	35	—	1	27	1	2	37	—	1
Fibula distal width	37	—	1	39	5	2	31	—	1
Astragalus length	72	2	8	58	1	9	68	4	6
Astragalus trochlear width	62	2	8	49	1	8	61	3	6
Calcaneum length	119	2	5	94	5	8	114	3	3
Calcaneum width @ sustentaculum	71	3	4	55	2	7	58	8	2
Mt2 length	165	3	6	139	5	5	156	7	2
Mt2 proximal width	35	1	6	27	1	5	27	9	2
Mt3 length	177	4	7	151	4	8	161	17	3
Mt3 proximal width	41	3	7	35	1	6	44	4	3
Mt4 length	163	6	2	142	7	7	147	—	1
Mt4 proximal width	38	1	2	33	5	6	28	—	1

Table 5.4. Postcranial measurements of *Menoceras* and *Floridaceras* (in mm)

MEASUREMENT	M. arikarense			M. barbouri			F. whitei		
	Mean	SD	N	Mean	SD	N	Mean	SD	N
Atlas length	54	4	7	66	4	7	86	—	1
Atlas width	85	4	7	99	4	7	132	—	1
Axis length	70	1	2	73	6	7	75	—	1
Axis anterior width	81	5	2	102	5	7	181	—	1
Axis height	—	—	—	94	11	5	115	—	1
Scapula length	259	21	6	321	16	4	420	42	2
Glenoid fossa width	45	1	7	56	2	6	91	10	2
Humerus length	240	7	6	278	8	5	430	12	8
Humerus midshaft width	32	2	7	44	4	5	63	3	7
Humerus distal width	54	3	6	63	1	5	104	9	6
Radius length	244	10	8	283	11	6	381	20	4
Radius midshaft width	29	2	8	37	2	8	47	5	4
Radius distal width	44	2	8	57	1	8	90	5	3
Ulna length	271	14	6	320	10	6	407	37	6
Ulna midshaft width	21	2	9	29	3	6	46	3	5
Mc2 length	118	3	8	137	6	6	154	—	1
Mc2 proximal width	20	1	7	25	2	6	35	—	1
Mc3 length	132	3	8	151	5	6	—	—	—
Mc3 proximal width	26	1	8	33	3	6	—	—	—
Mc4 length	114	4	8	132	7	6	155	5	4
Mc4 proximal width	17	1	8	22	1	6	362	2	4
Femur length	313	7	8	376	11	5	535	40	3
Femur midshaft width	33	3	9	40	1	6	75	1	2
Femur distal width	74	3	9	88	3	5	120	5	3
Patella length	56	3	8	71	3	6	101	—	1
Patella width	51	4	8	67	2	6	16	—	1
Tibia length	268	7	7	296	14	8	360	45	8
Tibia midshaft width	29	2	7	37	2	6	53	1	6
Tibia distal width	47	3	7	57	2	6	94	4	6
Fibula length	247	8	3	280	—	1	353	39	2
Fibula midshaft width	10	2	6	15	—	1	19	1	3
Astragalus length	48	1	7	56	2	8	79	2	2
Astragalus trochlear width	41	1	7	51	3	8	82	4	2
Calcaneum length	78	5	8	91	3	6	116	3	3
Calcaneum width @ sustentaculum	45	2	8	51	3	6	78	2	3
Mt2 length	115	4	7	126	4	6	142	7	3
Mt2 proximal width	15	1	7	20	5	6	30	2	3
Mt3 length	126	5	8	142	4	6	157	8	3
Mt3 proximal width	22	1	8	27	1	6	46	4	3
Mt4 length	116	3	10	131	6	6	144	5	4
Mt4 proximal width	17	1	10	22	1	6	32	1	4

Table 5.5. Postcranial measurements of *Peraceras* (in mm)

MEASUREMENT	P. profectum			P. superciliosum			P. hessei			Giant acerathere F:AM114400
	Mean	SD	N	Mean	SD	N	Mean	SD	N	
Atlas length	—	—	—	178	—	1	—	—	—	—
Atlas width	—	—	—	75	—	1	—	—	—	—
Axis length	—	—	—	124	—	1	—	—	—	—
Axis anterior width	—	—	—	191	—	1	—	—	—	—
Axis height	—	—	—	157	—	1	—	—	—	—
Scapula length	—	—	—	400	38	3	—	—	—	—
Glenoid fossa width	—	—	—	63	8	2	—	—	—	—
Humerus length	295	27	5	463	21	4	262	—	1	—
Humerus midshaft width	48	7	7	81	5	5	45	—	1	—
Humerus distal width	82	13	7	114	3	4	67	—	1	—
Radius length	233	21	4	356	20	5	217	—	1	—
Radius midshaft width	33	4	2	62	8	6	40	8	3	—
Radius distal width	58	4	2	95	8	7	62	2	3	—
Ulna length	375	26	4	425	5	3	263	—	1	—
Ulna midshaft width	31	3	5	47	6	3	30	1	2	—
Mc2 length	—	—	—	138	2	3	30	—	1	—
Mc2 proximal width	—	—	—	46	4	3	21	—	1	—
Mc3 length	101	—	1	167	8	4	—	—	—	—
Mc3 proximal width	33	—	11	55	5	4	—	—	—	—
Mc4 length	87	—	1	135	5	3	—	—	—	—
Mc4 proximal width	21	—	1	24	3	3	—	—	—	—
Femur length	358	12	4	570	8	5	346	10	6	640
Femur midshaft width	47	6	3	84	4	5	131	8	4	83
Femur distal width	84	11	3	134	7	6	91	9	6	150
Patella length	59	—	1	93	8	5	—	—	—	—
Patella width	55	—	1	103	4	5	—	—	—	—
Tibia length	252	9	6	340	14	4	264	8	3	410
Tibia midshaft width	38	2	6	60	5	4	36	7	3	67
Tibia distal width	53	9	6	103	10	4	66	2	2	119
Fibula length	220	—	1	345	21	1	—	—	—	400
Fibula midshaft width	14	—	1	27	1	2	—	—	—	27
Astragalus length	50	3	5	80	6	11	49	—	1	84
Astragalus trochlear width	50	4	5	87	4	8	66	2	2	94
Calcaneum length	83	6	2	126	1	2	—	—	—	139
Calcaneum width @ sustentaculum	52	4	2	86	2	2	—	—	—	88
Mt2 length	97	3	3	131	2	3	—	—	—	141
Mt2 proximal width	22	3	3	31	2	3	—	—	—	33
Mt3 length	108	8	3	156	4	7	101	5	2	—
Mt3 proximal width	28	2	3	50	3	7	26	1	2	—
Mt4 length	97	8	3	139	1	2	—	—	—	151
Mt4 proximal width	23	2	3	35	3	2	—	—	—	30

Table 5.6. Postcranial measurements of *Aphelops* (in mm)

MEASUREMENT	*A. megalodus*			*A. malacorhinus*			*A. mutilus*		
	Mean	SD	N	Mean	SD	N	Mean	SD	N
Atlas length	76	6	5	84	10	2	103	4	2
Atlas width	122	2	4	132	8	2	134	2	2
Axis length	—	—	—	101	—	1	98	—	1
Axis anterior width	—	—	—	159	—	1	139	—	1
Axis height	—	—	—	135	—	1	130	—	1
Scapula length	360	—	1	375	—	1	420	21	2
Glenoid fossa width	81	1	2	85	7	3	86	5	3
Humerus length	304	13	6	388	12	4	398	25	3
Humerus midshaft width	56	5	8	59	2	4	65	5	4
Humerus distal width	99	9	8	92	3	4	116	5	4
Radius length	274	19	8	342	17	7	326	13	5
Radius midshaft width	44	4	9	46	5	9	54	3	6
Radius distal width	75	4	8	82	5	9	88	2	6
Ulna length	288	26	5	360	—	1	378	18	2
Ulna midshaft width	73	4	6	43	—	1	52	8	2
Mc2 length	126	4	6	153	6	4	145	5	2
Mc2 proximal width	33	2	6	33	1	4	36	4	2
Mc3 length	140	12	3	177	7	5	153	5	6
Mc3 proximal width	41	3	3	42	2	5	48	3	6
Mc4 length	120	5	5	146	5	5	135	7	4
Mc4 proximal width	30	4	5	30	1	5	73	2	5
Femur length	424	29	8	443	14	5	488	24	5
Femur midshaft width	58	7	10	60	3	5	73	6	5
Femur distal width	110	12	10	106	3	4	133	13	5
Patella length	75	1	2	—	—	—	97	9	7
Patella width	81	3	2	—	—	—	88	8	6
Tibia length	292	11	9	339	13	5	331	32	6
Tibia midshaft width	46	4	10	50	3	5	57	5	6
Tibia distal width	78	6	8	85	6	4	93	5	6
Fibula length	258	6	4	—	—	—	—	—	—
Fibula midshaft width	20	2	4	25	2	3	—	—	—
Astragalus length	64	4	5	69	4	6	68	4	8
Astragalus trochlear width	64	3	5	67	2	6	72	3	8
Calcaneum length	101	5	7	86	3	6	111	6	6
Calcaneum width @ sustentaculum	65	3	7	71	2	6	76	3	6
Mt2 length	111	10	7	132	3	6	120	1	2
Mt2 proximal width	25	4	7	24	1	6	26	1	2
Mt3 length	139	9	7	145	4	6	128	4	5
Mt3 proximal width	36	2	7	36	3	6	42	4	5
Mt4 length	110	8	5	130	3	4	113	5	5
Mt4 proximal width	23	3	5	23	2	4	29	3	5

Table 5.7. Postcranial measurements of *Teleoceras* (in mm)

MEASUREMENT	*T. americanum*			*T. medicornutum*			*T. meridianum*		
	Mean	SD	N	Mean	SD	N	Mean	SD	N
Atlas length	79	9	3	86	4	3	—	—	—
Atlas width	137	9	3	131	3	3	—	—	—
Axis length	78	—	1	89	2	2	—	—	—
Axis anterior width	128	—	1	158	26	2	—	—	—
Axis height	—	—	—	120	26	2	—	—	—
Scapula length	—	—	—	335	13	3	240	—	1
Glenoid fossa width	—	—	—	84	3	3	—	—	—
Humerus length	—	—	—	333	21	7	264	10	3
Humerus midshaft width	—	—	—	56	5	8	48	3	6
Humerus distal width	—	—	—	101	6	7	88	3	6
Radius length	257	16	3	265	16	8	215	14	2
Radius midshaft width	40	2	3	48	2	8	42	2	2
Radius distal width	75	6	3	86	7	8	77	6	3
Ulna length	305	—	1	310	30	5	—	—	—
Ulna midshaft width	29	—	1	41	5	6	—	—	—
Mc2 length	—	—	—	107	11	4	93	1	2
Mc2 proximal width	—	—	—	43	9	5	32	4	2
Mc3 length	120	4	2	134	13	6	104	1	2
Mc3 proximal width	39	2	2	51	5	6	44	2	2
Mc4 length	100	4	2	101	3	3	—	—	—
Mc4 proximal width	30	3	2	41	2	3	—	—	—
Femur length	340	7	2	433	24	5	398	—	1
Femur midshaft width	51	2	3	67	6	7	64	—	1
Femur distal width	133	6	3	121	12	7	110	—	1
Patella length	68	6	3	89	6	4	—	—	—
Patella width	70	6	3	88	9	3	—	—	—
Tibia length	250	16	5	276	19	4	218	7	4
Tibia midshaft width	45	3	5	47	6	4	40	6	4
Tibia distal width	70	4	5	88	5	4	70	5	4
Fibula length	223	4	2	222	12	5	—	—	—
Fibula midshaft width	16	1	2	19	3	5	—	—	—
Astragalus length	57	8	2	70	6	7	—	—	—
Astragalus trochlear width	72	8	2	73	4	7	—	—	—
Calcaneum length	85	1	2	105	12	5	—	—	—
Calcaneum width @ sustentaculum	54	6	2	63	5	5	—	—	—
Mt2 length	—	—	—	96	7	4	—	—	—
Mt2 proximal width	—	—	—	34	3	4	—	—	—
Mt3 length	102	—	1	110	6	5	87	—	1
Mt3 proximal width	37	—	1	48	4	5	32	—	1
Mt4 length	—	—	—	93	4	4	—	—	—
Mt4 proximal width	—	—	—	34	2	5	—	—	—

Table 5.8. Postcranial measurements of *Teleoceras* (in mm)

MEASUREMENT	*T. major* Mean	SD	N	*T. brachyrhinum* Mean	SD	N	*T. fossiger* Mean	SD	N
Atlas length	101	6	9	84	9	2	101	6	6
Atlas width	142	7	8	132	6	2	143	10	6
Axis length	—	—	—	68	6	3	83	4	2
Axis anterior width	—	—	—	137	18	3	138	1	2
Axis height	—	—	—	115	7	2	111	4	2
Scapula length	315	34	6	280	—	1	246	19	6
Glenoid fossa width	90	10	5	74	—	1	88	9	6
Humerus length	314	8	7	311	16	6	307	28	7
Humerus midshaft width	67	11	7	54	7	7	61	8	7
Humerus distal width	97	6	7	83	7	8	99	10	7
Radius length	257	26	9	257	14	7	259	13	8
Radius midshaft width	48	3	9	41	2	6	50	5	8
Radius distal width	94	2	7	73	5	5	99	7	8
Ulna length	293	21	8	294	12	5	260	10	3
Ulna midshaft width	42	4	8	31	3	5	35	3	4
Mc2 length	107	10	4	100	8	4	101	7	3
Mc2 proximal width	46	3	4	33	6	4	43	4	3
Mc3 length	118	11	4	113	10	5	115	7	8
Mc3 proximal width	49	5	4	42	2	5	52	5	8
Mc4 length	93	7	4	96	3	3	89	5	6
Mc4 proximal width	40	3	4	29	4	3	38	2	6
Femur length	384	15	7	396	13	4	440	33	5
Femur midshaft width	69	4	7	53	4	4	76	7	7
Femur distal width	123	7	7	98	12	4	113	5	5
Patella length	67	5	7	69	1	2	78	5	8
Patella width	73	6	7	66	3	2	86	5	8
Tibia length	236	23	8	263	25	3	236	24	8
Tibia midshaft width	49	3	7	42	6	3	50	3	8
Tibia distal width	85	7	7	74	12	3	87	5	7
Fibula length	245	14	2	233	4	2	201	18	6
Fibula midshaft width	20	5	3	17	2	3	22	5	5
Astragalus length	61	4	8	61	4	5	61	10	8
Astragalus trochlear width	70	6	8	61	5	5	72	2	8
Calcaneum length	125	5	4	94	4	5	130	7	6
Calcaneum width @ sustentaculum	74	5	4	55	4	4	78	5	6
Mt2 length	83	6	2	80	8	4	73	8	2
Mt2 proximal width	30	1	2	28	6	4	37	10	2
Mt3 length	98	7	5	99	7	6	109	9	5
Mt3 proximal width	44	1	5	37	3	6	50	7	5
Mt4 length	71	4	5	83	7	5	80	6	8
Mt4 proximal width	28	1	5	28	3	5	37	3	8

Table 5.9. Postcranial measurements of *Teleoceras* (in mm)

MEASUREMENT	*T. proterum*			*T. hicksi*			*T. guymonense*		
	Mean	SD	N	Mean	SD	N	Mean	SD	N
Atlas length	103	8	56	94	—	1	—	—	—
Atlas width	147	9	5	160	—	1	—	—	—
Axis length	80	11	7	—	—	—	—	—	—
Axis anterior width	138	12	7	—	—	—	—	—	—
Axis height	117	18	7	—	—	—	—	—	—
Scapula length	293	11	2	290	14	2	—	—	—
Glenoid fossa width	91	4	7	86	—	1	—	—	—
Humerus length	284	35	8	315	12	4	—	—	—
Humerus midshaft width	61	11	9	65	5	4	—	—	—
Humerus distal width	91	5	8	99	5	4	—	—	—
Radius length	252	11	9	250	6	5	235	7	2
Radius midshaft width	50	4	9	53	2	5	52	4	2
Radius distal width	92	5	10	95	10	4	87	4	2
Ulna length	262	6	7	279	8	6	270	—	1
Ulna midshaft width	38	4	6	42	4	6	38	—	1
Mc2 length	79	4	8	87	4	4	98	—	1
Mc2 proximal width	39	5	8	38	2	4	41	—	1
Mc3 length	105	7	8	115	4	5	117	—	1
Mc3 proximal width	50	3	8	51	4	5	54	—	1
Mc4 length	84	7	8	84	4	6	89	—	1
Mc4 proximal width	40	4	8	36	3	6	37	—	1
Femur length	392	17	7	440	20	3	—	—	—
Femur midshaft width	61	5	7	71	12	3	—	—	—
Femur distal width	102	12	6	115	9	3	—	—	—
Patella length	74	8	8	82	15	7	71	14	3
Patella width	86	3	8	88	10	7	74	9	3
Tibia length	195	13	7	230	18	4	220	—	1
Tibia midshaft width	47	4	8	52	4	4	40	—	1
Tibia distal width	75	4	7	86	6	4	76	—	1
Fibula length	196	11	7	183	4	2	—	—	—
Fibula midshaft width	22	4	9	32	4	2	—	—	—
Astragalus length	53	2	8	57	7	6	77	—	1
Astragalus trochlear width	70	5	8	77	2	6	87	—	1
Calcaneum length	124	10	8	125	12	2	—	—	—
Calcaneum width @ sustentaculum	70	7	8	74	3	2	—	—	—
Mt2 length	74	9	8	71	6	3	—	—	—
Mt2 proximal width	34	5	8	73	1	3	—	—	—
Mt3 length	86	5	8	84	3	3	—	—	—
Mt3 proximal width	40	3	8	39	3	3	—	—	—
Mt4 length	67	4	8	69	1	2	—	—	—
Mt4 proximal width	38	3	8	42	1	2	—	—	—

6. Biogeography and Diversity Patterns

INTRODUCTION

As is apparent from the locality data in Chapter 4, the Rhinocerotidae was an extremely widespread and successful family of mammals in North America from the middle Eocene until the earliest Pliocene. During much of their history, they were among the largest land mammals on the continent, and competed only with brontotheres in the middle and late Eocene, and proboscideans after the middle Miocene, for occupation of the large herbivore (or "megaherbivore") niche. During the Oligocene and early Miocene (from 34 to 16 Ma), they had no competition from any other large land mammal, but were the largest mammals on the continent for those 18 million years. A comprehensive review of a large North American mammalian family like the Rhinocerotidae thus demands that their geographic patterns of distribution be discussed.

The first caveat of such an analysis is the completeness of the stratigraphic record for each time interval. As is apparent from the maps in this chapter, the actual outcrop area of a given age is very limited compared to the total surface area of the North American continent. Most of the area east of the one-hundredth meridian has no terrestrial mammal-bearing beds (except in Florida, and occasionally New Jersey or the Texas-Mississippi part of the Coastal Plain Tertiary passive margin wedge). And those maps are overestimates, because the outcrop indicated as "Eocene" or "Oligocene" may have only a tiny portion actually exposed at the surface, bearing fossils, and representing the smaller time slice of each land mammal age. The available outcrop is even further restricted because most of the rocks of the appropriate age (particularly on the Atlantic and Gulf coasts) are marine beds with limited potential for producing terrestrial mammals, or they are volcanics with little preservation potential for fossils. However, as noted below, some remarkable rhino finds have occurred in shallow marine beds, because rhino bones are more durable than those of most mammals.

Not only are outcrops of the suitable age and facies very rare, but during certain time intervals, they are restricted by the small number of fossiliferous formations. For example, during the Orellan and Whitneyan, there were few important fossil-bearing outcrops outside the White River Group of the northwestern High Plains, with minor exceptions in Montana, Florida, and Mississippi. Thus, we know almost nothing of land mammal biogeography west or east of the High Plains during a long interval of time (29–34 Ma).

Although the restrictions on a biogeographic analysis are severe, there is still much positive information to be obtained from such an analysis. In many places, the absence of evidence is not evidence of absence, since for most of North America east of the Mississippi River, we have no terrestrial-mammal bearing beds at all. However, the absence of a rhinoceros record in some regions cannot always be dismissed as sampling failure. In many areas (discussed in detail below), we often have excellent faunas with large samples including plenty of fossil mammals in the same size range of rhinos, yet the contemporary rhino species have never been found. This is clear evidence that these rhinos did not range into a given area, despite their presence elsewhere. Wherever possible, I have listed the localities that have produced large faunas but still no evidence that rhinos lived in the area.

EOCENE

Uintan (later middle Eocene, 40–47 Ma)

The earliest record of the rhinocerotid lineage is the occurrence of *Uintaceras radinskyi* (Holbrook and Lucas, 1997). Although Holbrook and Lucas placed it as a sister-taxon of the family because it lacked the defining synapomorphies of the I1 chisel and i2 tusk combination, nevertheless its presence indicates several things. First of all, it was originally referred to the Asian genus *Forstercooperia*, which it closely resembles. Clearly, the rhinocerotids originated in Eurasia in the early middle Eocene from a sister-group within the Hyracodontidae, such as *Forstercooperia*.

Secondly, the Uintan (Fig. 6.1) is a particularly well-sampled interval, with excellent faunas from southern California, the Rocky Mountain region (especially the Uinta basin of Utah, the Sand Wash basin of Colorado, the Washakie, Wind River, and East Fork basins of Wyoming), New Mexico, Trans-Pecos Texas, Saskatchewan, and even the Texas Gulf Coastal Plain (Krishtalka *et al.*, 1987; Prothero and Emry, 1996). These include numerous samples of large mammals, particularly hyracodont and amynodont rhinos, as well as brontotheres, uintatheres, and tapiroids.

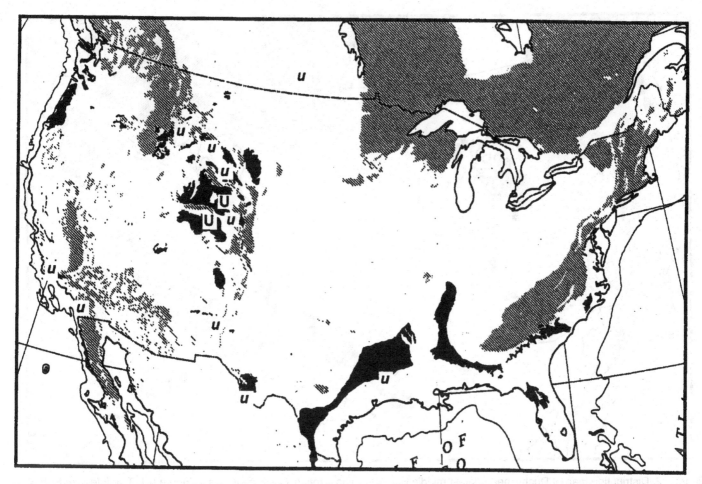

Figure 6.1. Map of Uintan (later middle Eocene) rhinoceros biogeography. Base map after Cook and Bally (1975). Light gray areas indicate uplift where all Cenozoic rocks have been stripped away. Black areas are outcrops of Eocene rocks (both marine and nonmarine). **U** = location of *Uintaceras radinskyi* specimens. *u* = Uintan localities that produce no rhinocerotids, but produce hyracodonts or amynodonts.

Thus, we can be confident that the sampling of large mammals is excellent in numerous regions, and that absence of *Uintaceras* outside its known distribution of the Uinta and Washakie basins is not an artifact of poor sampling. Instead, the sparse distribution of *Uintaceras* suggests that it was a rare immigrant, perhaps outcompeted by hyracodonts and amynodonts and other large perissodactyls during the Uintan. Apparently, the first invasion of the rhinocerotid lineage to North America was not a wave that swept across the entire continent, but a limited incursion that did not immediately take root. Comparable-aged faunas in Asia apparently had more representatives of the rhinocerotid lineage than did North America during the Uintan.

Duchesnean (latest middle Eocene, 37–40 Ma)
Although Duchesnean faunas (Fig. 6.2) are not as widespead and well known around the continent as the Uintan (Prothero and Emry, 1996), nevertheless there are important faunas in many locations, and more of them produce rhinocerotids than ever

before. In the Duchesnean, we see two genera of rhinocerotids (*Teletaceras* and *Penetrigonias*) distributed in several different localities. *Teletaceras* occurs in the ?late Uintan or Duchesnean Clarno l.f. in Oregon (its type locality), plus the late Duchesnean Porvenir l.f. of Trans-Pecos Texas and the ?Duchesnean or Chadronian Titus Canyon l.f. of Death Valley, California. *Penetrigonias* occurs in the late Duchesnean Lapoint l.f., Uinta Basin, Utah, and the Porvenir l.f. of Texas along with *Teletaceras*. Although this covers many of the classic Duchesnean localities, rhinocerotids are still missing from some localities, such as all the early Duchesnean faunas (especially those in southern California, which produce amynodonts and hyracodonts), as well as those in Saskatchewan, Wyoming, New Mexico, and Montana (Prothero and Emry, 1996). It is not clear that their exclusion from California, Saskatchewan, New Mexico and Montana can be attributed to poor sampling, because most of those regions are well sampled for large mammals. Nor can it be attributed to some sort of exclusion by competing perissodactyls, because

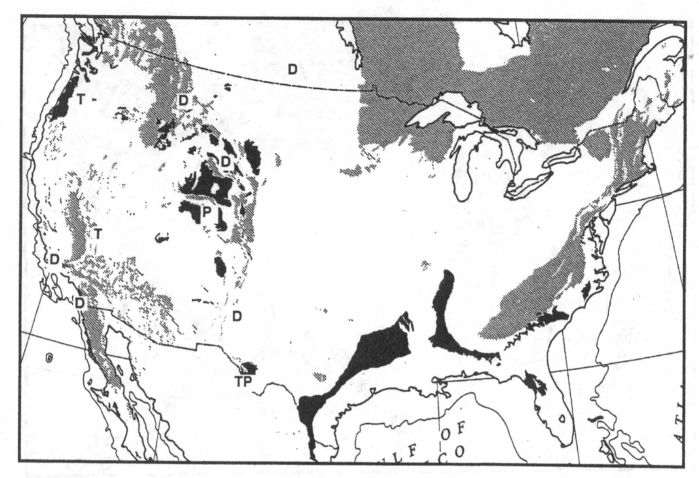

Figure 6.2. Distribution map of Duchesnean (latest middle Eocene) rhinocerotids. Conventions as in Figure 6.1. **T** = *Teletaceras*; **P** = *Penetrigonias*; **D** = Duchesnean localities that produce no rhinocerotids, but do yield amynodonts or hyracodonts. The Titus Canyon l.f. is here considered to be Duchesnean, rather than Chadronian, as traditionally correlated.

Teletaceras coexists with amynodonts in the Clarno l.f. (Hanson, 1996), as well as tapiroids and brontotheres, and the same is true in Trans-Pecos Texas. It seems that rhinocerotids, although successfully established in North America by the late Duchesnean, were simply not as dominant in the large herbivore niche as other groups (particularly hyracodonts and amynodonts) as they would become by the Oligocene.

Chadronian (late Eocene, 34–37 Ma)

In the Chadronian (Fig. 6.3), several phenomena occurred at once. By the late Chadronian, the diversity of rhinocerotids jumps up with the addition of *Trigonias osborni* and *T. wellsi*, *Amphicaenopus platycephalus*, and *Subhyracodon mitis* to the persistent *Penetrigonias dakotensis* (plus the addition of *P. sagittatus*). Diversity more than doubles, although *Teletaceras* apparently dies out (unless we count the Titus Canyon l.f. as earliest Chadronian). Yet the Chadronian also represents an interval of time with relatively restricted geographic coverage—mainly the

Chadron Formation and equivalent White River Group rocks in Wyoming, Nebraska, South Dakota, and Colorado, plus smaller faunas in Trans-Pecos Texas, New Mexico, Montana, and Saskatchewan. Nor can the diversity increase be attributed to the increased density of fossiliferous intervals in the Chadron Formation, because the localities in Trans-Pecos Texas, New Mexico, Montana, and Saskatchewan are comparable to their levels of Duchesnean and/or Uintan sampling. Nevertheless, the dramatic increase is real, despite fewer regions being represented (especially southern California), so there is no correlation between diversity and regions sampled. In addition, this is an interval when rhinos reached larger body sizes (especially *Trigonias*, *Subhyracodon*, and *Amphicaenopus*), and amynodonts and hyracodonts became more and more rare and lower in diversity. Many Chadronian faunas (such as those of Saskatchewan, Texas, and the Chadron Formation) have all three groups of rhinocerotoids found in the same faunas in about equal numbers, so clearly they were all established and successful. Some faunas (such as the Chadron

Formation and equivalents in Wyoming and South Dakota) produce all of the known rhinocerotid species (*P. dakotensis*, *T. osborni*, and *T. wellsi*, *S. mitis*, and *A. platycephalus*), plus numerous hyracodonts such as *Hyracodon priscidens* and *Triplopides rieli* (Prothero, 1996), and some *Metamynodon* as well. Along with the diversity of horses (five different co-existing species of *Mesohippus* and *Miohippus* in the late Chadronian, *fide* Prothero and Shubin, 1989), tapiroids (*Colodon*, *Protapirus*, and *Schizotheriodes*), and brontotheres (at least three genera according to Mader, 1989, despite the oversplitting of Osborn, 1929), the Chadronian was a peak in perissodactyl diversity and disparity, which was unmatched in the Cenozoic (except possibly the Uintan, when brontotheres were even more diverse, and eomoropid chalicotheres and other tapiroids were also common—although horses were low in diversity).

OLIGOCENE

Orellan (earliest Oligocene, 32–34 Ma)

Several striking changes occur in the Orellan (Fig. 6.4). The most important is that the available geographic sampling area is greatly reduced, becoming restricted to the White River Group of the High Plains (including the Cypress Hills of Saskatchewan—Storer, 1996) and equivalent rocks in Montana. There is no Orellan record for areas that had traditionally produced Uintan, Duchesnean, or Chadronian faunas, such as southern California, Trans-Pecos Texas, New Mexico, or western Wyoming, Colorado, and Utah. However, we do have the isolated occurrence of early Oligocene *Metamynodon* and *Subhyracodon* from the marine Byram Formation of the Mississippi Gulf Coastal Plain (Manning *et al.*, 1986; Manning, 1997).

The second striking change is the reduction in apparent

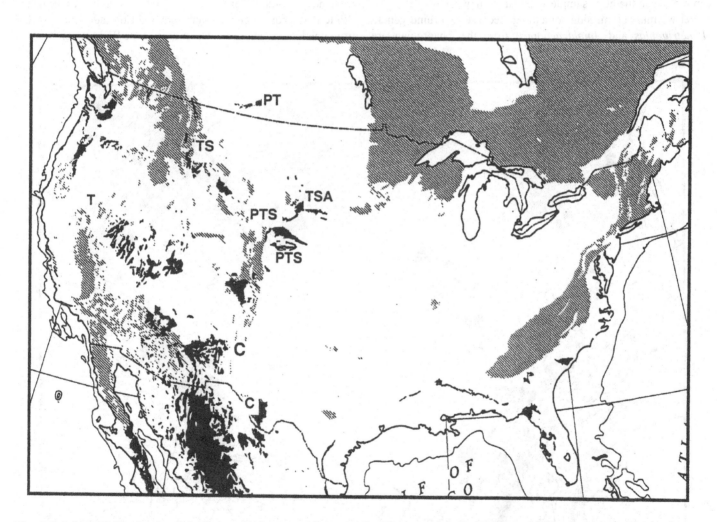

Figure 6.3. Map of Chadronian (late Eocene) rhinoceros biogeography. Black areas indicate areas of Oligocene (since Chadronian was once considered Oligocene) marine and non-marine outcrops (after Cook and Bally, 1975). Conventions as in Figure 6.1. **T** = *Trigonias*; **S** = *Subhyracodon*; **A** = *Amphicaenopus*; **P** = *Penetrigonias*; **C** = Chadronian localities with no rhinocerotids.

diversity of rhinocerotoids. Among hyracodonts and amynodonts, only one species of each (*Hyracodon nebraskensis* and *Metamynodon planifrons*) persists, and only one kind of rhinocerotid (*Subhyracodon occidentalis*) is common in the Orellan as well. It is tempting to dismiss this as a product of reduced sampling area, but even in areas far removed from the classic White River beds (such as Montana or Mississippi), we still get the same taxa of rhinocerotoids as are found in the Scenic, Orella, and Cedar Creek members of the Brule Formation. In addition, the sample size and quality in the Orellan increases dramatically, because the "lower Oreodon beds" of the Big Badlands and equivalent rocks in Wyoming and Nebraska are among the mostly densely fossiliferous mammal-bearing deposits in the world, with hundreds of skulls and skeletons of amazing quality (in contrast to the relatively poorly fossiliferous intervals that precede it). Thus, we have little reason to suspect that we are missing much of the Orellan sample, given the huge sample size that we now have.

Yet we must be missing something, because two rhino genera (*Penetrigonias* and *Amphicaenopus*) from the Chadronian are unknown from the incredibly large sample of the Orellan, yet reappear as "Lazarus taxa" during the Whitneyan. How this occurs is a mystery. As suggested in Chapter 4, some authors have attributed the disappearance of *Amphicaenopus* to competition from similarly large-bodied *Metamynodon*, but both taxa co-existed in the Chadronian and possibly even the Whitneyan as well—and this does not explain the apparent disappearance of the smaller species, *Penetrigonias dakotensis*. For whatever reason, we know that *Penetrigonias* and *Amphicaenopus* must have lived somewhere in North America in the Orellan, but not in the areas for which we have rich samples.

Whitneyan (late early Oligocene, 30–32 Ma)
Continuing the trend from the Chadronian and Orellan, Whitneyan outcrops (Fig. 6.5) are even more geographically restricted (Emry *et al.*, 1987). Over 90% of the faunas come from just the Big Badlands of South Dakota, with smaller assemblages from the White River equivalents in Nebraska and Colorado, and possibly from one locality in the Badwater area of the Wind River Basin of

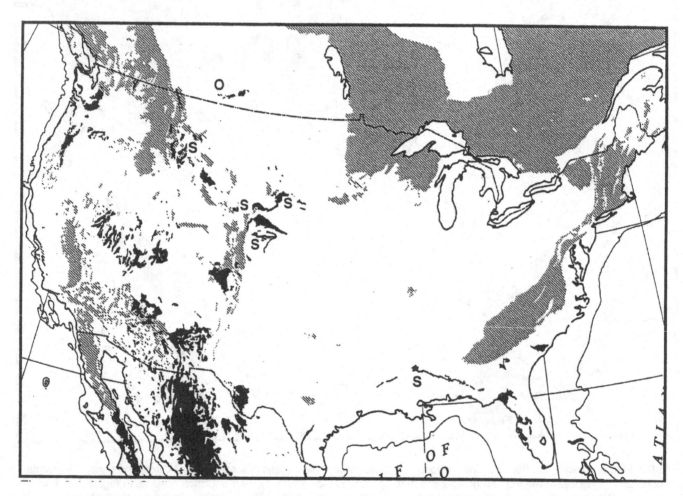

Figure 6.4. Map of Orellan (earliest Oligocene) rhinocerotid distribution. Conventions as in Figure 6.1. Black areas indicate marine and non-marine Oligocene outcrops (after Cook and Bally, 1975). **S** = *Subhyracodon*. **O** = Orellan localities without rhinocerotids.

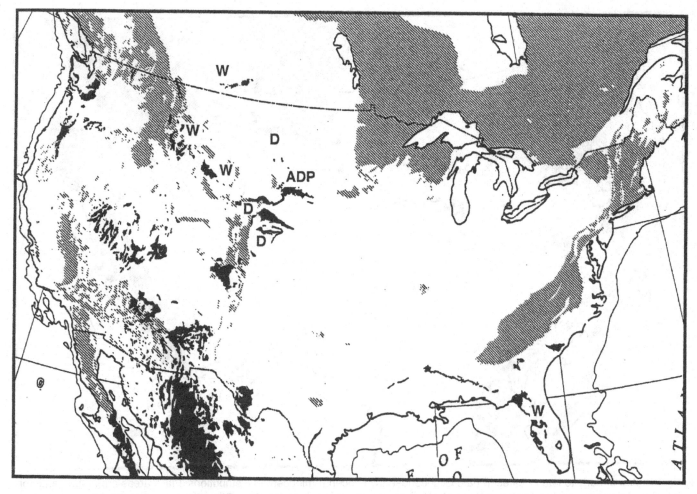

Figure 6.5. Map of Whitneyan (late early Oligocene) rhinocerotid distribution. Conventions after Figure 6.1 and 6.4. **A** = *Amphicaenopus*; **D** = *Diceratherium tridactylum*; **P** = *Penetrigonias dakotensis*; **W** = Whitneyan localities with no rhinocerotids.

Wyoming (Cedar Ridge l.f.—Setogouchi, 1978). Small scrappy Whitneyan faunas are known from Montana (Tabrum *et al.*, 1996) Saskatchewan (Storer, 1996), and the I-75 locality in Florida, which contains no rhinos (Patton, 1969), but not from other areas that yield Uintan, Duchesnean, or Chadronian faunas such as Texas, New Mexico, or California. Yet despite this geographic restriction, apparent diversity increases again. As mentioned above, *Penetrigonias* and *Amphicaenopus* reappear (although they are very rare), while *Subhyracodon occidentalis* is replaced by *Diceratherium tridactylum* (and yet the genus persists elsewhere, because *Subhyracodon kewi* is known from the ?Arikareean in California). Meanwhile, *Hyracodon nebraskensis* and *H. leidyanus* (Prothero, 1996) are both abundant, and *Metamynodon planifrons* persists although it is very rare. Thus, the apparent diversity of only three genera and three species of rhinocerotoids in the Orellan is replaced by six genera and seven species in the Whitneyan. Once again, the geographic spread of the relevant formations is no predictor of generic or specific diversity of rhinocerotoids during a given time interval.

Early to late Arikareean (late Oligocene to earliest Miocene, 19.2–30 Ma)

The Arikareean is a problematic interval. It is best known from only two regions (Fig. 6.6): the John Day beds of Oregon, and the Arikaree Group of the High Plains, with fragmentary faunas known from many other regions (primarily California, Montana, and the Texas Gulf Coastal Plain—Tedford *et al.*, 1987, 2004). Yet recent work on the Arikaree Group (Tedford *et al.*, 1996; MacFadden and Hunt, 1998) in the High Plains has shown that the classic tripartite division into equal intervals of the "Geringian," "Monroecreekian," and "Harrisonian" (e.g., Schultz and Stout, 1961) falls apart. Based on the recent chronostratigraphy of the Gering and Sharps formations and their included faunas, the "Geringian" is a very short but richly fossiliferous interval, spanning 2 m.y. from 28–30 Ma (Tedford *et al.*, 1996). The classic faunas of the Monroe Creek and Harrison formations, on the other hand, are much more poorly known with much less stratigraphic control, yet they span a very long interval of time: 19.2–28 Ma, over 8 m.y. (MacFadden and Hunt, 1998). The "typical"

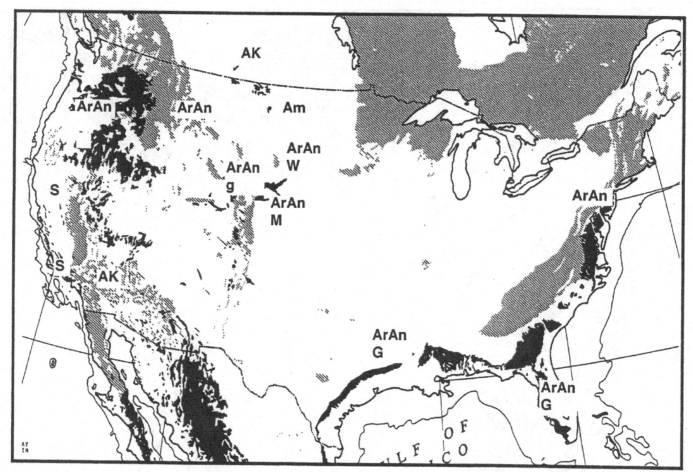

Figure 6.6. Map of early to late Arikareean (late Oligocene–early Miocene) rhinocerotid distributions. Black areas are marine and non-marine Miocene outcrops (after Cook and Bally, 1975). Conventions as in Figure 6.1. **Ar** = *Diceratherium armatum*; **An** = *Diceratherium annectens*; **Am** = *Amphicaenopus*; **M** = *Skinneroceras manningi*; **G** = *Gulfoceras westfalli*; **W** = *Woodoceras*; **g** = giant postcranials; **S** = *Subhyracodon kewi*. **AK** = Arikareean localities with no rhinocerotids.

Arikareean of the Wood Committee's (1941) original conception of the land mammal age is based on the Agate Springs fauna from the Anderson Ranch Formation of Hunt (2002), which is restricted to the last 0.5 m.y. of the Arikareean (18.7-19.2 Ma, *fide* MacFadden and Hunt, 1998). Thus, the "type" Arikareean of the High Plains is extremely long (over 11 m.y.), but only known well from its earliest (Gering-Sharps) and latest (Anderson Ranch) faunas, and poorly known through the 8 m.y. in the middle.

Some hope of better resolution is now occurring in the John Day Formation (Fremd *et al.*, 1994), where the sampling is more even through the entire Arikareean. Still, the collections of many taxa are sparse, or many of the old collections in the AMNH, YPM, and UCMP cannot be tied to modern localities. The best collections are still from the Arikaree Group in Nebraska, Wyoming, and South Dakota. This is especially true with the rhinocerotids, which are much more common and better preserved in the Arikaree Group than they are in the John Day Formation. In most of the other localities (Montana, California, Texas Gulf Coastal Plain, Florida), rhinos are much less common and more poorly preserved. Thus, our best knowledge of their biogeographic patterns comes primarily from the High Plains and central Oregon.

In those regions, diversity reaches an all-time low. Hyracodonts and amynodonts were gone, leaving only the rhinocerotids and horses (along with rare tapirids) surviving as the dominant perissodactyls. During most of the Arikareean, only two species in one genus of rhinocerotid (*Diceratherium annectens* and *D. armatum*) are common. Neither species shows much change in morphology or abundance through the entire early and middle Arikareean, unlike the rapidly changing record of other taxa, such as canids, nimravids, rodents, and oreodonts. Over 99% of the known rhinocerotid material from the Arikareean is from these two species. Similarly, the known Arikareean records in Montana (see Chapter 4) and Texas and Florida (Albright, 1998, 1999b) are also those two species.

Then there are the peculiar rare endemics that make up the last 1% of the Arikareean rhinocerotid faunas. Stone (1970) reported a possible early Arikareean *Amphicaenopus* (possibly the youngest specimen of its genus) from North Dakota, although I have not confirmed the veracity of this identification, nor its provenience. There is the unique endemic *Skinneroceras manningi* from the earliest Arikareean at Roundhouse Rock in the North Platte Valley of Nebraska, and the peculiar short-faced rhino *Woodoceras brachyops* from the early Arikareean Harris Ranch l.f. in South Dakota. Albright (1998, 1999b) described the tiny *Gulfoceras westfalli* from Toledo Bend and other localities in the Texas and Florida Gulf Coastal Plain. There is the peculiar relict survivor from the Orellan, *Subhyracodon kewi*, from the ?Arikareean Kew Quarry l.f. in the Sespe Formation of southern California, and possibly also from the Chili Gulch locality in northern California. Finally, there are the huge postcranials from two localities (77 Hill and Horse Creek) in Niobrara County, Wyoming (Chapters 4, 5), which may belong to an unusually large new species of diceratherine. Thus, the total diversity of the Arikareean appears unusually high, but the reality is that only two species in one genus dominate 99% of the localities for over 10 m.y.

MIOCENE

Latest Arikareean to Hemingfordian (early Miocene, 16–19.2 Ma)

Although Wood *et al.* (1941) used the latest Arikareean (Figs. 6.7, 6.8) Agate fauna (from the Anderson Ranch Formation of Hunt, 1998) to typify their concept of Arikareean, in most respects it has little in common with the fauna of the rest of the Arikaree Group, and is full of immigrant taxa that typify the Hemingfordian (Tedford *et al.*, 1987, 2004). This is particularly true of the rhinocerotids with the sudden and abundant appearance of *Menoceras arikarense* from Eurasian ancestry. Up until this point, there is little evidence of further immigration of Eurasian rhinos to North America once they were introduced in the Uintan and Duchesnean. Most of the Chadronian, Orellan, and Whitneyan rhinos (especially the *Subhyracodon–Diceratherium* lineage) are native groups that apparently speciated on this continent. However, from the latest Arikareean through the Hemingfordian, repeated waves of immigration of Eurasian rhinos occur, as well

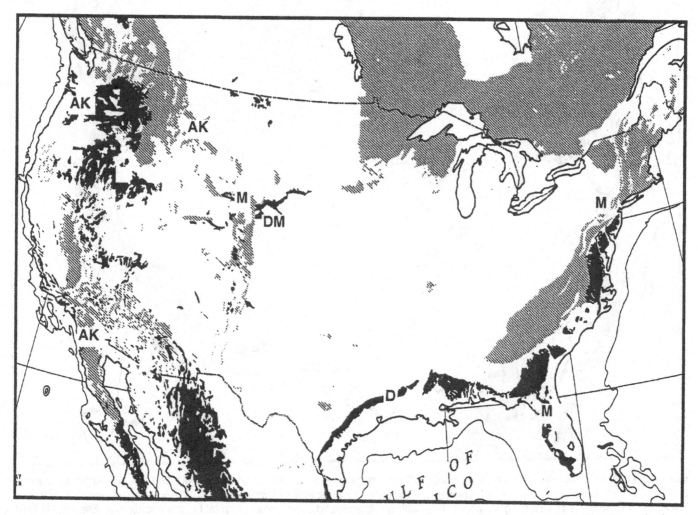

Figure 6.7. Latest Arikareean (early Miocene) distribution of rhinocerotids. Conventions as in Figures 6.1 and 6.6. **D** = *Diceratherium niobrarense*; **M** = *Menoceras arikarense*. **AK** = Arikareean localities with no rhinocerotids.

as many other immigration events (Woodburne and Swisher, 1995). After the late Hemingfordian, however, rhino diversity in North America stabilizes down to three lineages (*Teleoceras, Aphelops,* and *Peraceras*), with little evidence of further immigration.

One surviving species of the native *Diceratherium* lineage, *D. niobrarense*, is found in several late Arikareean and early Hemingfordian localities in Nebraska, and possibly persists until the late Hemingfordian (Railroad Canyon) in Idaho. However, the dominant lineage in the late Arikareean and early Hemingfordian is the immigrant *Menoceras*. During the late Arikareean, *M. arikarense* is known primarily from the High Plains (Nebraska, Wyoming), but specimens may also occur in the Farmingdale l.f., Monmouth County, New Jersey, and the Martin-Anthony l.f., Marion County, Florida (Tedford and Hunter, 1984). By the early Hemingfordian, *Menoceras barbouri* is found in nearly every locality that produces large mammal fossils, including the Thomas

Farm l.f. in Florida (its type locality), the Martin Canyon l.f. in Colorado, the Batesland Formation in South Dakota, the Runningwater Formation in Nebraska, the Zia Formation in New Mexico, the Delaho Formation in the Big Bend of Texas, the Garvin Gulley l.f. of the Texas Gulf Coastal Plain, and even from Mexico (Zoyatal l.f., Aguascalientes, Mexico) and Panama (Gaillard Cut l.f.).

Much less common are the earliest members of two other groups of immigrant rhinos that first appear in the early Hemingfordian: the aceratheriines and the teleoceratins. The oldest aceratheriine in North America appears to be *Floridaceras whitei*, known from only four places: its type locality in the Thomas Farm l.f. in Florida, the J.L. Ray Ranch in Dawes County, Nebraska, the Warm Springs l.f. in Oregon, and the Gaillard Cut l.f. in Panama. Although it is a large and distinctive taxon, diverse Hemingfordian samples from localities such as the Runningwater Formation, the Batesland Formation, the Martin Canyon

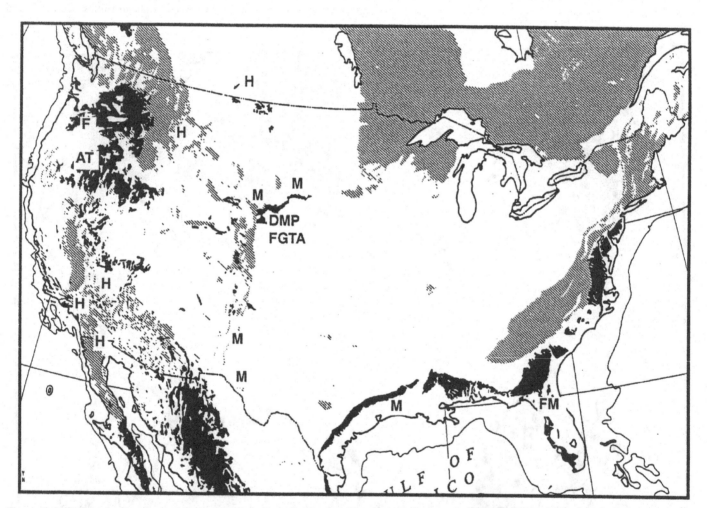

Figure 6.8. Hemingfordian (early Miocene) distribution of rhinocerotids. Conventions as in Figures 6.1 and 6.6. **A** = *Aphelops megalodus*; **D** = *Diceratherium niobrarense*; **F** = *Floridaceras*; **G** = *Galushaceras*; **M** = *Menoceras barbouri*; **P** = *Peraceras profectum*; **T** = *Teleoceras americanum*. **H** = Hemingfordian localities with no rhinocerotids. Not shown: Zoyatal l.f., Aguascalientes, Mexico; Gaillard Cut l.f., Panama, which both yield *Menoceras barbouri*; Panama yields *Floridaceras* as well.

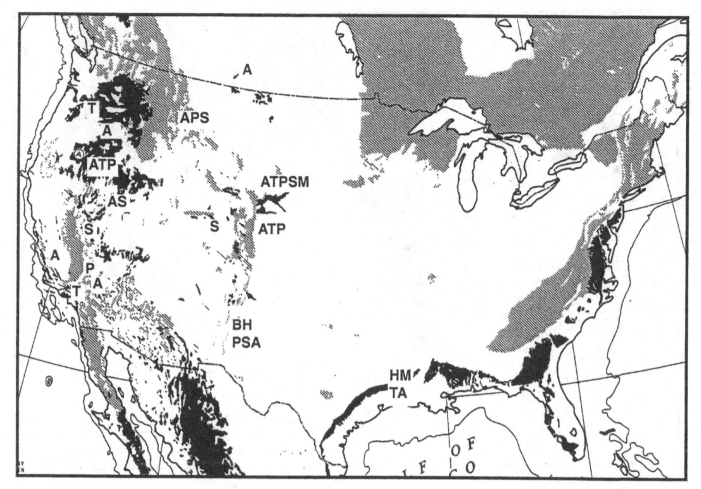

Figure 6.9. Barstovian (middle Miocene) distribution of rhinocerotids. Conventions as in Figures 6.1 and 6.6. **A** = *Aphelops megalodus*; **P** = *Peraceras profectum*; **H** = *P. hessei*; **S** = *Peraceras superciliosum*; **T** = *Teleoceras medicornutum*; **B** = *T. brachyrhinum*; **M** = *Teleoceras meridianum*.

Formation, the Garvin Gully fauna of Texas, and the Zia Formation of New Mexico, have so far not produced *Floridaceras*, so its distribution is very restricted. *Floridaceras* was followed shortly by another rare aceratheriine, *Galushaceras levellorum*, which so far is known only from the early middle Hemingfordian Box Butte Formation of Nebraska. Finally, the type specimen of *Teleoceras* ("*Brachypotherium*") *americanum* is also known from the early Hemingfordian Martin Canyon l.f. of Colorado, but it also occurs in the Runningwater and Box Butte formations of Nebraska. However, it is much more common in the late Hemingfordian Sheep Creek Formation of Nebraska, and it may also occur in the latest Hemingfordian Massacre Lake l.f. of Nevada (Morea, 1981).

The late Hemingfordian marks the dominance of aceratheriine and teleoceratin immigrants into North American faunas. *Teleoceras americanum* is abundant in the Sheep Creek Formation, as is *Peraceras profectum* and *Aphelops megalodus*. These three rhino genera will be the dominant ones through the rest of the Miocene. Meanwhile, *Menoceras barbouri, Floridace-*

ras whitei, and *Galushaceras levellorum*, which were dominant in the early Hemingfordian, had apparently become extinct. Only *Diceratherium niobrarense* remained as a late Hemingfordian relict in Idaho prior its extinction before the beginning of the Barstovian.

Barstovian (middle Miocene, 11–16 Ma)
The geographic distribution of fossil localities was at its minimum in the Whitneyan, and not much better in the Arikareean, but spreads to a number of regions by the Hemingfordian. In the Barstovian (Fig. 6.9) and Clarendonian, we begin to get extensive records from states that were poorly represented in the early Cenozoic, such as Nevada, Washington, Oregon, California, Arizona, and New Mexico, so the total geographic spread of localities, and the total number of localities, is much greater than in the Oligocene. In addition, we have good Barstovian samples from the Coastal Plain, especially in Texas, so that we are starting to see geographic provinciality as well (Tedford *et al.*, 1987).

Despite the wide range of localities in the Barstovian, the

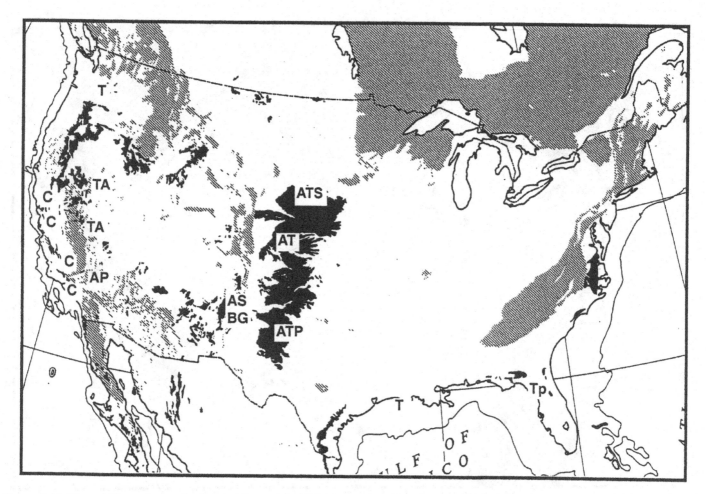

Figure 6.10. Map of Clarendonian (early late Miocene) localities producing rhinoceroses. Black areas show outcrops of Pliocene marine and non-marine beds (since the Clarendonian was once thought to be early Pliocene), after Cook and Bally (1975). Other conventions as in Figure 6.1. **A** = *Aphelops megalodus*; **P** = *Peraceras profectum*; **S** = *P. superciliosum*; **T** = *Teleoceras major*; **B** = *T. brachyrhinum*; **Tp** = *T. proterum*; **G** = giant aceratheriine. **C** = Clarendonian localities that produce no rhinoceroses.

diversity of rhinoceroses was not great. Only three genera are known, and they are never very diverse. For example, during the Barstovian there was only one species of *Aphelops* (*A. megalodus*) and three species each of *Peraceras* (*P. profectum*, *P. hessei*, and *P. superciliosum*) and three of *Teleoceras* (the widespread *T. medicornutum*, the short-nosed *T. brachyrhinum* in New Mexico, and the dwarf *T. meridianum* in New Mexico and the Gulf Coastal Plain). Because of the endemism of the Gulf Coastal Plain dwarfs, and the New Mexico *T. brachyrhinum*, the vast majority of Barstovian localities produce only one species each of one or two of the genera *Teleoceras*, *Aphelops*, and *Peraceras*, and very few (e.g., the Olcott and Valentine formations in Nebraska, or the Pawnee Creek Formation in Colorado) produce all three genera with the species *T. medicornutum*, *A. megalodus,* and *P. profectum.*

Prothero and Sereno (1982) and Prothero and Manning (1987) have already discussed the phenomenon of the dwarfing of *P. hessei* and *T. meridianum* in the Texas Gulf Coastal Plain faunas,

while they coexisted with their normal-sized relatives *T. medicornutum* and *A. megalodus*. Patton (1969) and Tedford *et al.* (1987) pointed out the high degree of endemism and the peculiarity of the Gulf Coastal Plain faunas of Texas and Florida, presumably in response to strikingly different, more humid and forested habitat. Prothero and Sereno (1982) suggested that the dwarfing in these rhinos may be analogous to the dwarfing in hippos or elephants that occurs on islands and in dense forests, where the resources are more patchy and restricted, so large body sizes are not as favored as they are in plains and grasslands. However, the occurrence of both dwarfs outside the Coastal Plain (*P. hessei* in New Mexico; *T. meridianum* in Nebraska) tempers this hypothesis somewhat, because they apparently did not remain restricted to the Coastal Plain forests, but returned to the plains in at least small numbers.

In addition to the endemism of the Gulf Coastal Plain dwarfs, and of *T. brachyrhinum* in New Mexico, there are other peculiar patterns as well. The large *Peraceras superciliosum* is most common in the northern Plains and Rockies (Montana, South Dakota,

Nebraska, northwestern Colorado, northern Nevada) plus New Mexico (which it shared with the other two species of *Peraceras*, plus *Teleoceras brachyrhinum* and *Aphelops megalodus*). However, it is not found in the Barstovian of the Pawnee Creek Formation of northeastern Colorado, or the Texas Gulf Coastal Plain, or in California (Barstow Formation, Ricardo Formation) or in Oregon (Mascall Formation), all of which produce *A. megalodus* and/or *T. medicornutum*. It became extremely rare by the early Clarendonian, when large species of *Teleoceras* (such as *T. major*) appear, so Prothero and Manning (1987) suggested that it was displaced by *Teleoceras*, and eventually driven to extinction in the late Clarendonian.

Another striking phenomenon in the Barstovian, Clarendonian, and Hemphillian is not apparent on the maps of localities. Even though a wide range of regions are now represented (particularly west of the Rockies) that were not represented in the Oligocene, rhinoceroses are not equally abundant in all those localities. In the High Plains Miocene (the classic "Clarendonian chronofauna"), rhinoceroses are among the most common large mammals other than horses, camels, and mastodonts. In some quarries, such as Quinn Rhino Quarry, Ashfall Fossil Beds, Hottell Ranch, Agate Springs Quarry (all in Nebraska), Wray in Colorado, and Long Island Rhino Quarry in Kansas, rhinoceros fossils (usually of one species only) make up 80-90% of the known specimens, with huge numbers of individuals represented. Such huge samples allow us to reconstruct the complete skeletons of many species (Chapter 4) and talk about population structure as well (Chapter 7). By contrast, although rhinoceroses are known from many localities in the Great Basin of Nevada, California, Oregon, and Washington, rhinoceros remains are relatively rare, compared to mastodonts, camels, and horses. In examining the large collections from these regions at the UCMP or LACM, it is striking how few rhino fossils occur in these places, and how scrappy these fossils are as well. Almost no skulls are known from these regions in the Miocene, let alone the multiple complete skeletons that we find in the High Plains.

This rarity in the Great Basin and West Coast cannot be simply attributed to poor sampling, because many of these localities yield hundreds of fossils with a diverse fauna and are well up the rarefaction curve. Nor can it be attributed to taphonomic factors, because rhino bones are robust to mechanical breakdown and easy to spot in surface collecting (so they should be overrepresented). In addition, other large mammals, such as mastodonts, horses, and camels, are well represented, so there is no size bias against large mammals in the collections. In many UCMP and LACM localities (such as the Caliente Formation, the Chanac Formation, the Black Hawk Ranch l.f., and the Modelo Formation, all in California), there are hundreds of specimens of horses, camels, and mastodonts, yet not a single rhino fossil has turned up despite decades of intense collecting. In other localities (such as the Dove Spring Formation in the Ricardo Group, or the Barstow Formation in California, or the Mascall and Rattlesnake formations in Oregon, or most localities in Nevada), rhinos are present but extremely rare compared to other large mammals. For example,

the Mascall Formation produces only a few scraps of *T. medicornutum* but no *Aphelops* or *Peraceras*. The Barstow Formation yields a few specimens of *A. megalodus*, but no *Teleoceras* or *Peraceras*. The Ricardo Group includes a few specimens of *P. profectum*, but no *Teleoceras* or *Aphelops*. Clearly, there must be something else at work here besides taphonomy or poor sampling. Given the overwhelming balance toward horses, antilocaprids, and camels in the Great Basin, perhaps these more upland, montane habitats favored long-limbed grazers, so that rhinos were relatively scarce. By contrast, the broad floodplains with huge river channels of the High Plains may have been better suited to large herds of semiaquatic mammals (particularly *Teleoceras*) that could not live in large numbers in regions without large rivers and floodplains, like the Great Basin. Alternatively, the taphonomy of large river channels in the High Plains may have favored the preservation of the rhinos living in and near the water, whereas such channel sandstones are rare in the montane uplands of Nevada, Oregon, or California. Further examination of the data is needed to test this hypothesis, but this is the overwhelming impression gained from examining the collections now available.

Clarendonian (early late Miocene, 8.8–11 Ma)

Many of the phenomena established in the Barstovian (with the beginning of the "Clarendonian chronofauna") are continued into the Clarendonian (Fig. 6.10). The wide geographic spread of Clarendonian localities from California, Nevada, Montana, and New Mexico, as far east as the localities in the High Plains, continues. In addition to these localities that were well represented in the Barstovian, there are the classic Clarendon beds of the northern Panhandle of Texas and Oklahoma, the Clarendonian Lapara Creek faunas of the Texas Gulf Coastal Plain, and several Clarendonian faunas (especially Love bone bed) in Florida. Hence, we have rhino records of much of North America except the northeast and midwest.

Other trends from the Barstovian continue as well. Only three genera of rhinos are known (*Teleoceras*, *Aphelops*, and *Peraceras*) with only one species of *Aphelops* (*A. megalodus* again), but only two species of *Teleoceras* (*T. major*, and the endemic New Mexican *T. brachyrhinum*), and only two species of *Peraceras* (*P. profectum* and the last of the *P. superciliosum*). Thus, rhinoceros generic diversity remained the same, but species-level diversity decreased, so that only three species (*T. major*, *A. megalodus*, or *P. profectum*) dominate the continent, and only one or two in most localities. Once again, the increase in geographic spread of localities has not resulted in increased diversity or disparity, but just the opposite. The raw diversity numbers or geographic ranges of taxa do not capture the actual pattern of species dominance. Most localities in the Clarendonian are dominated by one species of rhinoceros (such as the huge number of *T. major* in the Ashfall Fossil Beds). In localities that produce more than one species of rhino, typically there is a grazer (usually *Teleoceras*) and a browser (usually *Aphelops*), forming the "browser–grazer pairs" discussed by Prothero and Manning (1987).

In addition to the known species from the Clarendonian, there are also the huge aceratheriine bones (Chapter 4) from New Mexico whose taxonomic affinities are not yet known. Clearly, they represent yet another (as still unnamed) taxon which so far is only known from one locality in New Mexico.

As in the Barstovian, the tendency for huge numbers of rhino bones in the High Plains localities continues (especially the Clarendon beds of Texas, and in many localities in Nebraska), while most Clarendonian localities in the Great Basin and Pacific Coast have either very few rhinos (e.g., the Dove Spring Formation of the Ricardo Group), or none at all (Chanac Formation, Caliente Formation, Blackhawk Ranch l.f. in California). As discussed above, perhaps the montane habitats with few large rivers were inhospitable to the huge herds of water-loving semi-aquatic grazing rhinos that were so dominant in the High Plains.

Hemphillian (latest Miocene to earliest Pliocene, 4.5–8.8 Ma)

The Hemphillian represents the conclusion of the Clarendonian chronofauna, as well as the final stage of the Miocene savanna-grasslands faunas of North America before the great extinction

event at the end of the Miocene. However, rhino generic diversity declines to two genera, *Aphelops* and *Teleoceras*, with the extinction of *Peraceras profectum*. Within both of those genera, however, there is an increase in species, as there are two successive species of *Aphelops* in the Hemphillian (*A. malacorhinus* in the early Hemphillian, followed by *A. mutilus* in the late Hemphillian) and four species of *Teleoceras* (*T. fossiger* and *T. proterum* in the early Hemphillian, followed by *T. hicksi* and the dwarf *T. guymonense* in the late Hemphillian). Likewise, the geographic spread of Hemphillian localities is comparable to that of the Clarendonian, with excellent Florida records (Mixson's bone bed, Bone Valley Formation) along with extensive High Plains localities from Nebraska to Texas, and many other localities spread out from New Mexico to Nevada, California, Washington, and Oregon (plus isolated scraps from the Pipe Creek Sinkhole in Indiana, the only locality in the northeastern part of the United States—Farlow *et al.*, 2001; Gray Fossil Site in Tennessee may be another).

As before, the High Plains localities (especially Long Island Rhino Quarry and Edson in Kansas and Wray in Colorado, and several localities, such as Guymon, Box T Ranch Pit 1, and Coffee

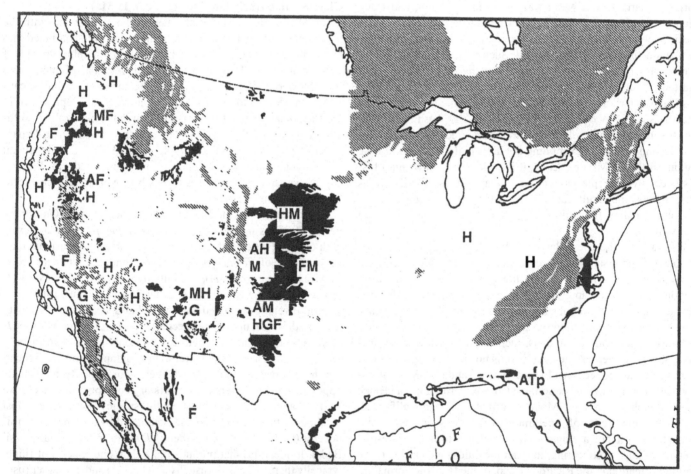

Figure 6.11. Hemphillian (latest Miocene–earliest Pliocene) distribution of rhinoceroses. Conventions as in Figures 6.1 and 6.9. **A** = *Aphelops malacorhinus*; **M** = *A. mutilus*; **Tp** = *Teleoceras proterum*; **F** = *T. fossiger*; **H** = *T. hicksi*; **G** = *T. guymonense*. Not shown: Gracias Formation, Honduras; Rancho el Ocote, Guanajuato, Mexico.

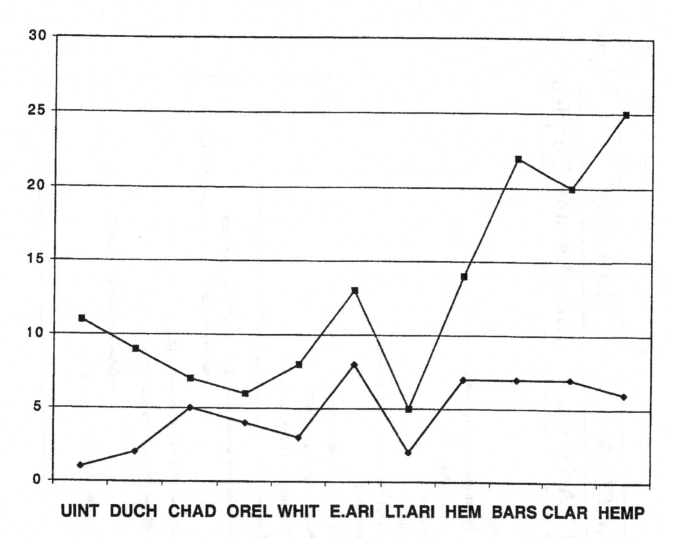

Figure 6.12. Temporal trends in number of species of rhinocerotids (lower curve with diamond symbols) and number of geographic regions/states during each North American land mammal age (upper curve with square symbols). Note that there is a a very weak trend, with low diversity during the Barstovian-Hemphillian despite great increase in geographic regions represented, and also increasing diversity in the Uintan through Chadronian despite a decreasing geographic spread. The two Arikareean data points are anomalous: the early Arikareean is distorted by the occurrence of a number of unique forms, while the latest Arikareean is only found in a few places (see text for discussion).

Ranch in the Texas–Oklahoma panhandle region) produce huge quantities of rhino bones, while those in the Great Basin and Pacific Coast produce only a few scraps of one taxon per locality. As discussed above, these differences may be due to the lack of large rivers and floodplains for the herds of semiaquatic rhinos in the Great Basin, and the abundance of such rivers and floodplains in the High Plains.

Although the endemic *T. brachyrhinum* is gone from New Mexico by the late Hemphillian, other endemics (such as *T. proterum* in Florida) do occur. By the late Hemphillian, the biogeographic and abundance patterns become very odd. As Prothero *et al.* (1989) and Prothero and Manning (1987) point out, the old notion that only *Aphelops* is present in the late Hemphillian is now contradicted by the collections. *Aphelops mutilus* is indeed more abundant in many classic late Hemphillian localities, such as Edson or Coffee Ranch, but rare specimens of *T. hicksi* and or *T. guymonense* are found in most late Hemphillian localities across the country. (By contrast, *T. hicksi* is hugely dominant over *Aphelops mutilus* in the Wray, Colorado, quarries). Actually, *A. mutilus* is rather restricted in its distribution (mainly the High Plains localities in Nebraska, Kansas, Oklahoma, and Texas, plus a few specimens in New Mexico and Oregon). *T. hicksi* is found in a wide variety of late Hemphillian localities, from Bone Valley Formation in Florida to the High Plains (again, mainly Colorado, Nebraska, Texas, and Oklahoma) to multiple localities in Nevada, Oregon, Washington, California, and even Arizona and New

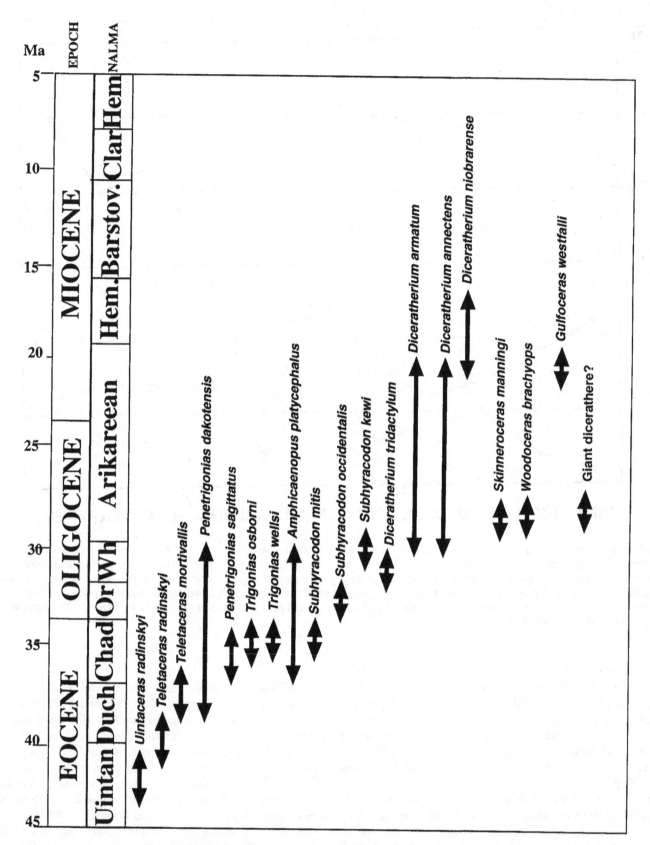

Figure 6.13. Chronologic ranges of the valid rhinoceros species discussed in this book. Time scale after Woodburne (2004) and Prothero and Emry (1996). "NALMA" = North American land mammal ages.

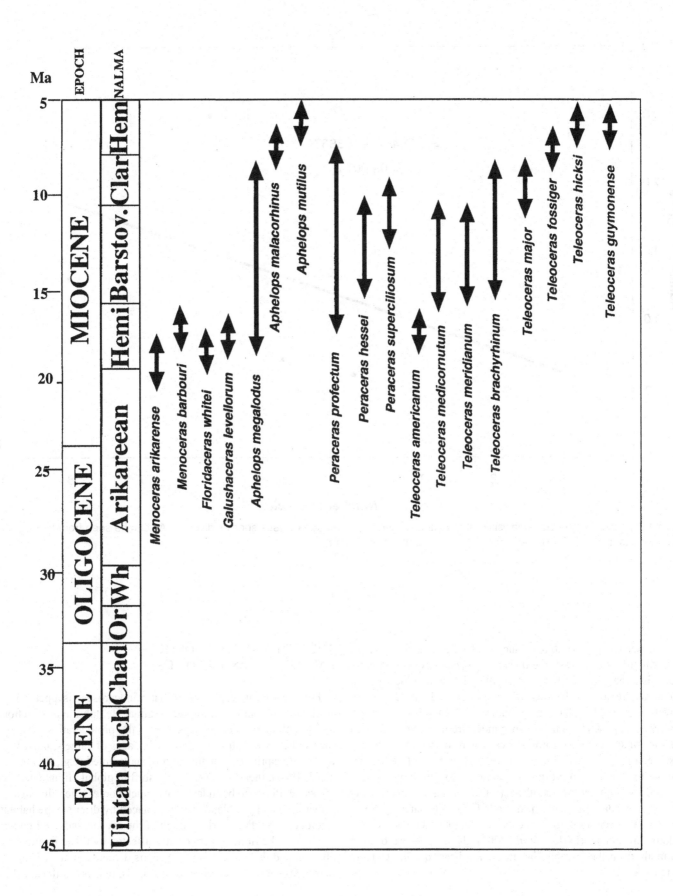

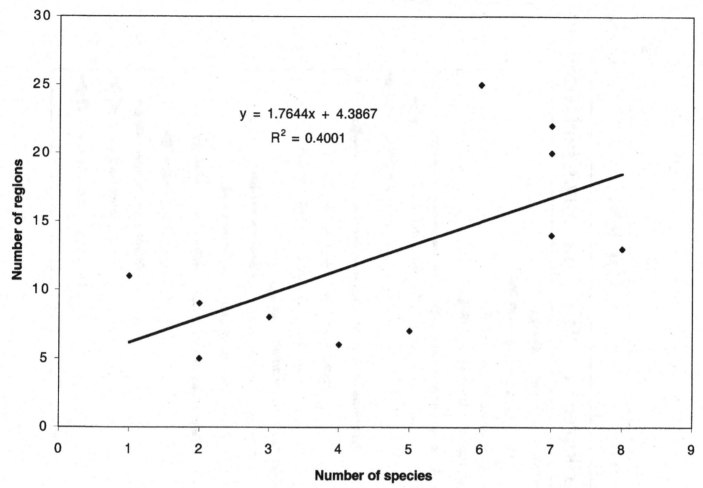

Figure 6.14. Linear regression analysis of the number of geographic regions versus species diversity. As can be seen from the regression statistics, there is a very weak correlation, with an r² value of 0.40.

Mexico (see Chapter 4), to the Rancho el Ocote l.f. and Yepomera l.f. in Mexico. In contrast, the dwarf *T. guymonense* is relatively rare. It is largely known from the Ogallala Formation localities in Oklahoma, Texas, and Kansas, where it coexisted with *T. hicksi,* as well as from the Chamita Formation in New Mexico (also a locality of *T. hicksi*), with rare specimens from the San Timoteo badlands of southern California. Thus, it is restricted to the southwest tier of the United States, whereas *T. hicksi* is found from Florida to Mexico to northern Nebraska to Washington and Oregon, as well as the southwest (California, Arizona, New Mexico, Nevada). In addition, a tooth fragment that may be *T. hicksi* is also reported from the latest Hemphillian Pipe Creek Sinkhole in Indiana (Farlow *et al.*, 2001), the only record of land mammals from the Hemphillian in the northeast quadrant of the United States.

THE EXTINCTION OF NORTH AMERICAN RHINOCEROTIDS

For decades, it seemed clear that rhinoceroses disappeared from North America during the terminal Hemphillian extinction, along with many other groups (especially protoceratids, dromomerycids, mylagaulids, plus many lineages of camels, horses, and antilocaprids). As many authors have noted (e.g., Webb, 1977, 1984), there was a dramatic shift in global climate near the Miocene–Pliocene boundary, with much cooler drier climates prevailing, and grasslands being replaced by drier steppe habitats in much of North America. Much of the savanna-adapted fauna (including the rhinos) were clearly affected by this climatic change, and the succeeding Blancan is depleted in large land mammals (only a few lineages of camels, horses, antilocaprids, and mastodonts make up the bulk of the early Blancan large land mammal fauna).

However, Madden and Dalquest (1990) reported a single scrap of probable rhino tooth enamel from the well-known Beck Ranch l.f., Scurry County, in the Texas Panhandle, which is of unquestioned Blancan age (3.4–3.6 Ma). In their paper, they argue that the specimen is found *in situ* among the other Blancan fossils and that it is not transported or reworked from the nearby beds, or part of the overlying Pleistocene gravels that later washed in. However, I am not convinced by their descriptions that the specimen wasn't reworked out of the nearby late Hemphillian deposits of the Ogallala Formation during the Blancan. Rhinoceros enamel is very durable to river transport and reworking, and as the recent experience with reworked Cretaceous shark and tyrannosaur teeth in the Paleocene of Montana has shown, it's easy to be fooled by fossils that have been reworked out of older beds.

Certainly, it is more parsimonious to assume that this one tooth scrap has been reworked than to assume that hundreds of Blancan localities all over North America have failed to fossilize one rhino specimen while they were still living on this continent. Thus, I regard the Beck Ranch specimen as questionable evidence of Blancan rhinos, and think it is more likely that rhinos died out in North America along with most of the rest of the terminal Hemphillian victims.

DIVERSITY VERSUS GEOGRAPHY THROUGH TIME

As discussed throughout this chapter, there is an ambiguous relationship between diversity and number of geographic regions through time. Range charts of the diversity of rhinoceroses through time are shown in Figure 6.13, giving the trend shown in Figure 6.12. As can be seen from this chart, there is only a weak correlation between number of species of rhinocerotids in each land mammal age, and the number of geographic regions sampled (as compiled from the hypodigms in Chapter 4 and the maps in this chapter, using each state or widely separated formation as a separate region). From the Uintan through Chadronian, the species diversity is increasing, while the geographic spread of regions is decreasing. In the Orellan and Whitneyan, the geographic spread of regions starts to increase, but the number of species decreases. The early Arikareean is slightly anomalous. As explained earlier in the chapter, it is dominated by two species, *Diceratherium armatum* and *D. annectens*. However, the unique rare species (*Gulfoceras, Skinneroceras, Woodoceras, Subhyracodon kewi*, and the giant postcranials from Wyoming) add five more species to the list, artificially inflating the apparent diversity even though only two species make up 99% of the specimens. Likewise, the latest Arikareean is also anomalous, because only a few regions are known of that age.

The most obvious trend is that of the Barstovian through Hemphillian, when the geographic coverage of the North American continent steadily increases, but the number of rhinocerotid species declines from a high in the Barstovian to a minimum in the Hemphillian (and there are actually only three species in the early Hemphillian and two in the late Hemphillian, but the compilation method plots the Hemphillian as 5 total species). Thus, there appears to be only a weak relationship between geographic spread of localities and species diversity through most of the Eocene, early Oligocene and Miocene.

Statistical analysis of these data (Fig. 6.14) suggests that a correlation exists, but it is a weak one. A linear regression through the data matrix yields a near-shotgun scatter of points that yields an r-squared value of 0.40, which indicates a very low level of correlation of the data. However, the regression is significant ($p = 0.0367$). Thus, for the Rhinocerotidae, there is a weak positive correlation between species diversity and geographic diversity.

7. Paleoecology and Evolutionary Patterns

PALEOECOLOGY

Although fossil rhinoceros skeletons have been known in North America for over a century, relatively little paleoecological and functional analysis has been conducted on them. With a few exceptions (like the extraordinary barrel-chested, short-legged *Teleoceras*), most genera have not stimulated much paleontological speculation about the life habits and functional morphology of most of the species.

However, over the years many individual clues have accumulated. For example, the earliest rhinocerotids (such as *Teletaceras* and *Penetrigonias*) were long-limbed, fairly gracile forms that closely resemble the long-limbed running rhino, *Hyracodon*. Wall and Hickerson (1995) conducted a biomechanical analysis of the long limbs of *Hyracodon*, and concluded that it was subcursorial (i.e., a runner, but not as specialized a runner as most modern high-speed ruminants). Apparently *Hyracodon*-like rhinocerotoids had running abilities comparable to those of pigs and peccaries. Matthew (1901) was the first to suggest that long-limbed forms like *Hyracodon* (and presumably *Teletaceras* and *Penetrigonias* as well) were open plains dwellers which may have lived in herds. The open plains hypothesis was corroborated by Clark *et al.* (1967), but no large quarry samples of most of these rhinos has ever been found to suggest they lived in herds (except for *Teletaceras radinskyi*, which occurs in large numbers in Hancock Quarry in the Clarno Formation of Oregon). Even though *Hyracodon nebraskensis* is very abundant in the Lower Nodular Zone of the Scenic Member, Brule Formation in the Big Badlands of South Dakota, it never occurs in quarry concentrations (to my knowledge), but always as isolated skulls and jaws (Prothero, 1996). This is also corroborated by trackways (Terry *et al.*, 1995) found in Toadstool Park in Nebraska. Mead and Wall (1998a) analyzed the skull shape (especially the long premaxilla and the well-developed anterior dentition) of *Hyracodon*, and suggested that it was a browser of nuts, fruits, twigs, and tougher vegetation growing on the floodplains far from the river banks. Mead and Wall (1998b) also analyzed lower jaw biomechanics, and suggested that a *Hyracodon*-like skull and jaws was indicative of a browser of low tough vegetation away from the stream banks and out in the open plains. According to Zeuner's (1945) methods, the occipital condyles suggest that *Hyracodon* held its snout downward, and fed mostly on low browse near the ground (Mead and Wall, 1998a).

By contrast, the larger-bodied rhinos such as *Trigonias* and *Subhyracodon* must have had a very different ecology. Clearly, they were heavier with more massive limbs, so they could not have been subcursorial runners like *Hyracodon, Teletaceras*, or *Penetrigonias*. Matthew (1901) first suggested that they might have lived in riparian habitats and river channels, and this hypothesis was corroborated by Clark *et al.* (1967). The fact that both of these genera are known from several large quarry samples (e.g., *Trigonias* Quarry in Colorado, plus the Harvard Fossil Reserve, Rockerville, and Big Pig Dig samples of *Subhyracodon*) suggests that they might have lived in herds, or at least congregated in large numbers near waterholes and rivers. In addition, their trackways in the Toadstool Park area (Terry *et al.*, 1995) suggest that they roamed in herds (Fig. 7.1), in contrast to the evidence of trackways of *Hyracodon* from the same area. Mead and Wall (1998a) analyzed the anterior skull of *Subhyracodon*, with its short premolar row, enlarged lateral incisors, narrow and more delicate premaxillae, and weaker cingula, and argued that it was a mixed feeder (both browsing and grazing), which used both high-fiber vegetation and succulent browse in the wooded habitats around streams. Mead and Wall (1998b) studied the lower jaw of *Subhyracodon* which also suggested that it was a mixed feeder on succulent vegetation and high browse in the riparian strip around streams and waterholes. According to Zeuner's (1945) methods, the occipital condyles suggest that *Subhyracodon* held its snout forward (Mead and Wall, 1998a), and fed mostly on mid-level and high browse well above the ground (as does the modern black rhino, *Diceros bicornis*).

Amphicaenopus was a heavier-bodied, broader-skulled form than *Subhyracodon*, and from the beginning paleontologists have thought that it was an aquatic form, more like a hippo than *Subhyracodon* (Osborn, 1898c). Although specimens are rare, the fact that they all come from river channel sandstones is suggestive of this hypothesis. So is the fact that they vanish from the "*Metamynodon* channel" sandstones of the Orellan, when there was apparent competition from the semi-aquatic amynodonts (and reappear in the Whitneyan when *Metamynodon* nearly disappears). However, no one has done a comprehensive analysis of the skull, jaws, and limbs of this taxon (even though we have much more material now) to test these hypotheses.

Diceratherium has not received a comparable level of ecological and functional analysis, even though it was described in detail by Peterson (1920). Once again, however, the fact that *D. arma-*

Figure 7.1. Trackways of *Subhyracodon* at Toadstool Park, Nebraska. Numerous parallel trackways of this taxon suggest that it traveled in small herds, while those of *Hyracodon* from the same layer are solitary.

tum and *D. annectens* do occur in large numbers in several quarries (such as Frick 77 Hill Quarry, Niobrara County, Wyoming), with subequal numbers of males and females, suggests that there was some preference for near-stream or waterhole habitats, and that they probably assembled in herds at least part of the time. The skull shape, occiput, and premaxillary dentition of *Diceratherium* is not much different from that of *Subhyracodon*, suggesting that they both fed on mid- and high-level succulent vegetation and high browse in the bushes and trees around watercourses.

Menoceras arikarense, on the other hand, has received extensive study, since it occurs in staggering numbers in the Agate Springs quarries. One slab of bone (Fig. 7.2) covering 44 square feet contained 4300 skulls and bones, suggesting that the Agate bone-bearing layer may have once contained at least 3 million bones from about 17,000 skeletons (most of which are *Menoceras arikarense*)! Hunt (1990, 1992) has examined the bone bed in detail using modern taphonomic methods. The population structure of the assemblage contains too few juveniles, and too many adults, to be a catastrophic death assemblage. Instead, it suggests an attritional assemblage, perhaps accumulating around a series of waterholes as they gradually dried up in severe droughts. The Agate bone bed was deposited in a broad river channel, with extensive volcanic ash in the matrix (as is typical of the highly volcaniclastic Arikaree Group). However, this is strictly fluvial and lacustrine water-laid ash; there is no evidence of an airfall tuff such as that which entombed the Ashfall Fossil Beds *Teleoceras major* (Voorhies, 1985, 1992).

One striking taphonomic occurrence of a fossil rhino is the famous Blue Lake "rhino cast cave" in the Columbia River flood basalts in the Grand Coulee area, Washington (Fig. 7.3). Contrary to the general rule that fossils do not occur in igneous rocks, in this instance they do. This particular find was made in 1935 (Chapell *et al.*, 1941), when three men hiking along the steep-walled canyons of Grand Coulee came upon the opening to a small cave. They crawled inside and recovered the jaw and a few leg bones of a rhinoceros, then realized that the cave was the natural body impression of a rhino carcass! In 1948, scientists from the University of Washington returned, and made a mold of the cave, and a cast of the rhino from that natural mold. From the cast, it is clear that the rhino carcass was bloated and floating upside down in a lake. The lava flows formed pillows as they hit the water, which molded around the rhino carcass before they cooled. Thus, the cast is not very detailed, but the shape of the rhino's head and prehensile lip, and the three blunt toes with the thick heel pad on each foot are well preserved. Unfortunately, the snout was between two pillows, so the condition of the nasal horns cannot be determined, and the available teeth are not diagnostic. They could belong to a late *Diceratherium*, a *Menoceras barbouri*, or a small *Peraceras* (although the 15–16 Ma date of the Columbia River basalts is more consistent with the latter).

Paleoecological hypotheses (e.g., Matthew, 1932) have seldom been suggested for the aceratheriines *Aphelops* and *Peraceras*, largely because their systematics was in chaos and complete skeletons were rare. However, some generalities have emerged. Most

Figure 7.2. Typical slab from Agate Springs Quarry, Nebraska, composed largely of disarticulated specimens of *Menoceras arikarense*.

aceratheriines have highly retracted nasals (and the nasal retraction increases from *A. megalodus* to *A. malacorhinus* to *A. mutilus*), suggesting that they have a much better developed prehensile lip (Fig. 4.28D) than even the modern black rhinoceros, so they could strip foliage even better than the modern black rhino does today. Their cheek teeth are generally the most brachydont of Miocene rhinos (becoming slightly hypsodont with *A. mutilus*), so they presumably were obligate browsers on middle- and low-level foliage. This has been confirmed by oxygen isotopic analysis of *A. malacorhinus* from Love and Mixson's bone beds in Florida (MacFadden, 1998). The teeth show that they at mostly C3 plants, with few or no C4 grasses. Only toward the late Hemphillian did they become more mixed feeders-grazers as their hypsodonty increased. As several people (e.g., Matthew, 1932; Webb, 1983) have noted, aceratheriines retained the longer limb proportions seen in living rhinos, so presumably they were more adapted for running than teleoceratins, and could sprint in short charges like modern rhinos do. All North American aceratheriines (except possibly *Peraceras superciliosum*) were apparently hornless, based on the lack of any rugosity on the tips of their nasals, so they probably did not engage in the intraspecific horn sparring seen in modern territorial horned rhinos such as *Ceratotherium simum*, and must have defended themselves and their territories with their i2 tusks (Penny, 1988; Dinerstein, 2003). In most aceratheriines, these tusks had no upper incisor to wear against (somewhat comparable to the living African rhinos, which have lost both upper

and lower incisors), so were never sharpened by thegosis. This suggests that most *Aphelops* and *Peraceras* were solitary browsers that roamed widely and did not defend a territory. The scarcity of *Aphelops* or *Peraceras* (compared to most teleoceratins) in the river channel deposits of the High Plains (with the notable exception of Coffee Ranch Quarry in Texas, and Edson Quarry in Kansas, which are dominated by *A. mutilus*) suggests that they did not congregate in large herds around water holes, but only occasionally visited rivers and water holes to drink. Mihlbachler (2003) studied the population structure of the extraordinary samples of *A. malacorhinus* from Love Bone Bed in Florida, and of *A. mutilus* from Coffee Ranch in Texas. The *Aphelops* assemblages were not significantly male-biased, nor do they show high mortality of young males, so there is no clear similarity to the ecology of modern rhinos. Mihlbachler (2003) suggests that the *Aphelops* social patterns were different from any living analogue, and that there was no tendency for young males to die preferentially (as happens in most living rhinos). Janis (1982) suggested that the aceratheriine rhinos fell into her "Category C2," which included ungulates living in woodland-savanna habitats that browse on trees and shrubs, living in isolation or in small monogamous or serially polygamous non-territorial mixed-sex herds.

The exception to these generalities about aceratheriines is *Peraceras superciliosum*. As shown in Chapter 4, it was a larger, more robust form than any other aceratheriine, with a thick rugose area on the male nasals that may have supported a small horn

Figure 7.3. The Blue Lake rhino mold. A. Panorama of the Grand Coulee where the mold was found. B. Shot inside the mold, showing the impressions of the feet. C. Reconstruction of the rhino mold cave and (D) cast of the rhino carcass made from the cave (formerly on display at the Burke Memorial Museum, University of Washington). E. Reconstruction of the Blue Lake rhino (after Chappell et al., 1941). (Photos B–D courtesy Burke Memorial Museum, University of Washington.)

(apparently, the only North American aceratheriine to bear horns). Its skull is broad and robust like that of a teleoceratin, with a broad muzzle suggesting that it was more of a grazer like *Teleoceras* or the living white rhino, *Ceratotherium*. As noted in Chapters 4 and 6, it had a peculiar distribution, being dominant mostly in the northern states (e.g., Nebraska, South Dakota, Montana, northern Colorado, and Nevada) and only rarely found in the same beds with *Teleoceras*. This suggests that it may have competed in some way with *Teleoceras* for the large-bodied grazer niche. However, no *P. superciliosum* became very hypsodont (unlike almost all teleoceratins), and the fact that it vanished in the early Clarendonian, while teleoceratins flourished all over North America until the late Hemphillian, suggests that *P. superciliosum* was not as effective a grazer as the teleoceratins.

In contrast to the lack of speculation about aceratheriines, the teleoceratins have generated a large literature concerning their ecology. Osborn (1898a, 1898b, 1910) and Scott (1913) first suggested that teleoceratins lived like modern-day hippos, spending their days in the rivers resting, and coming out on the banks at night to graze. This interpretation has been followed by most workers since that time (e.g., Webb, 1983; Voorhies, 1981, 1985, 1992; Prothero *et al.*, 1986, 1989; Prothero and Manning, 1987;

Prothero, 1992, 1998). In contrast, Matthew (1932) was not convinced that teleoceratins were necessarily aquatic, but simply lived on the rolling plains as open-field grazers. With their extremely hypsodont teeth and broad "lawnmower" snouts, there has been little question in the past that teleoceratins were grazers, and that was confirmed by the presence of grasses preserved in the throat regions of *Teleoceras major* from Ashfall Fossil Bed State Park (Voorhies and Thomasson, 1979).

However, the degree to which *Teleoceras* was a hippo analogue with aquatic habitats versus an open-field grazer has been disputed. The high number of *Teleoceras* specimens in river channel sandstones, and especially in the ash-filled pond deposit at Ashfall Fossil Bed State Park, seems to make a strong case for the hippo analogue. The short limbs and the barrel-shaped chest (first noted by Cope, 1879a, and Osborn, 1898a, 1898b) have always been considered *prima facie* evidence of this life style. But MacFadden (1998) found that the carbon isotopes in the teeth of *Teleoceras* indicated a mixed diet, not a strict grazer (at least in Florida). According to MacFadden (1998), there was no evidence in the oxygen isotope values of teeth from Florida that *Teleoceras* was primarily aquatic, but instead suggested that it was a grazer or mixed feeder, perhaps living like the modern terrestrial grazing

Figure 7.4. Reconstruction of the Ashfall water hole before the ash fell. A herd of female *Teleoceras major* and their calves hog the center of the water hole, with horses, camels, and birds nearby. (After Voorhies, 1992, by permission of the University of Nebraska State Museum.)

white rhino, *Ceratotherium*. Mihlbachler (pers. comm.) has re-examined these specimens, and found that the oxygen (but not carbon) isotopes were contaminated, so these conclusions are suspect. Clementz and Koch (2000) re-examined the oxygen isotopes of the teeth of *Teleoceras* from specimens outside Florida, and found that they showed very low variability typical of aquatic mammals, which are buffered in the oxygen isotope composition of the water they ingest by their aquatic surroundings.

Population structure has also been used as evidence. Wright (1980) argued that both *Teleoceras proterum* and *Aphelops malacorhinus* from Love Bone Bed in Florida had a mortality structure similar to that of modern rhinoceroses. Mihlbachler (2003) re-examined these same specimens, and also concluded that the mortality structure of *Teleoceras proterum* (but not *Aphelops malacorhinus*) samples from Love and Mixson's bone beds in Florida showed some high mortality among young males (comparable to that of *Diceros bicornis*, the black rhino), but not as low as that found in living *Hippopotamus*. Voorhies (1985) and Mead (2000) used the predominance of females (78% of the individuals) from the Ashfall Fossil Bed to argue that *Teleoceras major* formed large female-dominated herds like modern *Hippopotamus*, with most males living outside the herd as bachelors, and only a few dominant males in the herd. But Mihlbachler (2003) reported that the samples of *Teleoceras proterum* from both Love and Mixson's bone beds are largely male-dominated, and argued that this is more like modern rhinoceros analogues. In his view, the female-dominated assemblage at Ashfall might have more than one interpretation other than the hippo analogy, since the catastrophic instantaneous death assemblage at Ashfall does not clearly differ from the sex-specific distribution of individuals found in non-herding, non-territorial rhinos (e.g., *Rhinoceros unicornis*) in and around water. The attritional death assemblages in Florida, on the other hand, are a long-term accumulation, and apparently picked up samples of animals that lived all around the region for a longer span of time than the instantaneous slice of time represented by Ashfall Fossil Bed.

Ashfall Fossil Bed revealed many other things about *Teleoceras major* (Voorhies, 1981, 1985, 1992; Mead, 2000), most of which will be documented when Voorhies fully describes the specimens. The specimens are so exquisitely preserved that they even include the delicate hyoid bones of the throat and the tiny ear ossicles, neither of which is known from any other fossil rhino in North America. Many of the adult females had calves found in nursing position against their bodies, and several had fetal bones in the region of their pelvic cavities. The calves showed well-defined age classes, suggesting that there was a calving season about the same time each year, rather than year-round births. In addition, the skeletons (especially those of the males) showed healed fractures on their ribs, suggesting that there was intraspecific sparring and competition among males (as there is in the living rhinos). Much further detail is given in Voorhies (1985).

Clearly, much more can be learned by population analyses of many of the other large quarry samples of rhinos, isotopic studies of their teeth beyond the few studied so far, and functional analyses of taxa (especially the now well-known but seldom studied taxa such as *Amphicaenopus* and *Peraceras*). Hopefully, the systematic foundation presented in this volume will pinpoint the key collections and samples, highlight the important gaps in our knowledge, and encourage such studies in the future.

EVOLUTIONARY PATTERNS

With large databases of specimens and species now available, it is important to look at phenomena higher than the level of individual populations and species. The North American Rhinocerotidae now represent one of the largest, longest, and most complete records of a mammalian family in all of North America, comparable or superior in many ways to the overused evolution of the horse (Prothero, 1994). Several important patterns emerge when the entire rhino record is examined in detail.

Convergence and parallelism

Rhinocerotids are prime examples of patterns of evolutionary convergence and parallelism. As we have seen in Chapter 4, several lineages of rhinos (primarily *Teleoceras* and Hemphillian species of *Aphelops*) independently became hypsodont in the middle and late Miocene, and some of those (especially *Peraceras superciliosum* and the teleoceratins) converged on the large-bodied, hippo-like grazing body shape. The huge *Amphicaenopus* also converges on this body shape (or that of *Metamnyodon*) to some extent. The horn patterns of rhinocerotids are also highly convergent. As discussed before (Prothero *et al.*, 1986, 1989; Antoine, 2002), the character evidence clearly shows that the paired horns of *Diceratherium* and *Menoceras* are independently evolved, and apparently some of the early teleoceratins in Europe had paired horns as well (see Chapter 4.). This contradicts what earlier workers (e.g., Peterson, 1920) thought when they grouped all paired-horned rhinos in the "Diceratheriinae" and placed them all in the same genus, *Diceratherium*, as well. As discussed in Chapter 4 (especially Figure 4.16I-J) and reviewed in Prothero *et al.* (1986, 1989), the characters other than horns clearly show that *Diceratherium* and *Menoceras* are not closely related (nor is either close to the paired-horned European teleoceratins), and the detailed pattern of the horns (broad subterminal flanges in *Diceratherium* versus terminal knobs in *Menoceras*) corroborates the interpretation that the paired horns are not homologous in their detailed structure. Other examples of small-scale convergence are mentioned in many places in Chapter 4.

Dwarfing

As discussed by Prothero and Sereno (1982) and Prothero and Manning (1987), dwarfing appears to be a common phenomenon in the Rhinocerotidae. The most striking examples are the dwarf *Peraceras hessei* and *Teleoceras meridianum* from the Texas Gulf Coastal Plain, both of which are at least 20% smaller than their contemporary sister-species. As Prothero and Sereno (1982) pointed out, this dwarfing may be comparable to that seen in forest-

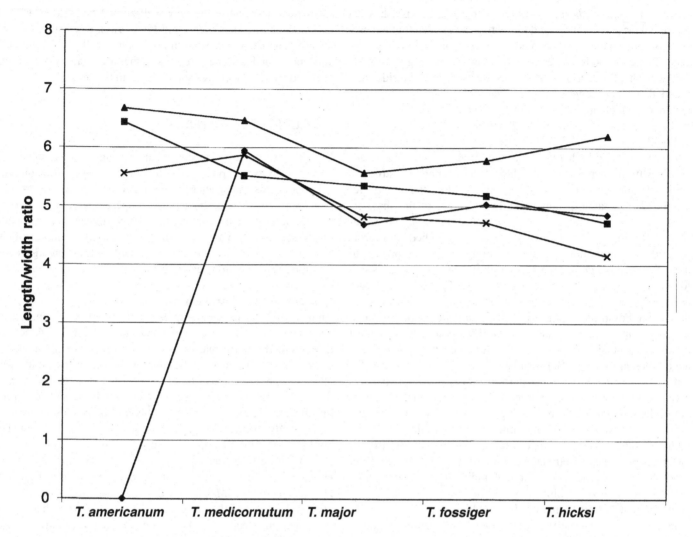

Figure 7.5. Length/width ratios of *Teleoceras* species through time. Triangles = femora; diamonds = humeri (no humerus is known for *T. americanum*); radii = squares; x = tibiae. See text for discussion.

dwelling or island-dwelling pygmy hippos and elephants, where the reduction in body size is driven by the more limited resource base, and the decreased need to roam large areas of grasslands to find food. However, there are multiple examples of dwarfing in the North American rhinocerotids. *Teleoceras guymonense* (Fig. 4.36) is almost 20% smaller than contemporary *T. hicksi,* and almost as much of a size reduction as *T. meridianum* was to its sister taxon, *T. medicornutum.* Unfortunately, *T. guymonense* is known from much less complete material than *T. meridianum,* so we are not able to analyze as many of the phenomena associated with dwarfing. *Skinneroceras manningi* is about 20% smaller (Table 4.4) than the next smallest dicerathere, *D. annectens,* and dramatically smaller than *D. armatum.* *Gulfoceras* is also an example of dwarfing, although its sister-group relationships are unclear (since the specimens are so incomplete).

In addition to dwarfing, there are some striking trends in body size through time. Some lineages (e.g., *Subhyracodon, Aphelops*)

steadily increase in size, whereas others (e.g., *Diceratherium, Peraceras*) split into both larger and smaller species. The *Teleoceras* lineage (Fig. 4.36) shows a mixed pattern: steady increase in size from *T. americanum* to *T. medicornutum* to *T. major* to the largest species, *T. fossiger,* followed by a slight size reduction with *T. hicksi.* There are also branching events for dwarfed *T. meridianum* and *T. guymonense,* and for the large (but short-nosed) *T. brachyrhinum.* Thus, the rhinocerotids demonstrate the full range of possible evolutionary trends through time, with no one pattern prevailing.

Limb bone proportions

Another macroevolutionary phenomenon well demonstrated by the rhinocerotids is the proportions of their limb bones. Although by and large they follow well established rules of scaling with increased and decreased body size, there are some striking exceptions. For example, Prothero and Sereno (1982) showed that when dwarfing occurred in *Peraceras hessei* and *Teleoceras meridianum*, their distal limb segments (radius, tibia) are much more robust than would be predicted from simple scaling. This is comparable to the robustness observed in living dwarfs, such as the pygmy hippopotamus, and the extinct dwarfed hippos and elephants from islands.

Stanley (1977) suggested that *Teleoceras* arose suddenly by chondrodystrophic dwarfing. But this completely ignores the fact that almost the entire tribe is characterized by these limb proportions (Prothero *et al.*, 1989). More importantly, the shortening and stumpiness of the limbs gets progressively more developed through the evolution of the genus. Figure 7.5 shows the trend in ratios through time. As measured by the ratio of the length of the limb bone divided by its midshaft width, the humerus, radius, and tibia all become more robust from *T. americanum* to *T. hicksi*; only the femur shows an inconsistent trend of becoming more robust up through *T. major*, then less so in *T. fossiger* and *T. hicksi*. Clearly, this is a progressive trend through almost 12 m.y. over at least five species, and not an instantaneous example of limb shortening by achondroplasia, which happens in a single generation.

Stasis versus gradualism

Ever since the proposal of punctuated equilibrium by Eldredge and Gould (1972), paleontologists have looked at their respective groups of fossils to see what they could contribute to the debate. The excellent records of the rhinocerotids in North America provides a fertile ground for testing such hypotheses.

Classically, paleontologists have looked at the transformation series of molarizing premolars seen in *Hyracodon, Trigonias*, and *Subhyracodon*, and interpreted this as a gradual change through time (Fig. 7.6). But as Prothero (1996, fig. 3) and Figure 2.8 in this volume show, there is no progressive change through time. Instead, almost the full range of variants is present in a single population at a single time horizon. There is no clear trend through time so that each successive population has only one possible premolar state, and the next population sample is a slightly more progressive state. Although there is a net change from unmolarized upper premolars at the beginning to fully molarized at the end, it is not a steady gradual progression, but a long period of stasis with high variability until the molarization process is concluded, and the upper premolars are static through the rest of their history.

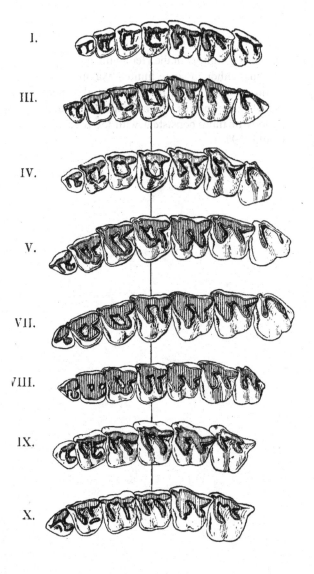

I.

III.

IV.

V.

VII.

VIII.

IX.

X.

Figure 7.6. "Trend" in molarization of premolars of *Hyracodon* through time (after Abel, 1910). Although the diagram purports to show a gradual change in molarizing upper premolars through successive populations, in fact most of the variants seen here occur in a single population (see Prothero, 1996). The same can be said for *Trigonias* and *Subhyracodon* (see Chapter 2).

On the other hand, there are some gradual size changes in lineages through time, although Gould and Eldredge (1977) excluded size change from the debate about evolutionary patterns. There are some other changes that might be touted as gradual. In the male nasals of *Subhyracodon occidentalis, Diceratherium tridactylum,* and *D. armatum,* there is a gradual increase in the number of individual specimens which show more robust nasals trending toward ridge-like rugosities and finally culminating in full-blown nasal ridges with rugose surfaces. Indeed, there are even specimens (Fig. 7.7) which show the intermediate condition between classic nasals of *D. tridactylum* and *D. armatum* in beds found between both species. The increase in hypsodonty through time in *Aphelops* and *Teleoceras* might also be considered gradual. So would the increase of robustness of limbs (Fig. 7.5), although this could be plotted as a series of static populations with rapid changes in between.

However, the most striking thing about the overall pattern of rhinocerotid evolution is that of stasis. Consistent with the predictions of the punctuated equilibrium model, most rhinocerotid species appear suddenly with few transitions between other species, and then are unchanged through most of their history. This is true of nearly every genus described in this volume (Fig. 6.13), and especially so of long-lived but static species such as *Amphicaenopus platycephalus* and *Penetrigonias dakotensis, Diceratherium armatum* and *D. annectens* (both of which range through almost 11 m.y. of the Arikareean with no visible change from beginning to end), and all three species of *Peraceras* (Fig. 4.31). As Figures 4.27 and 6.13 show, *Aphelops megalodus* is extremely stable through entire Hemingfordian, Barstovian, and Clarendonian before the transformation kicks in to *A. malacorhi-*

Figure 7.7. Specimen of *Diceratherium tridactylum* (F:AM 112171) from the Sharps Formation of South Dakota, which shows an intermediate level of nasal ridge development between typical *D. tridactylum* (Fig. 4.15C) and normal *D. armatum* (Fig. 4.16F).

nus and *A. mutilus* in the Hemphillian.

Thus, although some limited examples of gradual change can be documented in the rhinocerotids, the overwhelming pattern is one of stable species which show no measurable change over long periods of time, consistent with the predictions of Eldredge and Gould (1972).

References

Abel, O. 1910. Kritische Untersuchungen über die paläogenen Rhinocerotiden Europas. *Abhandlungen der Geologische Reichsanstall, Wien*, 20, 1-52.

Agusti, J., and Anton, M. 2002. *Mammoths, Sabertooths, and Hominids*. Columbia University Press, New York.

Albright, L.B. 1998. The Arikareean land mammal age in Texas and Florida: southern extension of the Great Plains faunas and Gulf Coast endemism. *Geological Society of America Special Paper*, 325, 167-184.

Albright, L.B. 1999a. Biostratigraphy and vertebrate paleontology of the San Timoteo badlands, southern California. *University of California Publications in Geological Sciences*, 144, 1-121.

Albright, L.B. 1999b. Ungulates from the Toledo Bend local fauna (late Arikareean, early Miocene), Texas Coastal Plain. *Bulletin of the Florida Museum of Natural History*, 42(1), 1-80.

Antoine, P.-O. 2002. Phylogenie et évolution des Elasmotheriina (Mammalia, Rhinocerotidae). *Mémoires du Muséum national d'Histoire naturelle*, 188, 1-359.

Antoine, P.-O., Duranthon, F., and Welcomme, J.-L. 2003. *Alicornops* (Mammalia, Rhinocerotidae) dans le Miocene supérieur des Collines Bugti (Balouchistan, Pakistan): implications phylogénétiques. *Geodiversitas,* 25(3), 575-603.

Archer, M. 1975. Abnormal dental development and its significance in dasyurids and other marsupials. *Memoirs of the Queensland Museum*, 17, 251-265.

Bales, G. 1995. A multivariate morphometric study of living and fossil rhinoceros skulls. Unpubl. Ph.D. Dissertation, University of Southern California, Los Angeles, California.

Barbour, E. H. 1906. Notice of a new Miocene rhinoceros, *Diceratherium arikarense*. Science (n.s.), 24, 780-781.

Beliacva, E. I. 1959. Sur la découverte de Rhinocéros tertiaires anciens dans la Province Maritime de l'U.R.S.S.. [in Russian summary]. *Vertebrata Palasiatica*, 3, 81-92.

Berger, J. 1994. Science, conservation, and black rhinos. *Journal of Mammalogy*, 75, 298-308.

Berggren, W.A., Kent, D. V., Swisher, C. C., III, and Aubry, M.-P. 1995. A revised Cenozoic geochronology and chronostratigraphy. *SEPM Special Publication*, 54, 129-212.

Bjork, P.R. 1978. The functional significance of a broken lower incisor in *Amphicaenopus*, an Oligocene rhinocerotid. *Proceedings of the South Dakota Academy of Science*, 57, 163-167.

Blumenbach, J.F. 1799. *Handbuch der Naturgeschichte*, 6th edn. Göttingen

Brandt, J.F. 1878. Tentamen synopseos rhinocerotidum viventium et fossilium. *Mémoires de l'Académie Imperiale de Science de St. Pétersbourg*, 7(5), 1-66.

Breyer, J. 1981. The Kimballian land mammal age: Mene, mene, tekel, upharsin (Dan. 5:25). *Journal of Paleontology*, 55, 1207-1216.

Butler, P.M. 1952. The milk-molars of Perissodactyla, with remarks on molar occlusion. *Proceedings of the Zoological Society of London*, 121, 777-817.

Carranza-Castañeda, O. 1989. Rhinocerontes de la fauna local Rancho el Ocote, Mioceno Tardio (Henfiliano Tardio) de Estado de Guanajuato. *Universidad Nacional Autónoma de Mexico, Instituto Geológia, Revista*, 8(1), 88-99.

Cerdeño, E. 1992. Spanish Neogene rhinoceroses. *Palaeontology* 35(2), 297-308.

Cerdeño, E. 1993. Etude sur *Diaceratherium aurelianense* et *Brachypotherium brachypus* (Rhinocerotidae, Mammalia) du Miocene moyen de France. *Bulletin du Muséum national d'Histoire naturelle*, fourth series, section C, 15(1-4). 25-77.

Cerdeño, E. 1995. Cladistic analysis of the family Rhinocerotidae (Perissodactyla). *American Museum Novitates*, 3143, 1-25.

Chappell, W.M., Durham, J.W., and Savage, D.E. 1941. Mold of a rhinoceros in basalt, Lower Grand Coulee, Washington, *Geological Society of America Bulletin*, 62. 907-918.

Chow, M.-C., and Chiu, C.-S. 1964. An Eocene giant rhinoceros. [in Chinese and English]. *Vertebrata Palasiatica* 8, 264-268.

Clark, J., Beerbower, J.R., and Kietzke, K.K. 1967. Oligocene sedimentation, stratigraphy, paleoecology, and paleoclimatology in the Big Badlands of South Dakota. *Fieldiana Geology Memoir*, 5, 1-158.

Clementz, M.T., and Koch, P.L. 2000. Assessment of the aquatic habits of the extinct rhinoceros, *Teleoceras*, using stable isotope analysis. *Geological Society of America Abstracts with Programs*, 32(7), A-221.

Cook, H.J. 1908. A new rhinoceros from the Lower Miocene of Nebraska. *American Naturalist*, 42, 543-545.

Cook, H.J. 1909. A new genus of rhinoceros from Sioux County, Nebraska. *Nebraska Geological Survey*, 3, 243-248.

Cook, H.J. 1912a. A new species of rhinoceros, *Diceratherium loomisi*, from the lower Miocene of Nebraska. *Nebraska Geological Survey*, 7, 29-32.

Cook, H.J. 1912b. Notice of a new genus of rhinoceros from the Lower Miocene. *Science* (n.s.), 35, 219-220.

Cook, H.J. 1912c. A new genus and species of rhinoceros, *Epiaphelops virgasectus*, from the lower Miocene of Nebraska. *Nebraska Geological Survey*, 7 21-22.

Cook, H.J. 1927. A new rhinoceros of the genus *Teleoceras* from Colorado. *Proceedings of the Colorado Museum of Natural History*, 7, 1-5.

Cook, H.J. 1930. New rhinoceroses from the Pliocene of Colorado and Nebraska. *Proceedings of the Colorado Museum of Natural History*, 9, 44-51.

Cook, T.D., and Bally, A.W. 1975. *Stratigraphic Atlas: North and Central America*. Princeton University Press, Princeton, N.J.

Cope, E. D. 1873a. On some new extinct Mammalia from the Tertiary of the Plains. *Palaeontological Bulletin*, 14, 1-2.

Cope, E. D. 1873b. Second notice of extinct Vertebrata from the Tertiary of the Plains. *Palaeontological Bulletin*, 15, 1-6.

Cope, E. D. 1873c. Third notice of extinct Vertebrata from the Tertiary of the Plains. *Palaeontological Bulletin*, 16, 1-8.

Cope, E. D. 1874a. Report on the vertebrate palaeontology of Colorado. In Hayden, F.V. (ed.), *Annual Report of the Geological and Geographical Survey of the Territories for 1873*. U.S. Government Printing Office, Washington, D. C. pp. 427-533.

Cope, E. D. 1874b. Report on the stratigraphy and Pliocene vertebrate palaeontology of northern Colorado. *Bulletin of the U.S. Geological and Geographical Survey of the Territories (Hayden)* 1 (1), 9-28.

Cope, E. D. 1875a. Report on the vertebrate palaeontology of Colorado. *Annual Report of the U.S. Geological and Geographical Survey of the Territories (Hayden)* 7, 427-533.

Cope, E. D. 1875b. On some new fossil Ungulata. *Proceedings of the Academy of Natural Sciences of Philadelphia*, 1875, 258-261.

Cope, E. D. 1875c. Report on the geology of that part of northwestern New Mexico examined during the field season of 1874. *Annual report upon the geographical explorations and surveys west of the 100th meridian*, 981-1017.

Cope, E. D. 1877. Report upon the extinct Vertebrata obtained in New Mexico by parties of the expedition of 1874: Chapter xiii. Fossils of the Loup, In *Geographical surveys west of the 100th meridian* (First Lieut. Geo. M. Wheeler, Corps of Engineers, U. S. Army, in charge). 4, 1-370.

Cope, E. D. 1878. Descriptions of new extinct Vertebrata from the upper Tertiary and Dakota formations. *Bulletin of the U.S. Geological and Geographical Survey of the Territories*, 4, 379-396.

Cope, E. D. 1879a. On the extinct species of Rhinoceridae of North America and their allies. *Bulletin of the U.S. Geological and Geographical Survey of the Territories* 5, 227-237.

Cope, E. D. 1879b. American Aceratheria. *American Naturalist*, 13, 333.

Cope, E. D. 1879c. On the extinct American rhinoceroses and their allies. *American Naturalist*, 13, 771a-771j.

Cope, E. D. 1880a. A new genus of Rhinocerotidae. *American Naturalist*, 14, 540.

Cope, E. D. 1880b. The genealogy of the American rhinoceroses. *American Naturalist*, 14, 610-611.

Cope, E. D. 1884. The Vertebrata of the Tertiary formations of the West. Book I. In Hayden, F.V., ed., *Report U. S. Geological Survey of the Territories*. Washington, 1-1009

Cope, E. D. 1885. The White River beds of Swift Current River, Northwest Territory. *American Naturalist*, 19, 163.

Cope, E. D. 1891a. On Vertebrata from the Tertiary and Cretaceous rocks of the Northwest Territory. I. The species from the Oligocene or Lower Miocene beds of the Cypress Hills. *Geological Survey of Canada, Contributions to Canadian Palaeontology* 3, 1-25.

Cope, E. D. 1891b. On two new Perissodactyls from the White River Neocene. *American Naturalist*, 25, 47-49.

Cope, E. D., and Matthew, W.D. 1915. *Hitherto unpublished plates of Tertiary Mammalia and Permian Vertebrata. Prepared under the direction of Edward Drinker Cope for the U. S. Geological Survey of the Territories, with descriptions of plates by William Diller Matthew.* American Museum of Natural History, Monograph Series, 2, New York.

Dalquest, W.W. 1983. Mammals of the Coffee Ranch local fauna, Hemphillian of Texas. *Pearce-Sellards Series, Texas Memorial Museum*, 38, 1-41.

Dalquest, W. W., and Hughes, J.T. 1966. A new mammalian local fauna from the Lower Pliocene of Texas. *Transactions of the Kansas Academy of Sciences*, 69, 79-87.

Dalquest, W.W., and Mooser, O. 1974. Miocene vertebrates from Aguascalientes, Mexico. *Pearce-Sellards Series, Texas Memorial Museum*, 21, 1-10.

Dalquest, W.W., and Mooser, O. 1980. Late Hemphillian mammals of the Ocote local fauna, Guanajuato, Mexico. *Pearce-Sellards Series, Texas Memorial Museum*, 32, 1-25.

Depéret, C., and Douxami, H. 1902. Les vertébrés oligocènes de Pyrimont-Challonges (Savoie). *Mémoires de la Société paléontologique de Suisse*, 29, 1-91.

Deussen, A. 1924. Geology of the Coastal Plain of Texas west of the Brazos River. *U.S. Geological Survey Professional Paper*, 126, 1-145.

Dinerstein, E. 1991 Sexual dimorphism in the greater one-horned rhinoceros (*Rhinoceros unicornis*). *Journal of Mammalogy*, 72, 450-457.

Dinerstein, E. 2003. *The Return of the Unicorns*. Columbia University Press, New York.

Dingus, L. 1990. Systematics, stratigraphy, and chronology for mammalian fossils (late Arikareean to Hemingfordian) from the uppermost John Day Formation, Warm Springs, Oregon. *PaleoBios*, 12 (47), 1-24.

Dollo, L. 1885. Rhinocéros vivants et fossiles. *Revue des Questions Scientifiques*, 17, 293-300.

Douglass, E. 1903. New vertebrates from the Montana Tertiary. *Annals of the Carnegie Museum*, 2, 145-199.

Douglass, E. 1908. Rhinoceroses from the Oligocene and Miocene deposits of North Dakota and Montana. *Annals of the Carnegie Museum*, 4, 256-266.

Duvernoy, G.-L. 1853. Nouvelles études sur les Rhinoceros fossiles. *Comptes Rendus de l'Académie des Sciences, Paris*, 36, 117-125, 169-176, 450-454.

Eldredge, N., and Gould, S.J. 1972. Punctuated equilibria: an alternative to phyletic gradualism. In Schopf, T.J.M. (ed.), *Models in Paleobiology*. Freeman Cooper, San Francisco, CA, pp. 82-115.

Emry, R.J. 1992. Mammalian range zones in the Chadronian White River Formation at Flagstaff Rim, Wyoming. In Prothero, D.R., and Berggren, W.A (eds.), *Eocene-Oligocene Climatic and Biotic Evolution*. Princeton University Press, Princeton, NJ, pp. 106-115.

Farlow, J.O., Sunderman, J.A., Havens, J.J., Swinehart, A.L., Holman, J.A., Richards, R.L., Miller, N.G., Martin, R.A., Hunt, R.M., Jr., Storrs, G.G., Curry, B.B., Fluegeman, R.H., Dawson, M.R., and Flint, M.E.T. 2001. The Pipe Creek Sinkhole Biota, a diverse late Tertiary continental fossil assemblage from Grant County, Indiana. *American Midland Naturalist*, 145, 367-378.

Felix, J., and Lenk, H. 1891. Übersicht über die geologischen Verhältnisse des mexicanischen Stäts Puebla. *Palaeontographica*, 37, 117-139.

Figgins, J.D. 1934a. The generic status of ?*Caenopus premitis*. *Proceedings of the Colorado Museum of Natural History*, 13, 1.

Figgins, J.D. 1934b. New material for the study of individual variation, from the Lower Oligocene of Colorado. *Proceedings of the Colorado Museum of Natural History*, 13, 7-14.

Freeman, G.H., and King, J.M. 1969. Relations amongst the various linear measurements and weight for black rhinoceroses in Kenya. *East African Wildlife Journal*, 7, 67-72.

Fremd, T., Bestland, E.A., and Retallack, G.J. 1994. *John Day Basin Paleontology Field Trip Guide and Road Log*. Society of Vertebrate Paleontology Field Trip Guide, pp. 1-56.

Freudenberg, W. 1922. Die Säugetierfauna des Pliocäns und Postpliocäns von Mexico. II Teil: Mastodonten und Elephanten. *Geologische und Paläontologische Abhandlungen, Jena* 14-18, 1-76.

Frick, C. 1937. Horned ruminants of North America. *Bulletin of the American Museum of Natural History*, 69, 1-669.

Frye, J. C., Leonard, A.B., and Swineford, A. 1956. Stratigraphy of the Ogallala formation (Neogene) of northern Kansas. *Bulletin of the Geological Survey of Kansas*, 118, 1-92.

Galbreath, E.C. 1953. A contribution to the Tertiary geology and paleontology of northeastern Colorado. *University of Kansas Paleontological Contributions*, 4, 1-120.

Galusha, T. 1975a. Childs Frick and the Frick Collection of fossil mammals. *Curator* 18(1), 5-38.

Galusha, T. 1975b. Stratigraphy of the Box Butte Formation. *Bulletin of the American Museum of Natural History*, 156, 1-68.

Galusha, T., and Blick, J.C. 1971. Stratigraphy of the Santa Fe Group, New Mexico. *Bulletin of the American Museum of Natural History*, 144(1), 1-127.

Ginsburg, L, Huin, J., and Locher, J.-P. 1981. Les Rhinocerotidae (Perissodactyla, Mammalia) du Miocene inférieur des Beilleaux à Savigne-sur-Lathan (Indre-et-Loire). [The Rhinocerotidae (Perissodactyla, Mammalia) of the Lower Miocene of Les Beilleaux at Savigne-sur-Lathan (Indre-et-loire).] [in French with English summ.]. *Bulletin du Muséum national d'Histoire naturelle*, section C, 345-361.

Gould, S.J., and Eldredge, N. 1977. Punctuated equilibria: the tempo and mode of evolution reconsidered. *Paleobiology*, 3, 115-151.

Green, M. 1958. Arikareean rhinoceroses from South Dakota. *Journal of Paleontology*, 32, 587-594.

Gregory, W. K., and Cook, H.J. 1928. New materials for the study of evolution: A series of primitive fossil rhinoceros skulls (*Trigonias*) from the Lower Oligocene of Colorado. *Proceedings of the Colorado Museum of Natural History*, 7(1), 1-32.

Groves, C.P. 1982. The skulls of Asian rhinoceroses: Wild and captive. *Zoo Biology*, 1, 251-261.

Groves, C.P. 1984. Phylogeny of the living species of rhinoceros. *Zeitschrift für zoologische Systematik und Evolutionforschung*, 21(4), 293-313.

Guérin, C. 1980. Les rhinoceros (Mammalia, Perissodactyla) du Miocene terminal au Pleistocene Supérieur en Europe Occidentale. *Documents des Laboratoires du Geologie, Lyon*, 79, 1-1185.

Gustafson, E. P. 1977. First record of *Teleoceras* (Rhinocerotidae) from the Ringold Formation, Pliocene of Washington. *PaleoBios*, 27, 1-3.

Hanson, C.B. 1989. *Teletaceras radinskyi*, a new primitive rhinocerotid from the late Eocene Clarno Formation, Oregon. In Prothero, D.R., and Schoch, R.M. (eds.), *The Evolution of Perissodactyls*. Oxford University Press, Oxford, pp. 379-398.

Hanson, C.B. 1996. Stratigraphy and vertebrate faunas of the Bridgerian-Duchesnean Clarno Formation, north-central Oregon. In Prothero, D.R., and Emry, R.J. (eds.), *The Terrestrial Eocene-Oligocene Transition in North America*. Cambridge University Press, Cambridge, pp. 206-239.

Harrison, J.A., and Manning, E.M. 1983. Extreme carpal variability in *Teleoceras* (Rhinocerotidae, Mammalia). *Journal of Vertebrate Paleontology*, 3, 58-64.

Hatcher, J. B. 1893. The *Titanotherium* beds. *American Naturalist*, 27, 204-221.

Hatcher, J. B. 1894a. A median-horned rhinoceros from the Loup Fork beds of Nebraska. *American Geologist*, 13, 149-150.

Hatcher, J. B. 1894b. Discovery of *Diceratherium*, the two-horned rhinoceros, in the White River beds of South Dakota. *American Geologist,* 13, 360-361.

Hatcher, J. B. 1894c. On a small collection of vertebrate fossils from the Loup Fork beds of northwestern Nebraska; with note on the geology of the region. *American Naturalist*, 28, 236-248.

Hay, O.P. 1902. Bibliography and catalogue of fossil Vertebrata of North America. *United States Geological Survey Bulletin*, 179, 1-868.

Hay, O.P. 1930. Second bibliography and catalogue of the fossil Vertebrata of North America. *Carnegie Institution Publication,* 390, 1-916.

Heissig, K. 1999. Family Rhinocerotidae. In Rössner, G.E., and Heissig, K. (eds.), *The Miocene Land Mammals of Europe*. Verlag Friedrich Pfeil, Munich, pp. 175-188.

Henshaw, P.C. 1942. A Tertiary mammalian fauna from the San Antonio Mountains near Tonopah, Nevada. *Publications of the Carnegie Institution of Washington*, 530, 77-168.

Hesse, C.J. 1935. A vertebrate fauna from the type locality of the Ogallala formation. *University of Kansas Science Bulletin*, 22, 79-117.

Hesse, C.J. 1943. A preliminary report on the Miocene vertebrate fauna of southeast Texas. *Proceedings of the Texas Academy of Science*, 26, 157-179.

Hickey, L.J., Johnson, K.R., and Dawson, M.R. 1988. The stratigraphy, sedimentology, and fossils of the Haughton Formation: a post-impact crater fill, Devon Island, N.W.T., Canada. *Meteoritics* 23, 221-231.

Holbrook, L.G., and Lucas, S.G. 1997. A new genus of rhinocerotoid from the Eocene of Utah and the status of North American "*Forstercooperia.*" *Journal of Vertebrate Paleontology* 17, 384-396.

Hunt, R.M., Jr. 1990. Taphonomy and sedimentology of Arikaree (Lower Miocene) fluvial, eolian, and lacustrine paleoenvironments, Nebraska and Wyoming; a paleobiota entombed in fine-grained volcaniclastic rocks. *Geological Society of America Special Paper*, 244, 69-111.

Hunt, R.M., Jr. 1992. Death at a 19-million-year-old waterhole: the bonebed at Agate Fossil Beds National Monument, western Nebraska. *Museum Notes, University of Nebraska State Museum,* 83, 1-6.

Hunt, R.M., Jr. 1998. Amphicyonidae. In Janis, C., Scott, K.M., and Jacobs, L.L. (eds.), *Evolution of Tertiary Mammals of North America, Vol. I: Terrestrial Carnivores, Ungulates, and Ungulatelike Mammals*. Cambridge University Press, Cambridge, 196-221.

Hunt, R.M., Jr. 2002. New amphicyonid carnivorans (Mammalia, Daphoeninae) from the early Miocene of southeastern Wyoming. *American Museum Novitates* 3385, 1-41.

International Commission on Zoological Nomenclature. 1999. *International Code of Zoological Nomenclature*, fourth ed.

International Trust for Zoological Nomenclature, London.

Janis, C.M. 1982. Evolution of horns in ungulates: ecology and paleoecology. *Biological Reviews*, 57, 261-318.

Janis, C.M., Scott, K.M., and Jacobs, L.L. (eds.). 1998. *Evolution of Tertiary Mammals of North America. Vol. I: Terrestrial Carnivores, Ungulates, and Ungulatelike Mammals*. Cambridge University Press, Cambridge.

Jepsen, G.L. 1966. Early Eocene bat from Wyoming. *Science* 154(3754), 1333-1339.

Kihm, A.J. 1987. Mammalian paleontology and geology of the Yoder Member, Chadron Formation, east-central Wyoming. *Dakoterra*, 3, 28-45.

Klaits, B. 1972. The moving mesaxonic manus: a comparison of tapirs and rhinoceroses. *Extrait de Mammalia,* 36, 126-145.

Klaits, B. 1973. Upper Miocene rhinoceroses form Sansan (Gers), France: the manus. *Journal of Paleontology*, 47, 315-326.

Krishtalka, L., Stucky, R.K., West, R.M., McKenna, M.C., Black, C.C., Bown, T.M., Dawson, M.R., Golz, D.J., Flynn, J.J., Lillegraven, J.A., and Turnbull, W.D. 1987. Eocene (Wasatchian through Duchesnean) biochronology of North America. In Woodburne, M.O. (ed.), *Cenozoic Mammals of North America: Geochronology and Biostratigraphy*. University of California Press, Berkeley, pp. 77-117.

Lambe, L.M. 1908. The Vertebrata of the Oligocene of the Cypress Hills, Saskatchewan. Contributions to Canadian Paleontology, 3, 1-65.

Lander, E.B. 1983. Continental vertebrate faunas from the Upper Member of the Sespe Formation, Simi Valley, California, and the Terminal Eocene Event. In Squires, R.L., and Filewicz, M.V. (eds.), *Cenozoic Geology of the Simi Valley area, Southern California*. Pacific Section SEPM, Book 35, p. 142-154.

Lane, H. H. 1927. A new rhinoceros from Kansas. *Kansas University Science Bulletin*, 17, 297-311.

Leidy, J. 1847. On a new genus and species of fossil Ruminantia: *Poëbrotherium wilsoni*. *Proceedings of the Academy of Natural Sciences of Philadelphia*, 3, 322-326.

Leidy, J. 1850a. [Remarks on *Rhinoceros occidentalis*]. *Proceedings of the Academy of Natural Sciences of Philadelphia,* 5, 119.

Leidy, J. 1850b. [Descriptions of *Rhinoceros nebrascensis, Agriochoerus antiquus, Palaeotherium proutii*, and *P. bairdii*]. *Proceedings of the Academy of Natural Sciences of Philadelphia*, 5, 121-122.

Leidy, J. 1851a. [Remarks on *Oreodon priscus* and *Rhinoceros occidentalis*.]. *Proceedings of the Academy of Natural Sciences of Philadelphia*, 5, 276.

Leidy, J. 1851b. [Remarks on *Aceratherium*.]. *Proceedings of the Academy of Natural Sciences of Philadelphia,* 5, 331.

Leidy, J. 1852a. Description of the remains of extinct Mammalia and Chelonia from Nebraska Territory, collected during the geological survey under the direction of Dr. D. D. Owen. In Owen, D.D. (ed.), *Report of a geological survey of Wisconsin, Iowa, and Minnesota and incidentally a portion of Nebraska*

Territory, U.S. Government Printing Office, Washington, D.C., pp. 534-572.

Leidy, J. 1852b. [Description of a new rhinoceros from Nebraska, *R. americanus.*]. *Proceedings of the Academy of Natural Sciences of Philadelphia*, 6, 2.

Leidy, J. 1853a. [Remarks on a collection of fossil Mammalia from Nebraska.]. *Proceedings of the Academy of Natural Sciences of Philadelphia*, 7, 392-394.

Leidy, J. 1854. The ancient fauna of Nebraska, or a description of remains of extinct Mammalia and Chelonia from the Mauvais Terres of Nebraska. *Smithsonian Contributions to Knowledge*, 6, article 7, 1-126.

Leidy, J. 1856. Notice of some remains of extinct Mammalia, recently discovered by Dr. F. V. Hayden in the Bad Lands of Nebraska. *American Journal of Science*, 2, 422-423.

Leidy, J. 1858. Notice of remains of extinct Vertebrata, from the valley of the Niobrara River, collected during the exploring expedition of 1857, in Nebraska, under the command of Lieut. G. K. Warren, U. S. Top. Eng., by Dr. F. V. Hayden. *Proceedings of the Academy of Natural Sciences of Philadelphia*, 1858, 20-29.

Leidy, J. 1865. [Descriptions of *Rhinoceros meridianus* and *R. hesperius.*]. *Proceedings of the Academy of Natural Sciences of Philadelphia*, 1865, 176-177.

Leidy, J. 1869. The extinct mammalian fauna of Dakota and Nebraska, including an account of some allied forms from other localities, together with a synopsis of the mammalian remains of North America. *Journal of the Academy of Natural Sciences of Philadelphia*, 2, 1-472.

Leidy, J. 1871. [Remarks on fossils from Oregon]. *Proceedings of the Academy of Natural Sciences of Philadelphia*, 1871, 247-248.

Leidy, J. 1872. On the fossil vertebrates of the early Tertiary formation of Wyoming. *U. S. Geological Survey of Montana and portions of adjacent Territories; F. V. Hayden, U. S. Geologist.* U.S. Government Printing Office, Washington, D. C., pp. 353-372.

Leidy, J. 1873. Contributions to the extinct vertebrate fauna of the Western Territories. *Report of the U. S. Geological Survey of the Territories, F. V. Hayden, U.S. geologist in charge.* U.S. Government Printing Office, Washington, D. C., pp. 14-358.

Leidy, J. 1884. Vertebrate fossils from Florida. *Proceedings of the Academy of Natural Sciences of Philadelphia*, 1884, 118-119.

Leidy, J. 1885. *Rhinoceros* and *Hippotherium* from Florida. *Proceedings of the Academy of Natural Sciences of Philadelphia*, 1885, 32-33.

Leidy, J. 1886. An extinct boar from Florida. *Proceedings of the Academy of Natural Sciences of Philadelphia*, 1886, 37-38.

Leidy, J. 1887. Fossil bones from Florida. *Proceedings of the Academy of Natural Sciences of Philadephia*, 1887, 309-310.

Leidy, J. 1890. *Hippotherium* and *Rhinoceros* from Florida. *Proceedings of the Academy of Natural Sciences of Philadelphia*, 1890, 182-183.

Leidy, J., and Lucas, F. 1896. Fossil vertebrates from the Alachua clays of Florida. *Transactions of the Wagner Free Institute of Science, Philadelphia*, 4, 15-61.

Lindsay, E.H., Mou, Y., Downs, W., Pederson, J., Kelly, T.S., Henry, C., and Trexler, J. 2002. Recognition of the Hemphillian/Blancan boundary in Nevada. *Journal of Vertebrate Paleontology*, 22, 429-442.

Loomis, F. B. 1908. Rhinocerotidae of the Lower Miocene. *American Journal of Science*, 4, 51-64.

Lucas, F.A. 1900. A new rhinoceros, *Trigonias osborni*, from the Miocene of South Dakota. *Proceedings of the U.S. National Museum*, 13m 221-223.

Lucas, S.G., and Sobus, J.C. 1989. The systematics of indricotheres. In Prothero, D.R., and Schoch, R.M. (eds.), *The Evolution of Perissodactyls.* Oxford University Press, Oxford, pp. 358-378.

Lucas, S.G., Foss, S.E., and Mihlbachler, M.C. 2004. *Achaenodon* (Mammalia, Artiodactyla) from the Clarno Formation, Oregon, and the age of the Hancock Quarry local fauna. *New Mexico Museum of Natural History and Science Bulletin*, 26, 89-96.

Lucas, S.G., Schoch, R.M., and Manning, E. 1981. The systematics of *Forstercooperia*, a middle to late Eocene hyracodontid (Perissodactyla: Rhinocerotidae) from Asia and western North America. *Journal of Paleontology*, 55, 826-841.

Macdonald, J.R. 1963. The Miocene faunas from the Wounded Knee area of western South Dakota. *Bulletin of the American Museum of Natural History*, 125, 139-238.

Macdonald, J.R. 1970. Review of the Miocene Wounded Knee faunas of southwestern South Dakota. *Los Angeles County Museum of Natural History Bulletin*, 8, 1-82.

MacFadden, B.J. 1998. Tale of two rhinos: isotopic ecology, paleodiet, and niche differentiation of *Aphelops* and *Teleoceras* from the Florida Neogene. *Paleobiology*, 24, 274-286.

MacFadden, B.J. 2004. Middle Miocene land mammals from the Cucaracha Formation (Hemingfordian-Barstovian) of Panama. *Bulletin of the Florida Museum of Natural History* (in press).

MacFadden, B.J., and Hunt, R.M., Jr. 1998. Magnetic polarity stratigraphy and correlation of the Arikaree Group, Arikareean (late Oligocene-early Miocene) of northwestern Nebraska. *Geological Society of America Special Paper*, 325, 143-166.

Madden, C.T., and Dalquest, W.W. 1990. The last rhinoceros in North America. *Journal of Vertebrate Paleontology*, 10, 266-267.

Manning, E.M. 1997. An early Oligocene rhinoceros jaw from the marine Byram Formation of Mississippi. *Mississippi Geology*, 18(2), 14-31.

Manning, E.M., Dockery, D.T., III, and Schiebout, J.A. 1986. Preliminary report of a *Metamynodon* skull from the Byram Formation (Lower Oligocene) of Mississippi. *Mississippi Geology*, 6(2), 1-16.

Marsh, O. C. 1870. Discovery of the Mauvaises Terres formation in Colorado. *American Journal of Science*, 2, 292.

Marsh, O. C. 1873. Notice of new Tertiary mammals. *American Journal of Science,* 3, 407-410, 485-488.

Marsh, O. C. 1875. Notice of new Tertiary mammals. IV. *American Journal of Science,* 3, 239-250

Marsh, O. C. 1887. Notice of new fossil mammals. *American Journal of Science,* 3, 323-331.

Matthew, W. D. 1899. A provisional classification of the freshwater Tertiary of the West. *Bulletin of the American Museum of Natural History,* 12, 19-75.

Matthew, W. D. 1901. Fossil mammals of the Tertiary of northeastern Colorado. *Memoirs of the American Museum of Natural History,* 1, 355-448.

Matthew, W. D. 1909. Faunal lists of the Tertiary Mammalia of the West. *Bulletin of the U.S. Geological Survey,* 341, 91-138.

Matthew, W. D. 1918. Contributions to the Snake Creek fauna; with notes upon the Pleistocene of western Nebraska; American Museum Expedition of 1916. *Bulletin of the American Museum of Natural History,* 38, 183-229.

Matthew, W. D. 1924. Third contribution to the Snake Creek fauna. *Bulletin of the American Museum of Natural History,* 50, 59-210.

Matthew, W.D. 1931. Critical observations on the phylogeny of the rhinoceroses. *University of California Publications, Bulletin of the Department of Geological Sciences,* 20, 1-9.

Matthew, W.D. 1932. A review of the rhinoceroses with a description of *Aphelops* material from the Pliocene of Texas. *University of California Publications, Bulletin of the Department of Geological Sciences,* 20, 411-480.

McKenna, M.C., and Bell, S.K. 1997. *Classification of Mammals Above the Species Level.* Columbia University Press, New York.

Mead, A.J. 2000. Sexual dimorphism and paleoecology in *Teleoceras,* a North American rhinoceros. *Paleobiology,* 26, 689-706.

Mead, A.J., and Wall, W.P. 1998a. Paleoecological implications of craniodental and premaxilla morphologies of two rhinocerotoids (Perissodactyla) from Badlands National Park, South Dakota. In Santucci, V.L., and McClelland, L. (eds.), National Park Service Paleontological Research. *Technical Report NPS/NRPO/NRTR-98/01,* pp. 18-22.

Mead, A.J., and Wall, W.P. 1998b. Dietary implications of jaw biomechanics in the rhinocerotoids *Hyracodon* and *Subhyracodon* from Badlands National Park, South Dakota. In Santucci, V.L., and McClelland, L. (eds.), National Park Service Paleontological Research. *Technical Report NPS/NRPO/-NRTR-98/01,* pp. 22-28.

Merriam, J.C., and Sinclair, W.J. 1907. Tertiary faunas of the John Day region. *University of California Publications, Bulletin of the Department of Geological Sciences,* 5, 171-205.

Mihlbachler, M.C. 2003. Demography of late Miocene rhinoceroses (*Teleoceras proterum* and *Aphelops malacorhinus*) from Florida: linking mortality and sociality in fossil assemblages. *Paleobiology,* 29, 412-428.

Mihlbachler, M.C. 2004. Sexual dimorphism in Miocene rhinoceroses, *Teleoceras proterum* and *Aphelops malacorhinus,* from Florida (in press).

Morea, M. 1981. The Massacre Lake local fauna (Mammalia, Hemingfordian) from northwestern Washoe County, Nevada. Ph.D. dissertation, University of California, Riverside.

Nichols, R. 1998. The Lemhi Valley Oligo-Miocene: an overview. *Idaho Museum of Natural History Occasional Papers,* 36, 10-12.

Olcott, T.F. 1909. A new species of *Teleoceras* from the Miocene of Nebraska. *American Journal of Science,* 4, 403-404.

Olson, E.C., and McGrew, P.O. 1941. Mammalian fauna from the Pliocene of Honduras. *Geological Society of America Bulletin,* 52, 1219-1244.

Osborn, H.F. 1890. The Mammalia of the Uinta formation. Part III. The Perissodactyla. Part IV. The evolution of the ungulate foot. *Transactions of the American Philosophical Society,* 16, 505-572.

Osborn, H. F. 1893. *Aceratherium tridactylum* from the Lower Miocene of Dakota. *Bulletin of the American Museum of Natural History,* 5, 85-86.

Osborn, H. F. 1898a. A complete skeleton of *Teleoceras,* the true rhinoceros from the upper Miocene of Kansas. *Science,* 2, 554-557.

Osborn, H. F. 1898b. A complete skeleton of *Teleoceras fossiger.* Notes upon the growth and sexual characters of this species. *Bulletin of the American Museum of Natural History,* 10, 51-59.

Osborn, H. F. 1898c. The extinct rhinoceroses. *Memoirs of the American Museum of Natural History,* 1, 75-164.

Osborn, H. F. 1900. Phylogeny of the rhinoceroses of Europe. *Bulletin of the American Museum of Natural History,* 13, 229-267.

Osborn, H. F. 1904. New Miocene rhinoceroses with revision of known species. *Bulletin of the American Museum of Natural History,* 20, 307-326.

Osborn, H.F. 1905. Ten years of progress in the mammalian paleontology of North America. *American Geologist,* 36, 199-229.

Osborn, H. F. 1910. *The Age of Mammals in Europe, Asia and North America.* Macmillan, New York.

Osborn, H.F. 1923. The extinct giant rhinoceros *Baluchitherium* of western and central Asia. *Natural History,* 23(3), 209-228.

Osborn, H.F. 1929. The titanotheres of ancient Wyoming, Dakota, and Nebraska. *U.S. Geological Survey Monograph,* 55, 1-953.

Osborn, H. F., and Matthew, W.D. 1909. Cenozoic mammal horizons of western North America. *Bulletin of the U.S. Geological Survey,* 341, 1-90.

Osborn, H. F., and Wortman, J.L. 1894. Fossil mammals of the Lower Miocene White River beds. Collection of 1892. *Bulletin of the American Museum of Natural History,* 7, 199-228.

Owen-Smith, N. 1988. *Megaherbivores: The Influence of Very Large Body Size on Ecology*. Cambridge University Press, Cambridge.

Patton, T.H. 1969. Miocene and Pliocene artiodactyls, Texas Gulf Coastal Plain. *Bulletin of the Florida State Museum, Biological Science,* 14, 115-226.

Penny, M. 1988. *Rhinos: Endangered Species*. Facts on File, New York.

Peterson, O.A. 1906a. The Agate Spring fossil quarry. *Annals of the Carnegie Museum,* 3, 487-494.

Peterson, O.A. 1906b. The Miocene beds of western Nebraska and eastern Wyoming and their vertebrate faunae. *Annals of the Carnegie Museum,* 4, 21-72.

Peterson, O.A. 1906c. Preliminary description of two new species of the genus *Diceratherium* Marsh, from the Agate Spring fossil quarry. *Science* (n.s.) 24, 281-283.

Peterson, O.A. 1920. The American diceratheres. *Memoirs of the Carnegie Museum,* 7, 399-456.

Pocock, R.I. 1945. A sexual difference in the skull of Asiatic rhinoceroses. *Proceedings of the Zoological Society of London,* 115, 319-322.

Prothero, D.R. 1993. Fifty million years of rhinoceros evolution. In Ryder, O.A. (ed.), *Proceedings of the International Rhino Conference*. San Diego Zoological Society, San Diego, CA, pp. 81-87.

Prothero, D.R. 1994. Mammalian evolution. In Prothero, D.R., and Schoch, R.M. (eds.), *Major Features of Vertebrate Evolution*. Paleontological Society Short Courses in Paleontology, 7, 238-270.

Prothero, D.R. 1996. Hyracodontidae. In Prothero, D.R., and Emry, R.J. (eds.), *The Terrestrial Eocene-Oligocene Transition in North America*. Cambridge University Press, Cambridge, pp. 634-645.

Prothero, D.R. 1998. Rhinocerotidae. In Janis, C., Scott, K.M., and Jacobs, L.L. (eds.), *Evolution of Tertiary Mammals of North America, Vol. I: Terrestrial Carnivores, Ungulates, and Ungulatelike Mammals*. Cambridge University Press, Cambridge, 595-605.

Prothero, D.R., and Emry, R.J. (eds.) 1996. *The Terrestrial Eocene-Oligocene Transition in North America*. Cambridge University Press, Cambridge.

Prothero, D.R., and Manning, E. 1987. Miocene rhinoceroses from the Texas Gulf Coastal Plain. *Journal of Paleontology,* 61(2), 388-423.

Prothero, D.R., and Schoch, R.M. 2003. *Horns, Tusks, and Flippers: The Evolution of Hoofed Mammals and their Relatives*. Johns Hopkins University Press, Baltimore.

Prothero, D.R., and Sereno, P.C. 1980. Allometry and paleoecology of medial Miocene dwarf rhinoceroses from the Texas Gulf Coastal Plain. *Geological Society of America Abstracts with Programs,* 12(7), 504.

Prothero, D.R., and Sereno, P.C. 1982. Allometry and paleoecology of medial Miocene dwarf rhinoceroses from the Texas Gulf Coastal Plain. *Paleobiology,* 8(1), 16-30.

Prothero, D.R., and Shubin, N. 1989. The evolution of Oligocene horses. In Prothero, D.R., and Schoch, R.M. (eds.), *The Evolution of Perissodactyls*. Oxford University Press, New York, pp. 142-175.

Prothero, D.R., Guérin, C., and Manning, E. 1989. The history of the Rhinocerotoidea. In Prothero, D.R., and Schoch, R.M. (eds.), *The Evolution of Perissodactyls*. Oxford University Press, New York, pp. 322-340.

Prothero, D.R., Howard, J., and Dozier, T.H.H. 1996. Stratigraphy and paleomagnetism of the upper middle Eocene to lower Miocene (Uintan-Arikareean) Sespe Formation, Ventura County, California. In Prothero, D.R., and Emry, R.J. (eds.), *The Terrestrial Eocene-Oligocene Transition in North America*. Cambridge University Press, Cambridge, pp. 156-173.

Prothero, D.R., Manning, E., and Hanson, C.B. 1986. The phylogeny of the Rhinocerotoidea (Mammalia, Perissodactyla). *Zoological Journal of the Linnean Society of London,* 87, 341-366.

Quinn, J.H. 1955. Miocene Equidae of the Texas Gulf Coastal Plain. *University Texas Bureau of Economic Geology Publication,* 5516, 1-102.

Rachlow J., and Berger, J. 1997. Conservation implications of patterns of horn regeneration in dehorned white rhinos. *Conservation Biology,* 11, 84-91.

Radinsky, L.B. 1965. Evolution of the tapiroid skeleton from *Heptodon* to *Tapirus*. *Bulletin of the Museum of Comparative Zoology,* 134, 69-106.

Radinsky, L.B. 1966. The families of the Rhinocerotoidea (Mammalia, Perissodactyla). *Journal of Mammalogy,* 47 631-639.

Radinsky, L.B. 1967. A review of the rhinocerotoid family Hyracodontidae (Perissodactyla). *Bulletin of the American Museum of Natural History,* 136 1-46.

Rich, T.H.V. 1981. Origin and history of the Erinaceinae and Brachyericinae (Mammalia, Insectivora) in North America. *Bulletin of the American Museum of Natural History.* 171, 1-116.

Ringström, T. 1924. Nashörner der Hipparion-Fauna Nord-Chinas. *Palaeontologia Sinica,* series C, 1, 1-156.

Rose, K.D., and Smith, B.H. 1979. Dental anomaly in the early Eocene condylarth *Ectocion*. *Journal of Paleontology,* 53, 756-760.

Russell, L.S. 1934. Revision of the Lower Oligocene vertebrate fauna of the Cypress Hills, Saskatchewan. *Transactions of the Royal Canadian Institute,* 20, 49-67.

Russell, L.S. 1982. Tertiary mammals of Saskatchewan. VI: The Oligocene rhinoceroses. *Life Science Contribution of the Royal Ontario Museum,* 133, 1-58.

Schlaikjer, E.M. 1935. Contributions to the stratigraphy and palaeontology of the Goshen Hole Area, Wyoming. III. A new basal Oligocene formation. *Bulletin of the Museum of Comparative Zoology,* 76, 72-93.

Schlaikjer, E.M. 1936. Contributions to the stratigraphy and palaeontology of the Goshen Hole area, Wyoming. IV. New vertebrates and the stratigraphy of the Oligocene and early Miocene. *Bulletin of the Museum of Comparative Zoology,* 76, 97-189.

Schlosser, M. 1902. Die fossilen Säugethiere Chinas. *Zentralblatt Mineralogie, Geologie, Paläontologie,* 1902, 529-535.

Schultz, C. B., and Stout, T.M. 1961. *Field conference on the Tertiary and Pleistocene of western Nebraska.* Special Publication of the University of Nebraska State Museum, 2, 1-10.

Schultz, G. (ed.) 1977. *Guidebook, Field Conference on Late Cenozoic Biostratigraphy of the Texas Panhandle and Adjacent Oklahoma, August 4-6, 1977.* Special Publication 1, West Texas State University, Canyon, TX.

Scott, W. B. 1913. *A History of Land Mammals in the Western Hemisphere.* Macmillan, New York.

Scott, W. B. 1941. The mammalian fauna of the White River Oligocene. Part V. Perissodactyla. *Transactions of the American Philosophical Society,* 28, 747-980.

Sellards, E. H. 1914. The relation between the Dunnellon formation and the Alachua clays of Florida. *Report of the Florida Geological Survey,* 6, 161-162.

Setogouchi, T. 1978. Paleontology and geology of the Badwater Creek area, central Wyoming. Part 16. The Cedar Ridge local fauna (late Oligocene). *Carnegie Museum of Natural History Bulletin,* 9.

Shotwell, J.A. 1956. Hemphillian mammal assemblage from northeastern Oregon. *Bulletin of the Geological Society of America,* 67, 717-738.

Shotwell, J.A. 1963. The Juntura Basin: studies in earth history and paleoecology. *Transactions of the American Philosophical Society,* 53(1), 1-77.

Shotwell, J.A. 1968. Miocene mammals of southeastern Oregon. *Bulletin of the Oregon Museum of Natural History,* 14.

Simpson, G.G. 1930. Tertiary land mammals of Florida. *Bulletin of the American Museum of Natural History,* 59, 149-211.

Simpson, G.G. 1945. The principles of classification and a classification of the mammals. *Bulletin of the American Museum of Natural History,* 85, 1-350.

Simpson, W.F. 1985. Geology and paleontology of the Oligocene Harris Ranch badlands, southwestern South Dakota. *Dakoterra,* 2(2), 303-333.

Sinclair, W.J. 1924. The faunas of the concretionary zones of the *Oreodon* beds, White River Oligocene. *Proceedings of the American Philosophical Society,* 63, 93-133,

Sisson, W., and Grossman, R. 1975. *The Anatomy of the Domestic Animal.* W.B. Saunders, Philadelphia, PA.

Skinner, M.F., and Johnson, F.W. 1984. Tertiary stratigraphy and the Frick Collections of fossil vertebrates from north-central Nebraska. *Bulletin of the American Museum of Natural History,* 178(3), 215-368.

Skinner, M.F., and Taylor, B.E. 1967. A revision of the geology and paleontology of the Bijou Hills, South Dakota. *American Museum Novitates,* 2300, 1-53.

Skinner, M.F., Skinner, S.M., and Gooris, R.J. 1968. Cenozoic rocks and faunas of Turtle Butte, south-central South Dakota. *Bulletin of the American Museum of Natural History,* 138(7), 379-436.

Skinner, M.F., Skinner, S.M., and Gooris, R.J. 1977. Stratigraphy and biostratigraphy of late Cenozoic deposits in central Sioux County, western Nebraska. *Bulletin of the American Museum of Natural History,* 158(5), 263-371.

Stanley, S.M. 1977. *Macroevolution.* W.H. Freeman, New York.

Stecher, R.M., Schultz, C.B., and Tanner, L.G. 1962. A Middle Miocene rhinoceros quarry in Morrill County, Nebraska (with notes on hip disease in *Diceratherium*). *Bulletin of the Nebraska State Museum,* 4(7), 101-111.

Stevens, M.S. 1977. Further study of the Castolon local fauna (early Miocene), Big Bend National Park, Texas. *The Pearce-Sellards Series, Texas Memorial Museum,* 28, 1-69.

Stevens, M.S., Stevens, J.B., and Dawson, M.R. 1969. New early Miocene formation and vertebrate local fauna, Big Bend National Park, Brewster County, Texas. *The Pearce-Sellards Series, Texas Memorial Museum,* 15, 3-52.

Stock, C. 1921. Later Cenozoic mammalian remains from the Meadow Valley region, southeastern Nevada. *American Journal of Science,* 5, 250-264.

Stock, C. 1933. Perissodactyla from the Sespe of the Las Posas Hills, California. *Publications of the Carnegie Institution of Washington,* 440, 15-28.

Stock, C. 1949. Mammalian fauna from the Titus Canyon formation, California. *Publications of the Carnegie Institution of Washington,* 584, 229-244.

Stock, C, and Furlong, E.L. 1926. New canid and rhinocerotid remains from the Ricardo Pliocene of the Mohave Desert, California. *Bulletin of the Department of Geology, University of California,* 16, 40-60.

Stone, W.J. 1970. Stratigraphic significance of fossil rhinoceros remains in Slope County, North Dakota. *Proceedings of the North Dakota Academy of Science,* 24(1), 32.

Storer, J.E. 1975. Tertiary mammals of Saskatchewan. Part III: The Miocene fauna. *Life Sciences Contribution, Royal Ontario Museum,* 103, 1-134.

Storer, J.E. 1996. Eocene-Oligocene faunas of the Cypress Hills Formation, Saskatchewan. In Prothero, D.R., and Emry, R.J. (eds.), *The Terrestrial Eocene-Oligocene Transition in North America.* Cambridge University Press, Cambridge, pp. 240-262.

Tabrum, A.R., Prothero, D.R., and Garcia, D. 1996. Magnetostratigraphy and biostratigraphy of the Eocene-Oligocene transition, southwestern Montana. In Prothero, D.R., and Emry, R.J. (eds.), *The Terrestrial Eocene-Oligocene Transition in North America.* Cambridge University Press, Cambridge, pp. 278-312.

Tanner, L.G. 1959. Notice of a new Pliocene rhinoceros, *Aphelops kimballensis. Proceedings of the Nebraska Academy of Sciences,* 69, 15.

Tanner, L.G. 1967. A new species of rhinoceros, *Aphelops kim-*

ballensis, from the latest Pliocene of Nebraska. *Bulletin of the University of Nebraska State Museum*, 6, 1-16.

Tanner, L.G. 1969. A new rhinoceros from the Nebraska Miocene. *Bulletin of the University of Nebraska State Museum*, 8, 395-412.

Tanner, L.G. 1972. A new species of *Menoceras* from the Marsland Formation of Nebraska. *Bulletin of the University of Nebraska State Museum*, 9, 205-213.

Tanner, L.G. 1975. The stratigraphic occurrence of *Teleoceras* species in Pliocene deposits of Nebraska and adjacent states. *Proceedings of the Nebraska Academy of Sciences and Affiliated Societies*, 82, 45-46.

Tanner, L.G. 1976. A new species of *Peraceras* from the Lower Pliocene of Boyd County, Nebraska. *Proceedings of the Nebraska Academy of Sciences and Affiliated Societies*, 86, 50.

Tanner, L.G. 1977. A new species of *Diceratherium* from the Lower Pliocene (Valentinian) of Boyd County, Nebraska. *Transactions of the Nebraska Academy of Sciences and Affiliated Societies*, 4, 121-128.

Tanner, L.G., and Martin, L.D. 1972. Notes on the deciduous and permanent dentition of the hyracodonts. *Transactions of the Nebraska Academy of Sciences, Earth Science*, 1, 61-72.

Tanner, L.G., and Martin, L.D. 1976. New rhinocerotids from the Oligocene of Nebraska. *Life Science Miscellaneous Publications of the Royal Ontario Museum*, pp. 210-219.

Taylor, R.M.S. 1982. Aberrant maxillary third molars, morphology and developmental relations. In Kurten, B. (ed.), *Teeth: Form, Function, and Evolution*. Columbia University Press, New York, pp. 64-74.

Tedford, R.H. 1981. Mammalian biochronology of the late Cenozoic basins of New Mexico. *Geological Society of America Bulletin*, 92, 1008-1022.

Tedford, R.H., and Barghoorn, S. 1997. Miocene mammals of the Española and Albuquerque basins, north-central New Mexico. *New Mexico Museum of Natural History and Science Bulletin*, 11, 77-96.

Tedford, R.H., and Hunter, M.E. 1984. Miocene marine-nonmarine correlations, Atlantic and Gulf Coastal Plains, North America. *Palaeogeography, Palaeoclimatology, Palaeoecology*, 47, 129-151.

Tedford, R.H., Albright, L.B., III, Barnosky, A.D., Ferrusquia-Villafranca, I., Hunt, R.M., Jr., Storer, J.E., Swisher, C.C., III, Voorhies, M.R., Webb, S.D., and Whistler, D.P., 2004. Mammalian biochronology of the Arikareean through Hemphillian interval (late Oligocene through early Pliocene epochs). In Woodburne, M.O., ed., *Late Cretaceous and Cenozoic Mammals of North America: Biostratigraphy and Geochronology*. Columbia University Press, New York, pp. 169-231.

Tedford, R.H., Galusha, T., Skinner, M.F., Taylor, B.E., Fields, R.W., Macdonald, J.R., Rensberger, J.M., Webb, S.D., and Whistler, D.P. 1987. Faunal succession and biochronology of the Arikareean through Hemphillian interval (Late Oligocene through earliest Pliocene epochs) in North America. In Woodburne, M.O. (ed.), *Cenozoic mammals of North America. Geochronology and biostratigraphy*. University of California Press, Berkeley, CA, pp. 153-210.

Tedford, R.H., Swinehart, J., Prothero, D.R., Swisher, C.C., III, King, S.A., and Tierney, T.E. 1996. The Whitneyan-Arikareean transition in the High Plains. In Prothero, D.R., and Emry, R.J. (eds.), *The Terrestrial Eocene-Oligocene Transition in North America*. Cambridge University Press, Cambridge, pp. 295-317.

Terry, D.O., LaGarry, H.E., and Wells, W.B. 1995. The White River Group revisited: vertebrate trackways, ecosystems, and lithostratigraphic revision, redefinition, and redescription. In Flowerday, C.A. (ed.) *Geologic field trips in Nebraska and adjacent parts of Kansas and South Dakota*. Conservation and Survey Division, University of Nebraska, Lincoln, NE, Guidebook, 10, 43-57.

Toula, F. 1902. Das Nashorn von Hundsheim. *Rhinoceros (Ceratorhinus* Osborn) *hundsheimensis* nov. form. mit Ausführungen über die Verhältnisse von elf Schädeln von *Rhinoceros (Ceratorhinus) sumatrensis*. *Abhandlung geologische Reichsanst Wien*, 19, 1-92.

Trouessart, E. L. 1904. *Catalogus Mammalium tam viventium quam fossilium. Quinquinale supplementum*. Berolini, Italy. 1-546.

Troxell, E.L. 1921a. New amynodonts in the Marsh collection. *American Journal of Science*, 5, 21-34.

Troxell, E.L. 1921b. New species of *Hyracodon*. *American Journal of Science*, 5, 34-40.

Troxell, E.L. 1921c. *Caenopus*, the ancestral rhinoceros. *American Journal of Science*, 5, 41-51.

Troxell, E.L. 1921d. A study of *Diceratherium* and the diceratheres. *American Journal of Science*, 5, 197-208.

Voorhies, M.R. 1981. Dwarfing the St. Helens eruption: ancient ashfall creates a Pompeii of prehistoric animals. *National Geographic*, 159, 66-75.

Voorhies, M.R. 1985. A Miocene rhinoceros herd buried in volcanic ash. *Research Report, National Geographic Society*, 19, 671-688.

Voorhies, M.R. 1990. *Vertebrate Paleontology of the Proposed Norden Reservoir Area, Brown, Cherry, and Keya Paha Counties, Nebraska*. Division of Archeological Research, University of Nebraska, Lincoln, Technical Report 82-09, 1-591.

Voorhies, M.R. 1992. Ashfall: Life and death at a Nebraska waterhole ten million years ago. *Museum Notes, University of Nebraska State Museum*, 81, 1-4.

Voorhies, M.R., and Stover, S.G. 1978. An articulated fossil skeleton of a pregnant rhinoceros, *Teleoceras major* Hatcher. *Proceedings of the Nebraska Academy of Science*, 88, 48.

Voorhies, M.R., and Thomasson, J.R. 1979. Fossil grass anthoecia with Miocene rhinoceros skeletons: diet of an extinct species. *Science*, 206, 331-333.

Voorhies, M.R., Holman, J.A., and Xue, X.-X. 1987. The Hottell

Ranch rhino quarries (basal Ogallala: medial Barstovian), Banner County, Nebraska. Part I: Geologic setting, fauna lists, lower vertebrates. *Contributions to Geology, University of Wyoming,* 25, 55-69.

Wall, W.P., and Hickerson, W. 1995. A biomechanical analysis of locomotion in the Oligocene rhinocerotoid *Hyracodon.* In Santucci, V.L., and McClelland, L. (eds.), *National Park Service Paleontological Research.* Technical Report NPS/NRPO/NRTR-95/16, pp. 19-26,

Warren, L. 1998. *Joseph Leidy: The Last Man Who Knew Everything.* Yale University Press, New Haven, CT.

Webb, S.D. 1977. A history of the savanna vertebrates in the New World. Part I: North America. *Annual Reviews of Ecology and Systematics.* 8, 355-380.

Webb, S.D. 1983. The rise and fall of the late Miocene ungulate fauna in North America. In Nitecki, M., ed., *Coevolution.* University of Chicago Press, Chicago, pp. 267-306.

Webb, S.D. 1984. Geology and vertebrate paleontology of the Late Miocene Gracias Formation in central Honduras. *Research Reports, National Geographic Society,* 17, 913-930.

Webb, S.D., and Perrigo, S.C. 1984. Late Cenozoic vertebrates from Honduras and El Salvador. *Journal of Vertebrate Paleontology,* 4, 237-254.

Webb, S.D, MacFadden, B.J., and Baskin, J.A. 1981. Geology and paleontology of the Love Bone Bed from the late Miocene of Florida. *American Journal of Science,* 281, 513-544.

Whitlock, C., and Dawson, M.R. 1990. Pollen and vertebrates of the early Neogene Haughton Formation, Devon Island, Arctic Canada. *Arctic,* 43(4), 324-330.

Whitmore, F. C., Jr., and R. H. Stewart. 1965. Miocene mammals and Central American seaways. *Science,* 148, 180-185.

Whitney, J. D. 1865. *Geological Survey of California.* Geology, 1, 1-498.

Wilson, J.A. and Schiebout, J.A. 1984. Early Tertiary vertebrate fauna, Trans-Pecos Texas: Ceratomorpha less Amynodon-tidae. *Pearce-Sellards Series, Texas Memorial Museum,* 39, 1-47.

Wood, H.E., II. 1926. *Hyracodon petersoni,* a new cursorial rhinoceros from the Lower Oligocene. *Annals of the Carnegie Museum,* 16, 315-318.

Wood, H.E., II. 1927. Some early Tertiary rhinoceroses and hyracodonts. *Bulletins of American Paleontology,* 13(50), 3-89.

Wood, H.E., II. 1929. American Oligocene rhinoceroses—a postscript. *Journal of Mammalogy,* 10, 63-75.

Wood, H.E., II. 1931. Lower Oligocene rhinoceroses of the genus *Trigonias. Journal of Mammalogy,* 12, 414-428.

Wood, H.E., II. 1933. A fossil rhinoceros (*Diceratherium armatum* Marsh) from Gallatin County, Montana. *Proceedings of the U.S. National Museum,* 82(7), 1-4.

Wood, H.E., II. 1934. Revision of the Hyrachyidae. *Bulletin of the American Museum of Natural History,* 67, 181-295.

Wood, H.E., II. 1939. Research and publications on the anatomy, stratigraphic distribution and phylogeny of the rhinoceroses and related groups of perissodactyls. *American Philosophical Society Yearbook,* 1938, 245.

Wood, H.E., II. 1941. Trends in rhinoceros evolution. *Transactions of the New York Academy of Sciences,* 3, 83-96.

Wood, H.E., II. 1960. Eight historic fossil mammal specimens in the Museum of Comparative Zoology. *Bulletin of the Museum of Comparative Zoology,* 123, 85-110.

Wood, H.E., II. 1964. Rhinoceroses from the Thomas Farm Miocene of Florida. *Bulletin of the Museum of Comparative Zoology,* 130, 361-386.

Wood, H. E., II, and Wood, A.E. 1937. Mid-Tertiary vertebrates from the Texas coastal plain: fact and fable. *American Midland Naturalist,* 18, 129-146.

Wood, H.E., II, Chaney, R.W., Clark, J., Colbert, E.H., Jepsen, G.L., Reeside, J.B., Jr., and Stock, C. 1941. Nomenclature and correlation of the North American continental Tertiary. *Bulletin of the Geological Society of America,* 52, 1-48.

Woodburne, M.O. (ed.) 2004. *Late Cretaceous and Cenozoic Mammals of North America.* Columbia University Press, New York.

Woodburne, M.O., and Swisher, C.C., III. 1995. Land mammal high-resolution geochronology, intercontinental overland dispersals, sea level, climate, and vicariance. *SEPM Special Publication,* 54, 335-364.

Wright, D.B. 1980. Preliminary investigation of population dynamics of *Teleoceras* and *Aphelops* from the late Miocene of Florida. *Geological Society of America Abstracts with Programs,* 12(4), 213.

Yalden, D.W. 1971. The functional morphology of the carpus in ungulate mammals. *Acta Anatomica,* 76(4), 461-487.

Yatkola, D., and Tanner, L.G. 1979. *Brachypotherium* from the Tertiary of North America. *Occasional Papers of the Museum of Natural History, University of Kansas,* 77, 1-11.

Zeuner, F.E. 1945. New reconstructions of the woolly rhinoceros and Merck's rhinoceros. *Proceedings of the Linnaean Society of London,* 156(3), 183-195.

Printed in the United States
By Bookmasters